Handbook of Neural Networks for Speech Processing

For a listing of recent titles in the *Artech House Signal Processing Library*,
turn to the back of this book.

Handbook of Neural Networks for Speech Processing

Shigeru Katagiri

Editor

Artech House
Boston • London
www.artechhouse.com

Library of Congress Cataloging-in-Publication Data
Handbook of neural networks for speech processing / Shigeru Katagiri, editor.
    p. cm. — (Artech House signal processing library)
    Includes bibliographical references and index.
    ISBN 0-89006-954-9 (alk. paper)
      1. Automatic speech recognition—Handbooks, manuals, etc. 2. Neural networks
(Computer science)—Handbooks, manuals, etc. 3. Speech processing
systems—Handbooks, manuals, etc. I. Katagiri, Shigeru. II. Series.

TK7895.S65 H36 2000   00-055883
006.4'54—dc21   CIP

British Library Cataloguing in Publication Data
Handbook of neural networks for speech processing
    (Artech House signal processing library)
    1. Neural networks (Computer science) 2. Speech processing
systems 3. Speech perception
I. Katagiri, Shigeru
006.5'4

    ISBN 0-89006-954-9

Cover design by Igor Valdman

© 2000 ARTECH HOUSE, INC.
685 Canton Street
Norwood, MA 02062

International Standard Book Number: 0-89006-954-9
Library of Congress Catalog Card Number: 00-055883

10 9 8 7 6 5 4 3 2 1

Contents

Part III   Current Issues in Speech Signal Processing

10   Networks for Speaker Recognition   357

Preface

Speech is one of the most important means of human communication. Human beings communicate with one another by speaking and hearing, and they have made great efforts to achieve freer and more convenient communication by developing speech telecommunications technology such as telephony. In recent years, such technology has greatly advanced. The dream of communication—speaking with anybody, anywhere, anytime—is now becoming a reality. The dream, however, continues to expand. A more economic telecommunications system is desired. Even language barriers are expected to be overcome. Morever, work is being carried out to develop human-machine interfaces that use speech as a communications medium. Indeed, satisfying these demands is of great significance. Global and borderless speech communications assist daily human life, promote mutual understanding among nations, and greatly contribute to the general progress of human society. In pursuing these challenges, a key goal is to develop a useful engineering model of the human speech processing mechanism. This handbook seeks to provide a comprehensive introduction to one of the most important approaches to this technological challenge, speech processing using artificial neural networks (ANNs).

ANNs and speech processing are both transdisciplinary research fields, each having a history of several decades and involving many disciplines such as physiology, physics, statistics, psychology, linguistics, and engineering. Accordingly, a single book cannot be expected to fully cover the entire subject of speech processing using ANNs.

Fortunately, many excellent textbooks have been published in both the ANNs and speech processing fields. Readers can find detailed expertise in such books. In this handbook, we shall concentrate on the results of emerging and challenging studies on using ANNs for speech processing. We shall strive to convey the excitement of developments on the research front. The handbook will guide its readers through the unfinished but promising technological strides made in this field.

The book's 13 chapters are divided into three parts. Part I, entitled "Fundamentals," consists of 4 chapters that introduce basic information about speech processing and ANN technologies. In this introductory section, however, our focus is on speech processing technologies; the fundamentals of ANNs are discussed in Parts II and III. Part II, "Current Issues in Speech Recognition," emphasizes the considerable vigor with which ANNs have been investigated in their application to speech recognition. This second part consists of Chapters 5 through 9. Finally, Part III, "Current Issues in Speech Signal Processing," discusses speech-related topics such as speaker recognition, voice conversion, speech coding, and speech enhancement.

The handbook is aimed at researchers, engineers, and graduate-level students who wish to study the fundamentals and practical applications of neural-network-based speech processing. For these readers, Part I provides the necessary basis for proceeding to the later chapters on applications; Parts II and III provide a comprehensive introduction to the research front of neural networks for speech processing, which can be useful for further study of topics elaborated in related technical books and journals. Despite the primary aim of providing an introductory text, the book can also be used by researchers who have experience in ANN and speech processing technologies. The two latter parts provide good archives of results of the ongoing research. Each chapter was written with the aim of being as self-contained as possible. The book can therefore be used for various types of study by various readers. However, we, the chapter contributors, have a common motivation: we hope that the handbook becomes a vehicle to stimulate research by scientists and engineers on the human mechanism of speech processing and its engineering embodiment, i.e., neural networks for speech processing.

The inspiration for this handbook came from the series editor, Professor Alexsander D. Poularikas, University of Alabama in Huntsville. I wish to express my sincere thanks to him for his careful planning and for asking me to serve as editor of this text. The value of the book derives from the considerable efforts of the contributors, each of whom brought extensive experience to the task of writing the individual chapters. I would like to express my considerable gratitude to these authors. Finally, it would not have been possible to produce the book without the patience and professional assistance provided by the crew at Artech House, especially Mark Walsh, Alexia Rosoff, Traci Beane, and Barbara Lovenvirth. I wish to thank them for help in all phases of making this book a reality.

—Shigeru Katagiri

Part I:
Fundamentals

1

Introduction

Shigeru Katagiri

1.1 Speech Processing

A speech signal is an acoustic waveform; it is also a principal medium for conveying various kinds of information useful for human communications. Based on this fact, speech has been a central channel of telecommunications since the advent of telephony. The advancement of speech processing technology is therefore obviously desirable in making human life more convenient and comfortable.

Figure 1.1 illustrates a fundamental scheme of human speech communications, the *speech chain* [1]. In this scheme, a speaker first encodes his or her intention into linguistic units such as phonemes, controls articulators such as the tongue and the lips through neural information paths, and then produces a speech signal. The speech sound produced is transmitted as air vibration. The sound arriving at the ears of listeners vibrates the eardrums and is then converted into its corresponding neural representation. This neural information travels up to the brain system, is decoded into linguistic units, and ultimately leads to an understanding of the speaker's intention.

Figure 1.1 shows elements of the information conveyed by the speech sound. Some of these are human factors, such as the physical sizes and shapes of the articulators involved (the speaker and the listeners); others are nonhuman factors, such as the transmission loss in the air and distortion in the telecommunication system.

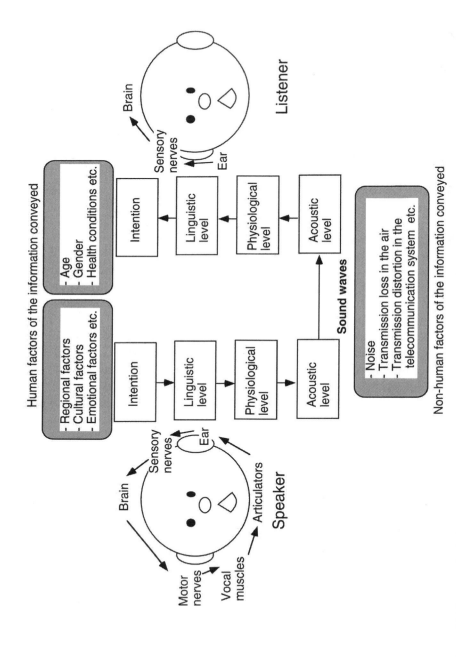

Figure 1.1 The speech chain.

Traditionally, the main purpose of speech processing technology has been to reduce interference between speaker and listener. Significant research efforts have been made to achieve high fidelity in microphones and (electrical) speakers. In addition, knowledge on a wide range of subjects in science and technology has been gathered with the aim of reducing the channel distortion of telephones and recording equipment. In this traditional environment, principal research disciplines have been electroacoustics and device materials engineering.

Extensive psychoacoustic research has also been done, but roughly speaking, its purpose has been to further the development of traditional disciplines by clarifying the perceptual limitations of human auditory mechanisms. Recent technological concerns, in contrast, have shifted to the discovery and emulation of the entire human process of speaking and hearing.

A goal of this new trend is to make machines speak and hear like humans do. With this advance, the world of human communications would be dramatically extended. Achieving the goal would also provide novel knowledge about cognition, understanding, and the mechanisms of speech production and hearing, certainly useful for various aspects of human life such as education and medical care. Research disciplines in this sphere include physiology of speech production, neurophysiology, linguistics, cognitive science, and information engineering such as pattern recognition and coding.

This book considers the very latest research environments, and mainly covers research issues in information engineering, especially speech recognition and understanding, speaker recognition, speech coding, and speech enhancement. Among the many factors determining the characteristics of the speech chain, the following three types of information are considered especially important for developing systems in the information engineering field: (1) phonological information, (2) syntactic and semantic information, and (3) speaker identity.

Phonological information indicates the linguistic difference between speech sounds, and forms a basis for speech pattern classification according to the most fundamental linguistic units, i.e., phonemes. Syntactic information basically corresponds to grammatical rules concerning words and sentences, and semantic information expresses the meanings of speech units. Accordingly, in order to achieve system development, technology appropriate for modeling the threefold nature of information is clearly desired.

1.2 Neural Networks

1.2.1 Fundamentals

Originally, the term *neural network* referred to the biological neural network system. In recent years, however, it has often been used to express an artificial neural network implemented on an electronic device such as a computer, and in fact, such artificial networks have been considered to be one of the most promising technological concepts for developing information systems such as pattern recognizers and function estimators. In this book, we adopt this current usage of the term.

Artificial neural networks are actually based on the biological neural system. They are comprised of many of the key features of the biological system, such as distributed computation mechanism, adaptivity (trainability), nonlinearity, and simplicity in the unit computation (see [2]). Despite great advances in neural science, however, the overall functionality of the neural system, especially the brain, has not yet been fully clarified, and consequently, artificial neural networks have typically been defined as systems that produce some output signal by transforming an input signal to a network through parallel (distributed) computation, conducted over computational units mutually connected and prepared to handle only rather simple computations.

Figure 1.2 illustrates the basic mechanism of artificial neural

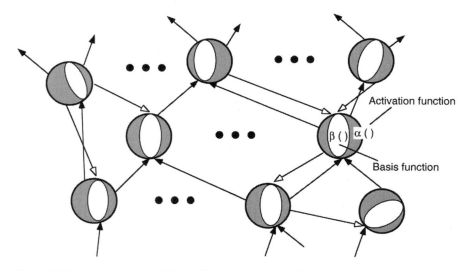

Figure 1.2 Basic mechanism of the artificial neural network.

networks. For simplicity, we will sometimes use the term *neural network* in place of *artificial neural network* in this book. The neural network consists of many nodes (circles in Figure 1.2), each of which conceptually corresponds to a neuron cell in the real biological neural system, and connections (arrows in Figure 1.2), each of which conceptually corresponds to an axon in the neural system. The nodes are mutually connected. Each node consists of two parts: one part for the computation of basis function $\beta(\ )$ (white ellipse) and the other part for the computation of activation function $\alpha(\ )$ (gray circle). The basis function computation unit receives the input signal, which can be an input to the network or an output of another node, and computes the input signal to the activation function unit; the activation function unit produces an output signal, which can be the final output of the network or an input to another node.

An arrow is used to indicate the direction of the information (data) flow. Note that arrows come out of the gray circles that compute the activation function values; they go into the white ellipses that compute the basis function values. There are two types of arrows: black arrows and white arrows. Assuming that an input signal to a network is given at

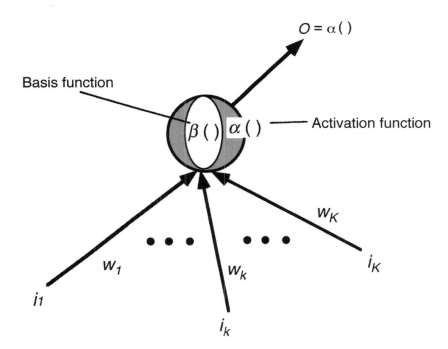

Figure 1.3 Computation mechanism of the neural network.

the bottom part of the network, a black arrow indicates information flow in the forward direction and a white arrow indicates information flow in the backward direction, in other words, the recurrence of information.

In Figure 1.3, we provide a detailed description of the computation mechanism of the neural network. Each connection is a data flow path that transfers an element of the input signal, and it is also assumed to be associated with a weight coefficient. The figure illustrates one neural node, to which K input connections and one output connection are assigned. It is assumed here that the k-th connection is associated with weight w_k and it transfers the k-th element of the input signal vector $\mathbf{i}$, i.e., i_k; $\mathbf{w} = (w_1, \cdots, w_k, \cdots, w_K)^T$ and $\mathbf{i} = (i_1, \cdots, i_k, \cdots, i_K)^T$, where T is the vector transpose. Then, the basic function $\beta(\ )$ computes an input signal to the activation function unit, e.g., $\beta(\mathbf{i}) = \sum_{k=1}^{K} w_k i_k$. The activation function computes the final node output O as $O = \alpha(\beta(\mathbf{i}))$. Actually, there are many possibilities for the computational selection of the basis function and of the activation function, and these selections are one of the main factors determining the mathematical characteristics of the artificial neural network.

1.2.2   Taxonomy of Neural Networks

1.2.2.1   Overview

In addition to the selection possibility for the computation of the basis and activation functions, various factors of the artificial neural network determine the functionality as well as the configuration of the network. This arbitrariness is due to the insufficiency of the clarification of the biological neural network, but it is in practice an important feature of artificial neural network technology that allows for the design of neural networks with high flexibility for a wide range of applications. Based on this arbitrariness or flexibility, many types of artificial neural networks have been developed, and various ways have been devised for the taxonomy of these networks (e.g., [3]). Among the many ways, we here introduce the taxonomy shown in Figure 1.4.

In the figure, we employ four main categorical issues, each characterizing the neural network: (1) structure, (2) measurement, (3) objective function, and (4) optimization method. The selection of the structure apparently determines the network configuration, and it accordingly controls the information flow in the network. It basically controls how flexibly the network represents information, or in other words, the representation capability of the network. The measurement

Category	Types		
Structure	Feedforward	Recurrent	
Connection	Partial	Full	
Measurement	Probability	Distance	Possibility
Basis function	Linear	Radial	
Activation function	Linear	Sigmoidal	Radial
Objective function	Squared error	Classification error	Mutual information
Supervision	Unsupervised	Supervised	
Optimization	Gradient search	Simulated annealing	Genetic algorithm

Figure 1.4 A taxonomy of neural networks.

selection determines how the network expresses the information contained in the input signal. It is mainly characterized by the selection of basis functions at neural nodes. The selection of the objective function determines for what purpose the network is designed, i.e., for data regression, for classification, or for coding. The final issue, i.e., the selection of the optimization method, determines how the network satisfies its design purpose.

1.2.2.2 Structure

There are two important types of structural selection: (1) the feedforward neural network and (2) the recurrent neural network. As illustrated in Figure 1.5, the feedforward neural network conveys node output signals in a one-way mode. The information goes up from the bottom layer of the network, which consists of input nodes, to the top layer. In contrast, the recurrent neural network, as its name indicates, describes the recurrence of information. The output signal of some neural nodes is allowed to go back to a neural node, excluding that which produced the input signal to the neural node. The output signal is sometimes allowed to return to the node that produced the signal itself. Fundamentally, the recurrent neural network has a larger capability of information representation than the feedforward neural network, and it is specially suited to the representation of time-series data such as speech signals. Recent research results have clearly shown that the feedforward neural network is a viable candidate for a powerful pattern classifier.

　　This point will be demonstrated in detail in later chapters.

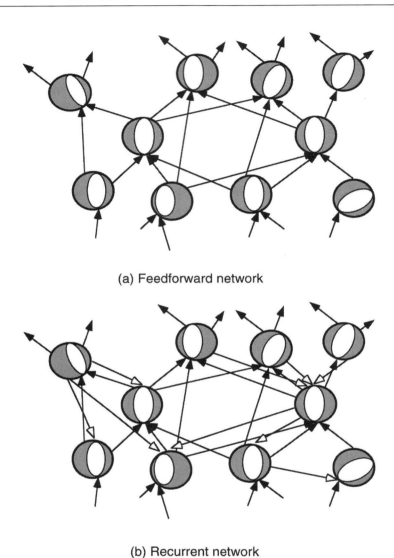

(a) Feedforward network

(b) Recurrent network

Figure 1.5 Feedforward network and recurrent network.

Compared with the feedforward neural network, the recurrent neural network requires basic analytic research efforts in order to embody its full capability in practical applications. The suitability of the recurrent structure, however, is clear, and discussions on this neural network form one of the main topics of this book.

There is another means of categorization in terms of the network structure, i.e., partial connection network versus full connection network. In a full connection network, all of the neural nodes are connected. Under the condition that the number of neural nodes is fixed, a fully connected network basically has a larger representation capability than a partially connected network because of a larger set of connection weights, which results in a high freedom of information representation. However, full connections are not always the best alternative. Accordingly, a careful design of the connections, i.e., the setting of partial connections, is often indispensable for neural network applications.

1.2.2.3 Measurement

The most popular selection for measurement in the recent information technology is probability. In particular, probability has formed a central mathematical framework for speech recognition. In neural network formalisms, too, the estimation problem of a posteriori probabilities, which play an important role in recognition decisions, has been vigorously investigated.

On the other hand, neural networks using distance measures, such as self-organizing feature maps and learning vector quantizers, have also constituted a major subarea of network research. Such measures, however, can have other more general interpretations than either probability or distance, e.g., possibility. This kind of interpretation is useful for discussing designs of neural network pattern classifiers from the perspective of the traditional Bayes decision theory.

The measurement is determined by the selection of the basis function and that of the activation functions. Figure 1.6 illustrates typical types of the basis function and the activation function. There are two main types of selection for the basis function: (1) linear basis and (2) radial basis. For the neural node of Figure 1.3, the linear basis is basically formulated as

$$I = \sum_{k=1}^{K} w_k i_k \qquad (1.1)$$

where I is the output signal of the basis function unit, which is also the input to the activation function unit. Essentially, this computation done over many neural nodes achieves a hyperplane-type mapping of the input signal. Similarly, the radial basis is formulated as

$$I = c \exp(-(\mathbf{i} - \mathbf{w})^T \mathbf{\Sigma}^{-1} (\mathbf{i} - \mathbf{w})) \qquad (1.2)$$

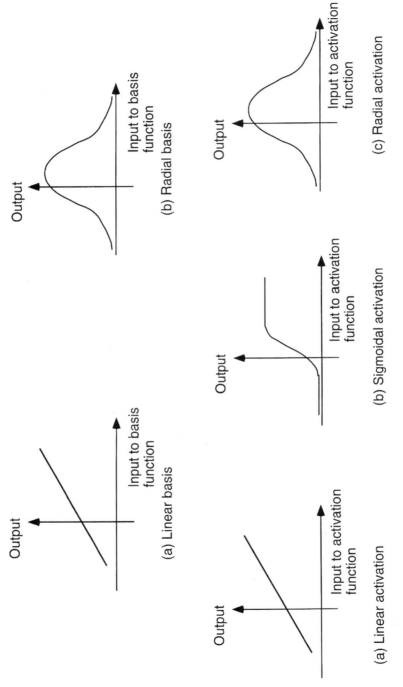

Figure 1.6 Basis functions and activation functions.

where c is a constant and Σ is a K-by-K matrix. This type of basis function achieves a hypersphere-type mapping of the input signal. Note here that the term inside exp( ) is of the form of the general distance measure.

The activation functions are generally grouped into three classes: linear basis, sigmoidal basis, and radial basis. The linear basis activation is simply a linear mapping, defined as

$$O = \alpha(I) = aI + b \qquad (1.3)$$

where a and b are constants. The sigmoidal basis is formulated as

$$O = \alpha(I) = \frac{1}{1 + \exp(-I/a)} \qquad (1.4)$$

where a is a constant, and the radial basis is formulated in the same way as in (1.2). Among these three types, the sigmoidal basis has been the most widely used. It has crucial links to many historic research results in studies involving the Bayes decision theory and in the field of artificial neural networks.

1.2.2.4   Objective Function

Even if a neural network has a significant representation capability, it must be trained, adapted, or designed appropriately, just as the human brain must be trained through education and experience. The design of a neural network is usually set by selecting an objective function, which is also often called a loss function, a risk function, and a criterion. A popular objective is the squared error, which is computed between the output signal of the network and its corresponding target signal. This is widely used for data regression, data coding, and even for pattern classification.

Another selection involves an objective function that closely emulates the classification error count. This is one of the emerging concepts in the neural network and pattern recognition fields, and will be highlighted in later pages in this book. To increase the classification power, mutual information has been used in pattern recognition literature. This can also be an objective function for designing neural network classifiers.

From an engineering standpoint, network design must be done by setting some clear objective (or goal). However, there are expectations for neural networks to learn automatically something important, such as the setting of the objective function, without explicit supervision. Automatic design mechanisms of this type are referred to as self-organization or unsupervised learning mechanisms; design using the objective function explicitly is called supervised learning. Most of the neural networks discussed in this book employ supervised learning. Nevertheless, it is worthwhile to introduce the following important neural networks trained in an unsupervised learning manner: self-organizing feature maps [4], ART [5], and competitive learning networks [6].

1.2.2.5 Optimization

An objective function, which is also called a loss function, computes a measure indicating how differently a neural network behaves from its ideal condition. Accordingly, computing the objective function over many samples (input signals) given for the design purpose produces a loss hypersurface that expresses the quality of the network behavior. The goal of the design clearly comes to searching for the lowest point of this loss surface, and various search algorithms, in other words, optimization algorithms, have been investigated. Among them, many have been heuristic algorithms which do not necessarily guarantee the discovery of the optimal lowest-loss point.

Mathematically supported algorithms are usually grouped into three classes: (1) gradient search, (2) simulated annealing, and (3) genetic algorithms. Gradient search algorithms are traditional optimization algorithms that use the gradient of the loss surface to find its lower positions. Algorithms of this type are fast and easy to handle, but they only guarantee the finding of local optimal points, which are not necessarily the lowest points of the loss surface. Based on their high practicality, however, they are used most widely in the design of neural networks. Most of the neural networks in this book use search algorithms of this type.

Simulated annealing algorithms were originally motivated by physical systems, and have a significant feature that enables them to basically achieve the global optimal (minimum loss) point of the surface. Genetic algorithms are based on biological evolution, and they aim at the global optimal point of the surface by repeating gene recombination operations.

The above global optimization algorithms often require a long training time and are not very popular in application-oriented neural network designs.

1.3 Neural Networks for Speech Processing

Speech processing had been an active research field even before the readvent of neural network research in the 1980s (e.g., see [7]). Many speech recognizers have been developed by using traditional prototype-based classifiers; speech recognition based on statistical methodologies such as the hidden Markov model is becoming a focus of research and development. Speech coding based on traditional vector quantization and autoregressive modeling techniques currently plays a crucial role in the development of speech synthesizers and cellular phone communications. Speech processing technology has clearly been advanced in this traditional research paradigm. Most traditional techniques used for speech processing, however, suffer from various limitations such as linearity in modeling and inconsistency in design. Unfortunately, most of these limitations are unavoidable in the traditional technological paradigm. Clearly, therefore, any paradigm shift is desired for alleviating the current limitations.

Artificial neural networks are neither a miracle nor a mutation. They have a close and important relationship with traditional technologies. At the same time, however, neural networks are clearly advanced with respect to several features, such as their nonlinearity and high classification capability, and can be expected to be involved in a new paradigm that promises further advancement of the current speech processing technology. The subject of neural networks for speech processing is therefore emerging as an important research field, causing people to anticipate the occurrence of a paradigm shift.

1.4 Handbook Overview

1.4.1 Part I: Fundamentals

This introductory chapter 1 has guided the reader through the world of neural networks for speech processing, providing an overview of neural network technologies.

Chapter 2 summarizes the acoustic nature of speech signals and the human process of handling this temporal carrier of communications, i.e., the speech production mechanism and the hearing mechanism.

Chapter 3 introduces the fundamentals of speech recognition, which involve a machine process of recognizing what a human speaker says by encoding an acoustic speech signal into linguistic units such as phonemes and words. By employing a general assumption that a recognition system, or recognizer, has a modular system structure, this chapter describes, in order, the most fundamental theoretical framework for classification decisions, i.e., the Bayes decision theory, acoustic feature extraction methods, and design issues for classification.

Chapter 4 introduces the fundamentals of speech coding, which compresses original speech signals for efficient data recording and for efficient data transmission over telecommunication channels. The chapter introduces techniques underlying coding technology, such as vector quantization and time-series modeling, and then overviews coding methodologies and their applications.

1.4.2   Part II: Current Issues in Speech Recognition

Chapter 5 focuses on prototype-based architectures, which are closely related to radial-basis, or kernel, function neural networks. The chapter reviews several approaches to prototype-based pattern recognition, such as learning vector quantization, from the perspective of discriminative training and the guidelines of the Bayes decision theory. It then discusses prototype-based speech recognizers that have been discriminatively trained in the recent minimum classification error design framework.

Chapter 6 discusses advantages and disadvantages in using recurrent neural networks instead of conventional techniques for speech recognition. The chapter highlights the concepts of uni- and bi-directional recurrency, and it shows the underlying theory, basic architectures, their problems, and merits with respect to speech recognition as well as procedures to train large networks with large amounts of data.

Chapter 7 presents neural-network-based speech recognition technologies, especially those using one of the historical neural networks for speech recognition, i.e., the time-delay neural network (TDNN). The chapter overviews TDNN, its extension, i.e., a multistate TDNN, and hybrid systems using the hidden Markov model. The chapter then elaborates design issues in using these TDNN-based speech recognizers, such as system modularity.

Every technique typically has multiple aspects. Neural networks closely linked to prototype-based systems, such as radial-basis neural networks, also have another face of probability estimators. Chapter 8 highlights the features of neural networks for probability estimation, and discusses their application for speech recognition.

Chapter 9 discusses an emerging approach, called the generalized probabilistic descent (GPD) method, for achieving minimum recognition error networks, suggesting that recognizer design considers the origin of most pattern recognition formalisms, i.e., the Bayes decision theory. The chapter first introduces the basics of GPD and then overviews its recent extensions and applications in the speech recognition field.

1.4.3 Part III: Current Issues in Speech Signal Processing

Chapter 10 covers speaker recognition, which is a machine process of recognizing who is speaking. Compared with traditional approaches, neural-network-based recognizers generally have a high discrimination capability. The chapter overviews the fundamentals of speaker recognition techniques and elaborates experimental evaluations of neural-network-based techniques.

Chapter 11 discusses networks for voice conversion that enable one to convert the speech signals of an original speaker to those of a different target speaker. The chapter highlights neural network models for transforming vocal tract information in the domain of formants, i.e., acoustic resonances based on the vocal tract shape.

Chapter 12 summarizes recent challenges in studies on neural networks for speech coding. The chapter first overviews the basics of neural-network-based quantizers and introduces their performance evaluations. Then it discusses coding application results with the introduction of a new modeling scheme of neural-network-based nonlinear prediction.

Chapter 13 guides the reader through the research world of neural networks for speech enhancement, which is a front-end process used for most speech processing techniques, such as recognition, to improve the quality and intelligibility of corrupted speech signals. The chapter elaborates on several approaches, such as time-domain filtering, transform-domain mapping, state-dependent model switching, and online iterative approaches.

References

[1]   Denes, P. B., and E. N. Pinson, *The Speech Chain: The Physics and Biology of Spoken Language*, New York: Freeman, 1993.

[2]   Haykin, S., *Neural Networks: A Comprehensive Foundation*, New York: MacMillan, 1994.

[3]   Kung, S.-Y., *Digital Neural Networks*, Englewood Cliffs, N.J.: Prentice-Hall, 1993.

[4]   Kohonen, T., *Self-Organizing Feature Maps*, New York: Springer-Verlag, 1995.

[5]   Carpenter, G. A., and S. Grossberg, "ART2: Self-Organization of Stable Category Recognition Codes for Analog Input Patterns," *Applied Optics*, Vol. 26, 1987, pp. 4919–4930.

[6]   Kosko, B., ed., *Neural Networks for Signal Processing*, Englewood Cliffs, N.J.: Prentice-Hall, 1992.

[7]   Rabiner, L., and B.-H. Juang, *Fundamentals of Speech Recognition*, Englewood Cliffs, N.J.: Prentice-Hall, 1993.

2

The Speech Signal and Its Production Model

Shinobu Masaki

2.1 Introduction

Studies of multimodal communication have revealed that a combination of verbal and nonverbal behaviors provides efficient human-to-human communication, and some reports on audiovisual speech processing have been published (e.g., [1]). Speech communication is a crucial channel for conveying various kinds of information, including linguistic, paralinguistic, and emotional information, from person to person [2]. Ever since the concept of the speech chain was presented [3], the production, perception, and feedback of speech by humans has been investigated as a basic research field. Simultaneously, applications of speech signal processing technology, such as recording (or encoding), accumulation, transmission, and reproduction (or decoding) have developed, along with accelerated electronics technology and computational power. Unfortunately, extensive collaboration, exchange, and discussion have not occurred between the research and application fields. This dissociation is remarkable given that neural net and stochastic techniques have been introduced into automatic speech recognition, and editing of speech waveforms is now a practical technique for text-to-speech synthesis applications.

This chapter deals with the human speech production process as a basis of speech signal processing. There are two reasons the chapter is located early in the book. First, it serves as an introduction to speech technology by reviewing the basic speech production studies, including recently reported findings. Second, and more importantly, it seeks to bridge the gap between basic (e.g., speech production and perception) studies and application (e.g., speech recognition and synthesis) studies.

In the following sections, the relationship between information transmitted by speech and its generation mechanisms is addressed. Section 2.2 describes the information included in speech. Then, the mechanism of speech production for transferring the information is discussed. Section 2.3 focuses on anatomical constraints of speech organs in producing various speech sounds. Section 2.4 describes the characteristics and modeling of the speech production mechanism. The last section, Section 2.5, discusses the orientation of future research in basic speech science, and the mutual interaction between basic speech science studies and application technologies for speech.

Table 2.1
Characteristics of linguistic, paralinguistic, and nonlinguistic information

Category of information	Content of the information	Characteristics of transferred information	Features to transfer the information
Linguistic	Lexical Syntactic Semantic Pragmatic	Discrete and Categorical	Mainly transferred by segmental features. (Suprasegmental features are also used; e.g., accent pattern)
Paralinguistic	Intentional Attitudinal Stylistic	Discrete and Categorical (Some attitudinal information is represented continuously)	Mainly transferred by suprasegmental features. (Segmental features—e.g., formant—are also affected)
Nonlinguistic	Physical Idiosyncratic (e.g., sex and age) Emotional	Continuous	Mainly transferred by suprasegmental features. (Segmental features—e.g., formant—are also affected)

2.2 Information Conveyed by Speech

The information embedded in speech sounds can be divided into three categories in terms of its content: linguistic, paralinguistic, and non-linguistic [2]. The relationship between these categories, and their physically observable features as carriers of information, are the main topics of this section (see Table 2.1).

2.2.1 Linguistic Information

The primary objective of human speech communication is to transfer linguistic information. Linguistic information can be defined as "symbolic information that is represented by a set of discrete symbols and rules for their combination" [2]. Therefore, linguistic information is discrete and categorical. An important characteristic of linguistic information, in contrast with nonlinguistic information, is that linguistic information can be controlled by the speaker. As a conveyer of linguistic information, a *segmental feature* plays an important role. A *suprasegmental feature* is also used to distinguish between the discrete meanings of words. In the following subsections, the characteristics of segmental and suprasegmental features and their roles in transferring linguistic information are presented.

2.2.1.1 Segmental Features

Suppose there is a man who wishes to convey his love to a woman, and both are native speakers of English. In order to compose the sentence "I love you," he will select three words: "I," "love," and "you." In this stage, each word is selected as a unit that has a unique (or at least restricted) correspondence to meaning and function. However, each word can be divided hierarchically into smaller segments: A word is divided into *syllables* (/lʌv/), and each syllable can be divided into *phonemes* (/l/, /ʌ/, and /v/); note that a phoneme is enclosed in slashes (/ /).) The phoneme is the smallest segment of sound that functions in a language to signal a difference in meaning. In this sense, phonemes play an important role in transferring linguistic information in human speech communication.

The *phone* is also used to determine the smallest segment of sound. A phone is a particular speech sound categorized discretely according to several acoustical features (e.g., formant frequencies for vowels) and temporal features (e.g., voice onset time [VOT] for stop consonants) of the sound. These features are called segmental features and are determined by a unique combination of independent speech production processes, such as the place and manner of articulation, voiced versus voiceless contrast, and the characteristics of the sound source. These speech production processes will be presented in Section 2.3. According to the criteria of these speech production processes, phones are symbolized by the International Phonetic Alphabet (IPA).

The difference between a phone and a phoneme is that a phoneme is a family of phones which has an identical function in a certain language. For example, the aspirated voiceless plosive sound [t^h] and the nonaspirated [t] are different phones, but they are allophones of the same phoneme /t/ in English. (Note that brackets are used for phone symbols in the IPA.)

2.2.1.2   Suprasegmental Features

Suprasegmental, or prosodic, features are acoustical characteristics distributed over a sequence of several segments, such as phonemes, syllables, words, phrases, and sentences. Acoustical characteristics that transfer the suprasegmental features include the fundamental frequency of glottal vibration for vowels and voiced consonants, the intensity of speech, and the temporal characteristics of segments. Suprasegmental features play a main role in transferring paralinguistic and nonlinguistic information. However, the features listed above also play important roles in transferring linguistic information.

The *word accent* is an example of suprasegmental features used to alter the meaning of words by controlling the fundamental frequency (F0), which is the frequency of vocal fold vibration, or intensity. In Japanese, for example, /ame/ ("rain") and /ame/ ("candy") have the same phoneme sequence. These words can be distinguished by different F0 patterns: For "rain," the F0 for the first mora is higher than that for the second mora, while the F0 pattern is in the opposite direction for "candy" in the Tokyo dialect. This is a typical example of *pitch accent*, since a talker intends to control the F0 to change the meaning, and this intention is perceived by detecting the change in pitch (a psychological measure corresponding to F0).

In English, the act of placing strong intensity on different syllables is used to differentiate the meaning and the function of the word. This kind of accent is called a *stress accent* in opposition to the pitch accent. For example, when stress is placed on the first syllable, "permit" becomes a noun, meaning license or permission. On the other hand, when stress is placed on the second syllable, it becomes a verb whose meaning is to allow or admit.

It is important to note here that physiological constraints of speech production play an important role in maintaining a robust channel for transferring accent information. Since the frequency and intensity of vocal fold vibration have high correlation, sound intensity can be changed as the F0 is adjusted for the pitch accent, and the F0 can be influenced by the stress alteration for the stress accent [4, 5]. These mutual influences between correlated acoustical features can be used to preserve the reliability of transferring accent information. The distinction between pitch accent and stress accent is dependent on what parameter the speaker intends to change.

Control of the segmental duration is also used to distinguish word meanings. For example, the distinction of the Japanese words /ita/ ("was present") and /itta/ ("went") is achieved by the single/geminate contrast at the word-medial voiceless stop. This contrast makes a significantly longer word-medial silence period for /itta/ than that for /ita/, resulting in a ratio of 3:2 for the total duration of /itta/:/ita/ [6]. As a result, native Japanese speakers recognize /itta/ and /ita/ as three-mora and two-mora words, respectively.

The Japanese mora, which is basically composed of a consonant-vowel (CV) sequence, has a subjectively equal duration. Any mora that does not have the CV sequence is called a *special mora*. The geminate is an example of a special mora [7]. A special mora does not form a syllable but occupies a time span roughly equal to an independent mora. Therefore, it is often difficult for those who learn Japanese as a second language to distinguish words with single/geminate contrast.

2.2.2 Paralinguistic Information

Paralinguistic information is defined as "information that is not inferable from a written counterpart but is deliberately added by the speaker to modify or supplement linguistic information" [2]. Paralinguistic information can have both discrete and continuous characteristics. For example, a speaker can control a sentence so as to categorize it as a declarative sentence or an interrogative sentence. Such qualities as the speaker's degree of suspicion can be represented continuously in the utterance of a question. To achieve these functions of paralinguistic information, the speaker mainly controls suprasegmental features but segmental features also come into play.

For example, when you are talking to friends, you might speak quickly and not articulate your words very clearly. On the other hand, when you are speaking to an elderly person who might have hearing

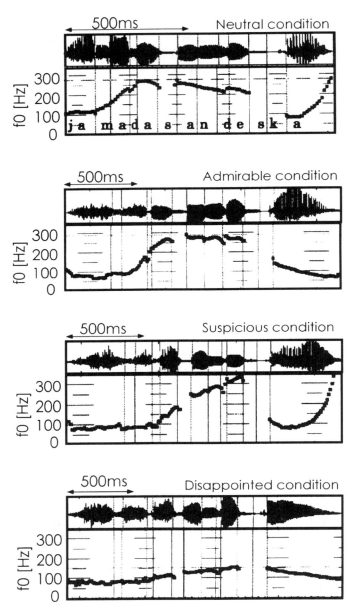

Figure 2.1 Effects of paralinguistic information on F0. Speech wave (upper panel) and F0 contour (lower panel) are shown for neutral (N), admirable (A), suspicious (S), and disappointed (D) conditions. These sentences are spoken by an identical male speaker. Vertical lines are boundaries between segments. (Reprinted with permission of Dr. Kikuo Maekawa.)

difficulties or to a person who does not speak your native language very well, you will speak slowly and articulate clearly. During conversation, you might raise the F0 and/or put stress on a segment of a word in order to make the topic of conversation clear or to answer a question that was asked.

Manipulation of the F0 pattern is also used to transfer the intention of spoken sentences. For example, the F0 of an interrogative sentence increases at the final part of the utterance. In such a sentence, the talker's intention and purpose can vary depending on the F0 pattern. Maekawa shows that the contrast of the F0 pattern of a sentence produces different paralinguistic effects [8]. Figure 2.1 shows the speech waves and corresponding F0 patterns during the utterance of the short Japanese question, /yamada saN desuka/ ("Are you Mr. Yamada?"), in neutral (N), admirable (A), suspicious (S), and disappointed (D) conditions. The situation of these conditions might be as follows. In the N condition, the speaker is simply asking to confirm the name of a person standing in front of him or her. In the A condition, the speaker expresses his or her gladness or respect in meeting (the famous or admired) Mr. Yamada. In the S condition, the speaker reveals doubt that the person in front of him or her is truly Mr. Yamada. In the D condition, the speaker responds to the unexpected and unwelcome appearance of Mr. Yamada.

As shown in the figure, the F0 pattern of the utterance in each

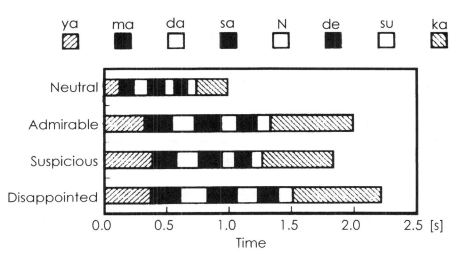

Figure 2.2 Effects of paralinguistic information on segmental duration. (Reprinted with permission of Dr. Kikuo Maekawa.)

particular condition is different from that of the neutral condition. The range of the F0 for the D condition is small and almost flat. In the S condition, the magnitude in which the F0 increases for the last vowel is larger than that of the N condition. In addition, the instance of the F0 increase for the initial word accent (observed during /m/ production in /yamada/) is delayed as compared with that of the N condition. This delayed F0 increase can also be observed in the A condition.

Figure 2.2 shows a comparison of segmental duration for the utterances in these four conditions. For all segments, the duration is prolonged as compared with the N condition. The magnitude of the lengthening for the first syllable /ya/ and the last syllable /ka/ is especially larger than that of the other segments.

In addition to the alteration of suprasegmental features (F0 and duration), segmental features are also modified to convey paralinguistic information. Figure 2.3 shows the distribution of the first two formant frequencies (F1 and F2) for the vowel /a/ in the final mora of the utterance. N represents the averaged plot for the N condition, while the measured formant frequencies for all ten trials are shown for each

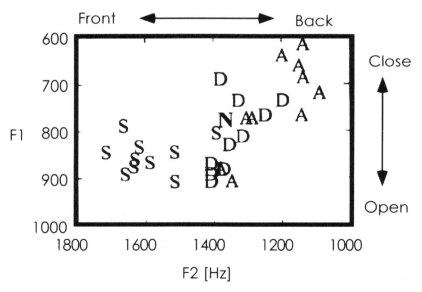

Figure 2.3 Effects of paralinguistic information on the formant frequencies for /a/ in the utterance-final mora /ka/. N indicates the average value of F1 and F2 for the natural condition of the utterance. F1 and F2 values for ten repetitions are shown for admirable (A), suspicious (S), and disappointed (D) conditions. (Reprinted with permission of Dr. Kikuo Maekawa.)

condition. The figure shows that the F1 and F2 for /a/ in the last mora for the S condition is higher than those for the N condition, and that the F1 and F2 of the same vowel for the A condition are lower than those for the N condition. In vowel production, F1 and F2 roughly correspond to vertical and horizontal (anterior-posterior) tongue positions, respectively. Therefore, the /a/ in the last mora for the S condition is supposedly produced with a larger jaw opening and an anterior tongue position. On the other hand, the /a/ should be produced with a posterior tongue position in the A condition. This kind of modification of articulation, which influences the segmental features, helps to convey paralinguistic information properly in accordance with the suprasegmental features.

2.2.3 Nonlinguistic Information

Speech contains nonlinguistic information in addition to linguistic and paralinguistic information. Nonlinguistic information concerns idiosyncratic factors and emotional states of the speaker [2]. Generally, the speaker cannot control these factors, although it is possible for a speaker to imitate some characteristics of these factors as actors do. Like paralinguistic information, nonlinguistic information can have continuous characteristics as well as discrete characteristics. It is also the suprasegmental features that play the most important role in conveying nonlinguistic information.

2.2.3.1 Idiosyncratic Factors

The acoustical characteristics of speech are influenced by idiosyncratic features such as age, gender, individual morphological characteristics, condition of health, and, in some cases, physical handicaps. For example, the gender of the adult speaker is identifiable by listening to spoken sounds. The one-octave difference in the F0 (approximately 100 Hz for males vs. 200 Hz for females) is an important cue for the gender identification of speakers. Since the F0 range for children is often similar to that for female adults, it does not always work to identify age. However, the difference in resonance characteristics (such as the F1–F2 space for vowels) due to the different scale of the vocal tract plays an important role in differentiating between adults and children.

The acoustical characteristics of speech also vary due to aging. Kasuya and Yang investigated the influence of aging on voice quality using the /e/ production of 854 males and 1,352 females whose ages ranged from twenties to eighties [9]. Their data showed that, as an effect of aging, the F0 for the female voice and that for the male voice converges in a similar F0 range (see Figure 2.4). In addition, their data shows an increase of hoarseness in the population due to aging. They concluded that the hoarseness of the voice is an important cue in identifying the age of speakers. These changes in voice quality can be mainly attributed to morphological changes in the vocal folds.

The identification of speakers is one of the most important roles of speech sounds. The individual morphological differences of the oral and nasal cavities are factors that can influence the acoustical characteristics of speakers' voices. For example, the size and shape of the oral cavity affects the range of articulatory movements resulting in a difference in acoustical characteristics in the same phoneme [10].

The acoustical characteristics of the nasal and the paranasal cavities play an important role in speaker identification and in an evaluation of the speaker's health. It is well known that voice quality changes when a person has a cold. This is partly because of the inflammation of the mucous membrane in the nasal cavity. The effects of the membrane in the nasal and paranasal cavity on the acoustical characteristics of

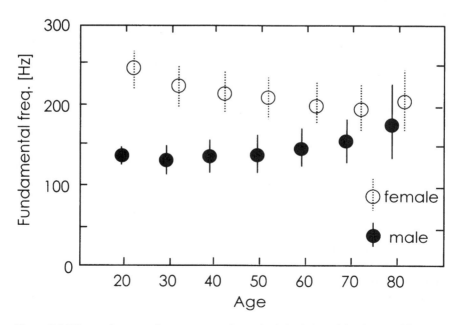

Figure 2.4 Effects of age on the average and standard deviation of fundamental frequency. (Reprinted with permission of the Acoustical Society of Japan and Dr. Hideki Kasuya.)

spoken sounds have been evaluated by Dang. A systematic difference in the pole and zero frequency of the vowel spectrum was observed between the absence and the existence of swelling in the mucous membrane [11].

More severe physical and physiological problems can cause a change in speech quality. Surgical operation of the speech organs directly affects several speech production processes. The surgical removal of the larynx and the tongue of a patient with cancer causes disorders in the generating source of speech sound (vocal fold vibration) and in the modifying transfer function of the oral cavity, respectively.

In addition to surgical changes of the speech organs, damage to the brain can cause problems in speech quality. For example, patients with dysarthria exhibit problems in both the segmental and suprasegmental features of speech sounds. The main symptom for patients with pseudobulber palsy or amyotrophic lateral sclerosis (ALS) is a weakness of the muscles. Therefore, the primary problem in speech production is the difficulty in realizing segmental (phonemic) features such as an incomplete closure for stop consonants; the secondary problem is a slowed speech rate. Patients with Parkinson's disease exhibit difficulty in keeping a consistent rhythm and range of intonation. Therefore, the low intelligibility of patients with severe dysarthria can be attributed to the co-occurrence of problems in the segmental and suprasegmental features.

2.2.3.2 Emotional Factors

It is well known that emotional conditions such as anger, sadness, and delight can have an effect on speech sound. These effects can be observed mainly in the suprasegmental features, such as the F0, intensity, and temporal characteristics of speech. Since muscle tension may be raised in some emotional conditions, there is a possibility that some segmental features are also influenced by the speaker's emotional condition.

Studies of signal processing of emotion in speech or speech processing with emotion have only recently been undertaken. Some studies have sought to extract the acoustical characteristics of speech containing emotional features and their effects on perception. However, the speech materials used in these studies have been imitated speech designed to replicate emotional conditions. In order to investigate the characteristics of emotional factors that usually cannot be controlled by the speaker, data must be collected from natural conversations, outside experimental environments.

2.2.4 Hierarchical Speech Production Processes

Figure 2.5 shows the hierarchy of speech production processes to convert linguistic, paralinguistic, and nonlinguistic information into segmental and suprasegmental features [2]. As shown in this figure, the process of spontaneous speech production is composed of four cascade functions: (1) message planning, (2) utterance planning, (3) motor command generation, and (4) speech sound production. Although these stages are represented by separate boxes, it is important to note that the processes at each stage can overlap in time and some processes can probably be performed simultaneously in spontaneous speech production.

At the first stage of message planning, like in the example of making the sentence "I love you" in section 2.2.1.1, linguistic units (words) and structures (word order) are selected on the basis of lexical and linguistic information following the rules and constraints of the grammar of spoken language. Then, at the utterance planning stage, the organization of the corresponding utterance is determined in terms of a sequence of phonemes as a process for generating linguistic information. In this stage, paralinguistic and nonlinguistic information are also processed to constrain the structure of the output speech sound by rules of suprasegmental (or prosodic) features (such as accentuation, phrasing, and pausing) for each spoken language. The coproduction of segmental and suprasegmental features at this stage was taken into account in a recently proposed speech production model [12].

At the command generation stage, commands to drive the movements of articulators and control the vocal folds are generated. In this process, linguistic, paralinguistic, and nonlinguistic information is implemented as a set of neuronal commands within the physiological constraints.

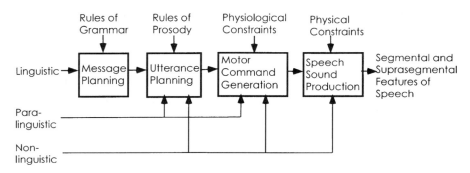

Figure 2.5 Hierchy of speech production processes.

The output of this stage is then transferred to the speech sound production stage in the form of the magnitude and timing of the muscle contractions to control each articulator's movement. In this last stage, the phonatory and articulatory systems are activated according to the neuronal commands generated in the previous stage. The phonatory system controls the lung and larynx for adjusting the amplitude and frequency of vocal fold vibration as well as voicing (the on-off of vibration). The articulatory system changes the position or shape of the articulators for varying the acoustical transfer function and generating fricative or burst noise as the sound source of consonants. The individual morphological difference of these systems and its effects on the dynamic characteristics (such as the range and the speed) of the articulatory movements are taken into account when representing nonlinguistic information such as idiosyncrasy.

As the final output of these processes, speech sounds are produced and transferred to the listener by sound (wave) propagation in the air. In addition to the listener (or perceiver) of the speech, the talker also monitors the speech sounds. During speech production, the talker is continuously listening to the speech sounds to evaluate the linguistic information (such as the sound quality of the phonemes), paralinguistic information (such as intonation), and nonlinguistic information (such as emotional and health conditions).

Kawahara evaluated the dynamic characteristics of a feedback loop of self-produced sound by the transformed auditory feedback (TAF) technique [13–15], where F0 perturbation applied to self-produced speech was fed back to the speaker. The results suggested that the delay between the presented and produced F0 changes was about 100 ms to 250 ms.

2.2.5 Summary

Linguistic, paralinguistic, and nonlinguistic information is transferred through the segmental and suprasegmental features. All of the information conveyed by speech sound issue from the stages of the speech production process is shown in Figure 2.5.

The techniques for extracting information from speech sounds are discussed elsewhere in this handbook. However, the information included in the speech sound has not been treated equally. Especially in the automatic speech recognition (ASR) technique, the focus of recent studies has been on the extraction of the sequence of phonemes as linguistic information. Paralinguistic and nonlinguistic information is treated as an item that should be excluded. Therefore, such information is not utilized effectively in increasing the recognition rate of the phoneme sequence and in detecting paralinguistic and nonlinguistic information.

However, it will be important to take the speaker's individual speech sound characteristics into account and detect the emotional state of the speaker for a future ASR system as a man-machine interface for use in our daily lives. As a basis for the future development of this ASR system, it will be important to accumulate knowledge of the human speech production system and the effects of idiosyncratic and emotional factors on it.

2.3 Physical and Physiological Processes in Speech Production

This section concerns the processes by which the speech organs convert speech motor commands to speech sounds, and their physical and physiological constraints. These processes are performed at the last stage of the speech production process as shown in Figure 2.5. The speech sound generation stage is divided into a *source* and a *filter* in the acoustic theory of speech production [16]. For vowel production, the source generator is the phonatory system, which serves as a transducer from the lung's airflow to form audible sound oscillation. The filter is an acoustic tube composed of the pharyngeal, oral, and nasal cavities that modify the quality of the sound generated by the phonatory system. For consonant production, noise generated in the middle of the vocal tract can be used as a sound source with or without vocal fold vibration. In the following subsections, the speech production mechanism is discussed with respect to the separate respiratory, laryngeal, and articulatory components.

2.3.1 Respiration System

2.3.1.1 Normal Breathing Without Speech Production

Respiration is an activity that brings oxygen into the body and expels carbon dioxide from it. Speech sound is generated using the expiratory airflow. For vowel and voiced consonant production, the airflow passing through the vocal folds is used to generate a quasi-periodic vibration as the source sound. For fricative and stop consonants, the airflow is also important for producing the fricative and burst noise and the pulse.

During normal breathing, inspiration is performed by the contraction of the external intercostal muscles to lift the ribs upward and outward in order to enlarge the lung volume with the help of the downward movement of the diaphragm. In expiration, on the other hand, the internal intercostal muscles contract to move the ribs downward and inward in order to decrease the lung volume with the help of abdominal muscle contraction and the elastic recoil of the lungs, the abdominal viscera, and the expanded rib cage. Breathing activity is achieved by the reciprocal control of activation and relaxation of these two groups of muscles for inspiration and expiration [17].

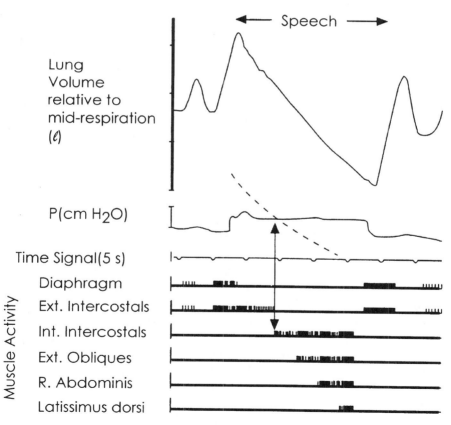

Figure 2.6 Measured lung volume and subglottal pressure and muscle activity. The dashed line indicates the relaxation pressure. (Reprinted with permission of the American Speech-Language-Hearing Association and Dr. Peter Ladefoged.)

2.3.1.2   Expiration in Speech Production

Speech production is performed during the expiration phase. This does not mean, however, that only the expiratory muscles are involved in speech production; both the muscles for expiration and inspiration are cooperatively activated during expiration to maintain stable subglottal pressure. Figure 2.6 shows the lung volume, subglottal pressure, and muscle activity as a subject counts from 1 to 32 at conversational volume [18]. As shown in this figure, the lung volume drops gradually while maintaining almost the same airflow rate, and the subglottal pressure is kept at almost the same level throughout the speech. However, to achieve this constant airflow, the muscles for inspiration are activated until the time that the relaxation pressure reaches the subglottal pressure level. Then the expiratory muscles are gradually activated to further reduce lung volume, extending the exhalation.

Recent kinematic and dynamic studies to monitor the anterior-posterior and vertical movements of the chest wall (the visible outer surface of the thorax) demonstrated that the abdominal muscle activity continues throughout the expiratory phase of speech breathing [19–20]. These cooperative activities of the muscles play an important role in controlling the transglottal pressure difference as well as the airflow rate, enabling the appropriate intensity of the speech sound.

2.3.2   Phonatory System

2.3.2.1   Framework of the Larynx

The larynx is located at the top end of the trachea in the path of the airflow from the lung to the vocal tract. Figure 2.7 shows the framework of the larynx, which is made up of four different cartilages: the epiglottis, thyroid, cricoid, and arytenoid cartilages. Although the epiglottis is fixed to the thyroid cartilage and the hyoid bone by ligaments, the other cartilages are connected by joints. These joints are the cricothyroid joint (between the cricoid and thyroid cartilages) and the cricoarytenoid joint (between the cricoid and arytenoid cartilages). In this framework, a pair of vocal folds (thyroarytenoid muscles) extend from the thyroid cartilage (below the thyroid notch) to the arytenoid cartilages. The main functions of the vocal folds in speech production are the on-off control of vibration and F0 control.

2.3.2.2   Abduction Versus Adduction

The vocal folds open and close the glottis in speech. The opening and closing movements are called abduction and adduction, respectively.

The abduction-adduction movements are controlled by the rotation of the pair of arytenoid cartilages on the cricoarytenoid joints. This control is important for the voiced versus voiceless distinction in consonant production. The mechanism underlying abduction vs. adduction of the vocal folds is the reciprocal activation of the abductor and adductor muscle groups. Figure 2.8 shows an example of the top view of the

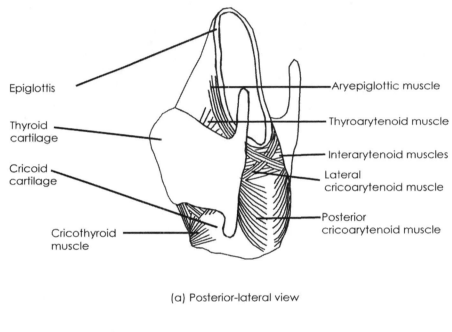

Epiglottis — Aryepiglottic muscle

Thyroid cartilage — Thyroarytenoid muscle

Cricoid cartilage — Interarytenoid muscles

Lateral cricoarytenoid muscle

Cricothyroid muscle — Posterior cricoarytenoid muscle

(a) Posterior-lateral view

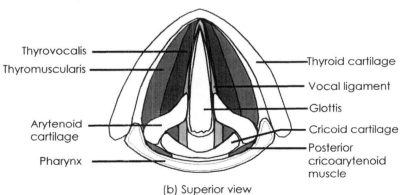

Thyrovocalis — Thyroid cartilage
Thyromuscularis — Vocal ligament

Glottis

Arytenoid cartilage — Cricoid cartilage

Posterior cricoarytenoid muscle

Pharynx —

(b) Superior view

Figure 2.7 Structure of the larynx.

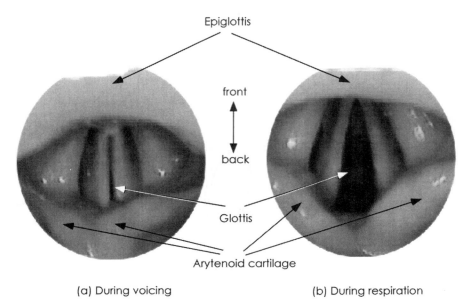

Epiglottis

front

back

Glottis

Arytenoid cartilage

(a) During voicing   (b) During respiration

Figure 2.8 Superior view of larynx.

larynx during phonation (adducted condition) and deep inspiration (abducted condition) observed using a tele-endoscope.

Figure 2.9 shows an example of ensemble-averaged EMG curves of the interarytenoid (INT, upper panel) and the posterior crico-arytenoid muscle (PCA, lower panel), for the test words abup (/əb'ʌp/) and apup (/əp'ʌp/) produced by an American English speaker [21]. The lineup point for averaging (zero on the abscissa) was identified by the voice offset of the stressed vowel in the second syllable. The horizontal bar graphs show the averaged segmental period to indicate the correspondence to the electromyographic (EMG) curves. The vocal folds are adducted during the production of the vowels (/ə/ and /ʌ/) and the voiced consonant /b/. However, the vocal folds are abducted for the word-final voiceless stop /p/ for both of the words and the intervocalic /p/. As shown in this figure, PCA activity is suppressed for the voiced portion (including the voiced stop /b/) when the vocal folds are adducted, whereas it increases for the production of the intervocalic /p/ as well as for the word-final /p/ when the vocal folds are abducted. On the other hand, INT shows a reciprocal pattern to that of PCA; the activity of INT increases for the voiced period when the vocal folds are adducted, whereas it decreases for the voiceless period for /p/ when the vocal folds are abducted.

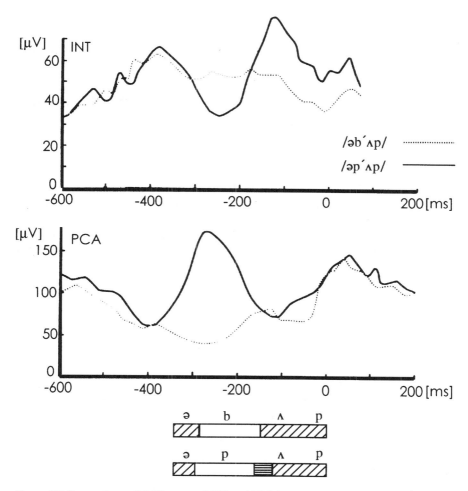

Figure 2.9 Comparison of EMG curve of INT and PCA between the utterances /əp'ʌp/ (solid line) and /əb'ʌp/ (dotted line). The origin of the time scale indicates the voice offset of the stressed vowel. Line bars at the bottom show the averaged period of each phoneme in the utterances. (Reprinted with permission of Blackwell Publishers and Dr. Hajime Hirose.)

2.3.2.3 F0 Control During Speech

The F0, or fundamental frequency, is the frequency of the vibration of the vocal folds. F0 control is achieved mainly by adjusting the effective mass and stiffness of the vocal fold. The cricothyroid muscle (CT) is the main contributor to raising the F0. The activity of the CT elevates the cricoid arch, thereby shortening the distance between the vocal processes of the arytenoid cartilages and the thyroid cartilages. This change elongates the vocal folds, resulting in a decrease in the effective mass and an increase in the stiffness of the vocal folds. Therefore, the activity of the CT increases the F0.

To achieve a word accent in Japanese or word stress in English, a unique relationship has been observed between CT activity and F0 change. Figure 2.10 shows the effects of CT activity on the F0 change during the production of an English accent [21].

The thyroarytenoid (TA) also participates in F0 control. The TA draws the arytenoid cartilages forward, which shortens and thickens the vocal folds, resulting in an increase of the effective mass of the vocal folds. The contribution of the TA is found to be reflected in the mode of the vocal fold vibration [22]; in the head register there is a marked decrease in TA activity compared with the chest register when raising the F0.

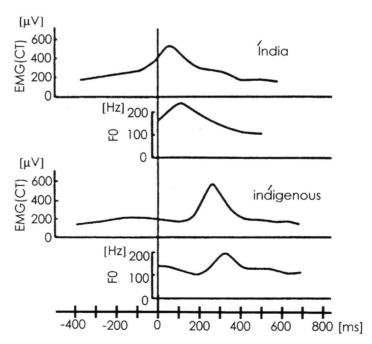

Figure 2.10 Examples of the time course between the averaged EMG activity of CT and F0 contours with different stress positions. (Reprinted with permission of Blackwell Publishers and Dr. Hajime Hirose)

2.3.3 Articulatory System

2.3.3.1 Morphology of Articulators

Figure 2.11 shows the articulators that are used for speech production. The hard palate and the upper teeth form a rigid framework for the articulators. The nasal cavity, which is located above the hard palate, is bounded by a rigid structure and plays an important role in generating nasal sounds. The connection between the oral cavity and the nasal cavity is controlled by the upward and downward movements of the soft palate. The jaw is a rigid structure that has some freedom of rotation and translation during speech [23]. The tongue is the most important organ for modifying the vocal tract area function when producing various phones (see sections 2.3.3.2 and 2.3.3.3). The shape of the tongue is modified by the intrinsic and extrinsic muscles. In addition, the position of the tongue is also changed by the jaw position since the tongue itself is placed on the oral floor muscles connected to the movable jaw. The upper and lower lips are also important articulators for speech production. Protrusion and rounding are factors in distinguishing vowels in some languages. The dynamic control of the aperture between the two lips or that between the lower lip and the upper teeth enable the production of some plosive and fricative consonants.

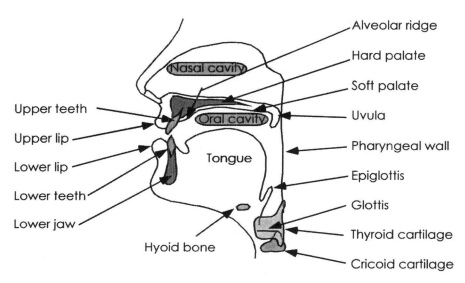

Figure 2.11 Articulators used for speech production.

2.3.3.2   Vowel Production

The principle of vowel production can be explained by the sound source and the modification of the sound quality by changes made to the vocal tract shape. The sound source is the periodical airflow generated by vocal fold vibration as described in Section 2.3.2. The spectral envelope of each vowel is determined by the transfer function of the vocal tract, which varies according to the position and shape of several articulators.

Figure 2.12 shows the midsagittal section of the articulators recorded by magnetic resonance imaging (MRI) during the production of five Japanese vowels. MRI is a powerful tool for visualizing the human body, but it is at a disadvantage in imaging teeth. To visualize the shape of the teeth, dental imaging plates with a contrast medium (Gadrinium) are used [24]. For this data acquisition, the subject maintained each vowel for 2.5 sec. Therefore, the shapes of the tongue may not be the same as those used during running speech. However, this figure gives the basic idea of the tongue position and shape for each vowel.

For [ɑ] production, the tongue is located at the posterior position, forming a constriction between the posterior part of the tongue and the pharyngeal wall. For [i] production, in contrast to [ɑ], the tongue body is located at the anterior and upper part of the oral cavity. The tongue dorsum is close to the highest point of the hard palate where the narrowest constriction is produced. For [e] production, the aperture of the lips is almost the same as that for [ɑ], and the position of the tongue is lower than that for [i], so that no characteristic constriction is made. For [o] and [ɯ] production, the aperture between the upper and lower lips is closer than that for [ɑ]. Between [o] and [ɯ], the distance between the upper and lower lips is closer for [ɯ] than for [o]. The tongue position for [o] is higher than that for [ɑ] so that a constriction is made between the posterior part of the tongue and the soft palate. The tongue position for [ɯ] is almost the same as that for [e] in this subject.

During running speech, the articulatory posture for vowels varies as an effect of the adjacent phones. This factor can be identified as one of the "coarticulation" effects. Figure 2.13 shows a comparison of the tongue shape for the /a/ production in the first syllable between /kaka/ and /tata/ utterances. These images are obtained by a dynamic MRI data acquisition method [25]. The difference in tongue position between the two contexts demonstrates the effects of the place of articulation in the adjacent consonants.

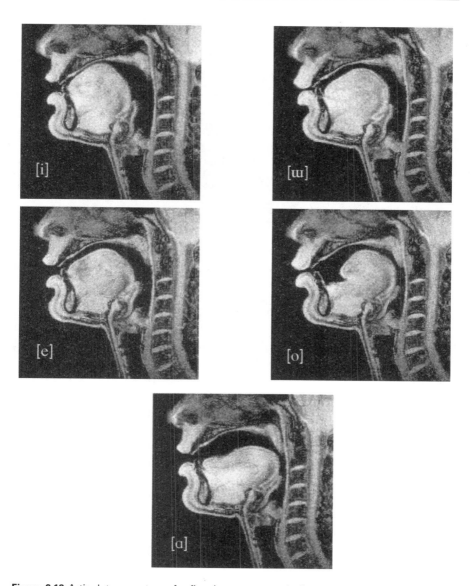

Figure 2.12 Articulatory postures for five Japanese vowels. Images are midsagittal sections taken by the MRI technique.

2.3.3.3   Consonant Production

Consonant production is categorized by the following three criteria: (1) manner of articulation, (2) place of articulation, and (3) voiceless versus voiced contrast. Table 2.2 shows how each consonant is classified by the

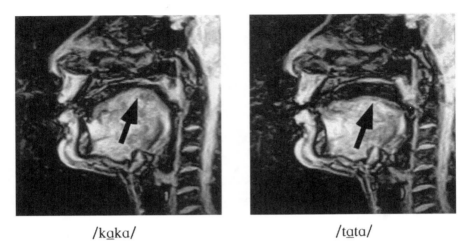

/kɑ̠kɑ/   /tɑ̠tɑ/

Figure 2.13 Effects of the consonant environment on tongue shape for /ɑ/. Tongue shape for /ɑ/ adjacent to /k/ shows the influence of the place of articulation for /k/.

IPA according to these criteria. In this table, each row and column indicates the manner and the place of articulation, respectively. In each cell, the consonant shown on the left side is a voiceless consonant and that on the right side is a voiced consonant.

A criterion of the voiceless versus voiced contrast is the existence or absence of vocal fold vibration during consonant production. For example, the contrast between [p] (voiceless) and [b] (voiced) can be distinguished by this criterion. The manner of articulation is categorized in terms of several criteria: The existence or absence of (1) closure or constriction to produce fricative noise or pulse sources in the vocal tract (stop or fricative), (2) the constriction without generating noise source (approximant), (3) a branch for sound propagation into nasal cavity (nasal), (4) a branch within the oral cavity (lateral approximant), and (5) dynamic movement for consonant production (trill, tap, and flap). The place of articulation is defined by the place where the closure or the constriction is located in the framework of the articulatory space.

Figure 2.14 shows articulatory postures for selected consonants in Japanese. This is the midsagittal section of the articulators during consonant production recorded by MRI using the same protocol as for the vowels shown in Figure 2.12. In order to get the articulatory posture, the subject produced /aCa/ (C: consonant) while sustaining the production of the intervocalic consonant during the data acquisition.

In Figure 2.14, the manner of articulation for the first row is "stop." In this manner, there is a closure to stop the airflow at the place of

Table 2.2

Characteristics of consonants defined by the International Phonetic Alphabet (IPA)

	labial		labiodental	dental		alveolar		postalveolar		palatal		velar		uvular		pharyngeal		glottal	
stop	p	b				t	d			c	ɟ	k	g	q	ɢ			ʔ	
nasal		m	ɱ				n				ɲ		ŋ		ɴ				
trill		ʙ					r								ʀ				
tap, flap							ɾ				ɽ								
fricative	ɸ	β	f v	θ	ð	s	z	ʃ	ʒ	ç	ʝ	x	ɣ	χ	ʁ	ħ	ʕ	h	ɦ
approximant			ʋ				ɹ				j		ɰ						
lateral approximant							l				ʎ		ʟ						

In each cell, the consonant shown on the left side is a voiceless consonant and that on the right side is a voiced consonant.

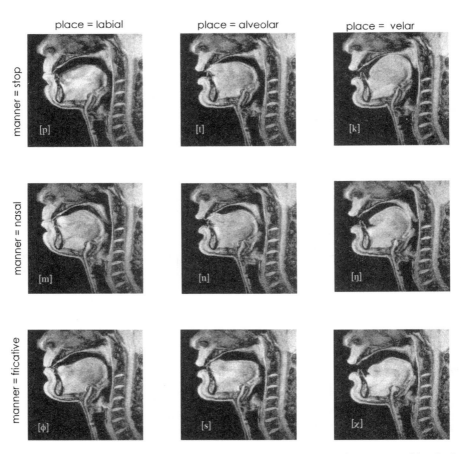

Figure 2.14 Articulatory postures for selected Japanese consonants. Images are midsagittal sections taken by the MRI technique.

articulation in the oral cavity. The places for these examples are the lips for [p], the alveolar ridge for [t], and the velum for [k]. A tight contact between the tongue and the corresponding places can be observed in the figures in this row. In the second row, articulatory postures for nasal consonants are aligned in the same order. The closure can be observed at the lips for [m], at the alveolar ridge for [n], and at the velum for [N]. The main difference between stop and nasal consonants is the position of the velum: For stop consonants, the velum contacts with the pharyngeal wall tightly so as not to let the airflow into the nasal cavity, while for nasal consonants, the velum is lowered to allow the airflow to branch into the nasal cavity. In the bottom row, the articulatory postures for fricative consonants are aligned. The velum position is high so as to close the channel to the nasal cavity. However, a thin air channel is observed at the place of articulation for each consonant: at the lips for [F], at the alveolar ridge for [s], and at the uvula for [X].

In addition to the MRI technique, there are several methods for visualizing articulatory movements, especially for distinguishing tongue properties. One of the conventional techniques for monitoring tongue activity is electropalatography (EPG) (e.g., [26]). EPG shows the tongue-palate contact pattern during dynamic speech production. In order to visualize the contact pattern in a 3-D view, a computer graphic technique was applied to the EPG data during Japanese fricative consonants [s] and

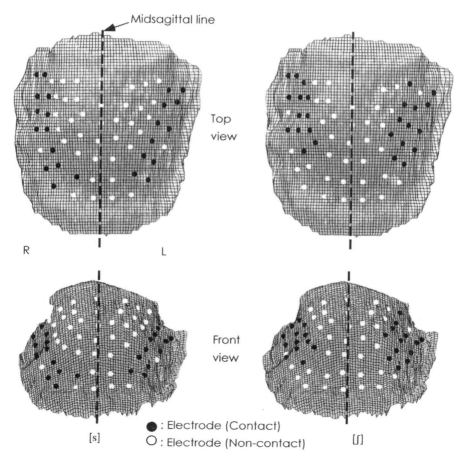

Figure 2.15 Three-dimensional representation of electropalatogram for [s] and [ʃ] production.

[ʃ] [27]. Figure 2.15 shows the contrast of the tongue-palate contact pattern between [s] in [asa] and [ʃ] in [aʃa] by the same Japanese speaker as in Figures 2.12–2.14. The contrast of the contact pattern between [s] and [ʃ] can be observed in the different height of the narrowest constriction.

Recently developed sensor technology enables us to obtain not only the contact pattern information but also the contact pressure information of the tongue to the hard palate [28]. Figure 2.16(a) shows the alignment of the sheet sensors on an artificial palate. Figure 2.16(b) shows a comparison of the highest tongue pressure measured at the alveolar ridge between [t] in [ata] and [d] in [ada] for nine Japanese subjects. For all the subjects, this comparison shows consistently higher pressure for [t] than for [d] [29]. The difference in the contact pressure can be attributed to the different intraoral pressure during the occlusion period between voiceless and voiced stop consonants [30].

With regard to the different intraoral pressure and tongue contact pressure between voiceless and voiced stop consonants, the dynamic MRI technique was applied to investigate laryngeal movements during [t] and [d] production [31]. Figure 2.17 shows a comparison of larynx height between the production of meaningful Japanese words /tata/ ("many") and /tada/ ("merely"), with high-low pitch accent patterns. In this dynamic MRI technique, a midsagittally-oriented moving image sequence (40 frames/second; FOV = 25 mm × 25 mm) is obtained from the scan data for 128 repetitions of each word. As shown in this figure, the height of the larynx during the occlusive period before the release of the voiced stop /d/ is lower than that of the voiceless stop /t/. In addition, an abrupt lowering movement of the larynx can be observed near the release of the voiced stop /d/. This lowering of the larynx for the voiced obstruent can help the expansion of the vocal tract to maintain glottal airflow during voiced stop production. Therefore, the intraoral pressure during the closure period until the release for the voiced stop /d/ should be lower than that for the voiceless stop /t/. This pressure difference can be linked to the tongue contact pressure control as observed by pressure palatography.

2.3.4 Summary

In this section, basic speech production processes were discussed. They are divided into three main systems: the subglottal respiration system

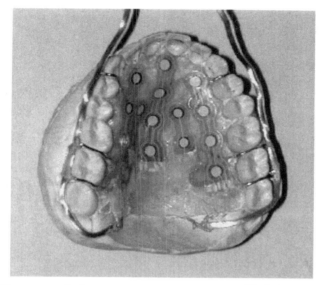

(a) Placement of pressure sensors on an artificial palate.

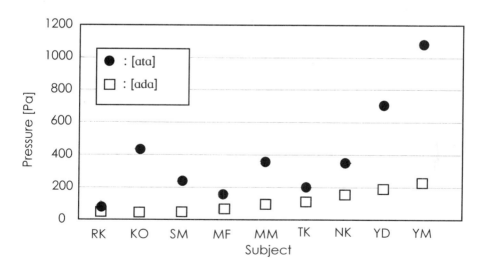

(b) Measured pressure at the period of stop consonants for 9 subjects.

Figure 2.16 Comparison of tongue pressure of alveolar ridge (center) between [t] and [d]. (Reprinted with permission of Pierre Fauchard Academy and Dr. Masahiko Wakumoto.)

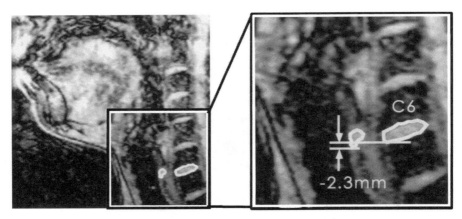

(a) Measurement of the vertical displacement of the larynx

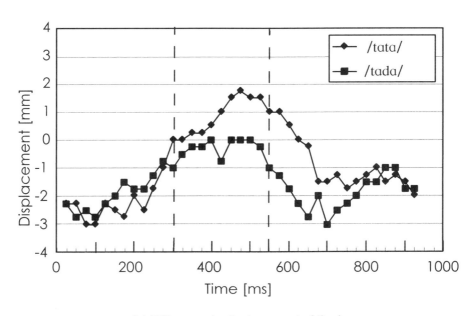

(b) Difference in displacement of the larynx
between /tɑtɑ/ and /tɑdɑ/ productions

Figure 2.17 Difference in the larynx height during the production of plosive consonants due to
voiceless and voiced distinction.

(section 2.3.1), the phonatory system (section 2.3.2), and the articulatory system (section 2.3.3). In each section, the basic ideas of the functions for each system have been presented. In addition, there are some new findings observed by using newly developed techniques for observing speech production processes. It is important to note that the functions as presented in each section must be linked with each other to produce phonemes, words, and sentence sequences. Section 2.4 explains how these systems and their linkages are modeled in speech production studies.

2.4 Models and Theories of Speech Production

Several models have been developed to simulate and investigate the behavior mechanism of the larynx and articulators and their interaction. The focus of this chapter is to introduce the method of modeling and the mechanisms to be clarified. As an application of this modeling, speech synthesis based on the speech production mechanism will also be presented.

2.4.1 Laryngeal System

Models have been proposed to demonstrate several aspects of laryngeal activities, including the vocal fold vibration mechanism, a description of the glottal airflow waveform, and functional control of the F0 during speech production.

2.4.1.1 Vocal Fold Vibration

Two models were proposed to simulate the self-oscillation mechanism of the vocal folds. The one-mass model [32] revealed the role of Bernoulli's law in self-oscillation and the effects of acoustic coupling on the oscillation. The two-mass model [33] represents the upper and lower portions of each vocal fold as two separated masses connected by elastic components (Figure 2.18). This model demonstrates additional important characteristics of vocal fold vibration, such as the phase difference between the upper and lower parts of the vocal folds resulting in a vertical wave in the vocal fold vibration. In order to explain the finer aspects of laryngeal kinematics, models with multiple components (e.g., the 16-mass model [34]) have been proposed.

The shape of the glottal airflow has been modeled as an LF-model based on the inverse filtering technique at the Royal Institute of

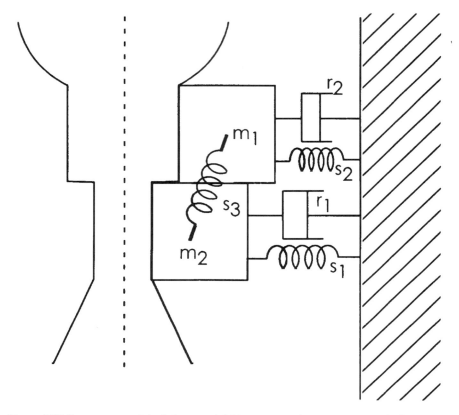

Figure 2.18 Two-mass model of the vocal folds; m, s, and r represent equivalent mass, stiffness, and viscous resistance, respectively.

Technology (KTH) [35]. The advantage of this model is that a small number of parameters can explain the characteristics of the glottal airflow. In addition, this model has been continuously revised in light of the experimental data to make it more suitable for rule-based speech analysis and synthesis [36].

2.4.1.2   F0 Control in Running Speech Production

Fujisaki proposed a functional model explaining various aspects of F0 patterns during speech (e.g., [2], [37]). In this model, as shown in Figure 2.19, the baseline of F0 function is defined by the impulse response of the second order system, which explains the gradual declination of the F0 in a phrase. Accent or stress is represented by the appropriately timed step responses which are added to the

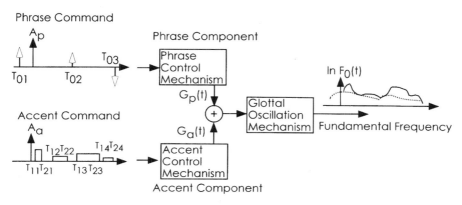

Figure 2.19 Functional model for generating F0 contours of sentences [2]. (Reprinted with permission of Springer-Verlag.)

previously defined baseline. This model has been applied to the analysis of F0 patterns for several languages, and the applicability and robustness of the model have been validated (in English [38], German [39], and Chinese [40]). In addition to this functional model approach, some other approaches have been applied in an attempt to understand the underlying mechanism of F0 pattern control during speech [41].

2.4.1.3 Vertical Movements of the Larynx

As presented in Section 2.3.3.3, vertical movements of the larynx play a crucial role in distinguishing voiced and voiceless consonants during running speech [31]. In addition, recent investigation of laryngeal movements during vowel production using the MRI technique revealed the importance of the vertical movements of the larynx for F0 control. Hirai et al. [42] and Honda [43] proposed a new F0 control mechanism (especially for lowering the F0) during vowel production. In their experiment, the subjects produced a sustained vowel /a/ with an F0 from 262 Hz to 87 Hz. The subjects showed the following changes during a descending musical scale: (1) vertical laryngeal movement takes place along the curvature of the cervical spine; (2) the posterior plate of the cricoid cartilage moves down along this curvature; and (3) the vertical larynx movement results in the rotation of the cricoid cartilage. This rotation of the cricoid cartilage causes the angle between the cricoid and thyroid cartilages to change. Finally, this change results in a shortening of the vocal folds and a lowering of the F0 (Figure 2.20). Since there are individual differences in the curvature of the cervical

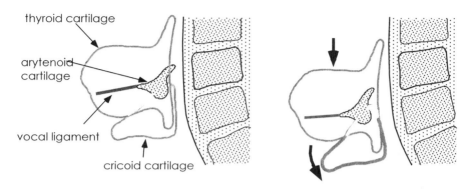

thyroid cartilage

arytenoid cartilage

vocal ligament

cricoid cartilage

Figure 2.20 A mechanism of F0 control by vertical movements of the larynx [43]. (Reprinted with permission of Dr. Kiyoshi Honda.)

spine, it is possible that not all speakers use this mechanism. However, this mechanism can explain why vertical movements occur along with F0 changes.

2.4.2 Dynamic Characteristics of Articulators

The dynamic characteristics of the articulators and their interaction play important roles in transmitting phonetic information. In the following subsections, unit models that are designed to simulate the dynamic characteristics of individual articulators (tongue, lip, and jaw) and composite models for understanding the overall characteristics of the articulatory system will be presented. In addition, trials of speech synthesis using these articulatory models will be discussed.

2.4.2.1 Models of Individual Articulators

Among the articulators, the tongue plays the most important role during speech production. A recent trend in articulatory modeling is to establish a three-dimensional (3-D) dynamic model of the articulators. Wilhelms-Tricarico proposed a dynamic model of the tongue using the finite element method based on the morphology of the tongue [44, 45] (Figure 2.21). This model is composed of 42 large-strain finite elements that incorporate fiber directions, a nonlinear muscle model, inertial properties, and a volume constancy constraint for simulating realistic dynamical behavior of the tongue. It considers the styloglossus (the two upward projections in the figure), the hyoglossus, and the genioglossus (anterior and posterior). The feasibility of this method has

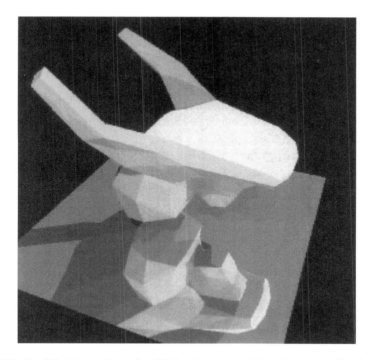

Figure 2.21 Simplified three-dimensional finite element model of the tongue. (Reprinted with permission of Dr. Reiner Wilhelms-Tricarico.)

been demonstrated by the simulation of typical tongue movements driven by EMG input obtained from previously reported data [46].

Lip and jaw movements are also important articulators for speech production. There are some 3-D dynamic models of these movements for analysis and synthesis during speech. For lip movements, 3-D analysis was conducted from simultaneously recorded video frames of frontal and sagittal views [47]. These data were used to model lip movements during speech and to create a text-to-facial movement synthesizer at the Institut de la Communication Parlée (ICP) [48].

As for the 3-D jaw movement analysis, for example, Vatikiotis-Bateson and Ostry used an optical device with an infrared LED [23]. They showed that the principle components of jaw motion were within the midsagittal plane. Generally, the relation between sagittal plane rotation and horizontal translation is linear. However, instances of pure rotation and pure translation are observed, supporting the previous claim that jaw rotation and translation are independently controlled [49]. They also showed quantitatively that the magnitude of displacement and rotation in the other plane is smaller than that in midsagittal plane.

2.4.2.2 Articulatory Models of Speech Production

Several models have been proposed to explain the activity of articulators in speech production. They have been called articulatory models [50–52]. In these models, parameters describing the vocal tract configuration, which are usually called articulatory parameters, are obtained by statistical analysis from the vocal tract shape obtained from midsagittal X-ray data during speech. Usually, articulatory parameters represent the position of the tongue (two to three parameters),

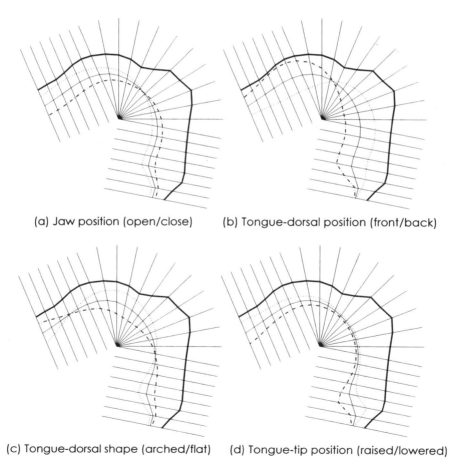

(a) Jaw position (open/close)   (b) Tongue-dorsal position (front/back)

(c) Tongue-dorsal shape (arched/flat)   (d) Tongue-tip position (raised/lowered)

Figure 2.22 Characteristics of the four articulatory parameters upon the tongue shape in Maeda's articulatory model. (Reprinted with permission of Dr. Shinji Maeda.)

the aperture of the lips (one or two parameters), the jaw opening (one parameter) and the velum height (one parameter). Figure 2.22 shows the characteristics of the parameters that were used for articulation of the tongue in Maeda's articulatory model [52]. Figure 2.23 shows an example of the potential of this model to describe the characteristics of articulation. This figure shows the ability to separate the components of the jaw and tongue activities to reveal the tongue height for a certain phoneme. As seen in this figure, the jaw and tongue are activated in a compensatory manner to produce the phoneme.

With regard to articulatory models, Yehia demonstrated an effective method for estimating the vocal tract area function from the acoustical characteristics (formant frequencies) of spoken sound [53]. He used the morphological restriction of the human vocal tract shape based on X-ray film data and simultaneously recorded spoken sound. He showed successful mapping from the sound to the vocal tract shape (or area function) for vowel sounds in French sentences.

In addition to the relationship between the sound and the articulator's activity, the relationship between the articulatory movements and the muscle activity used to make these movements has also been investigated. Honda discussed the relationship between the contribution of the muscles used to position the tongue for each vowel and the resulting acoustical characteristics [54]. Figure 2.24(a) shows a plot of the activity levels of muscles during the production

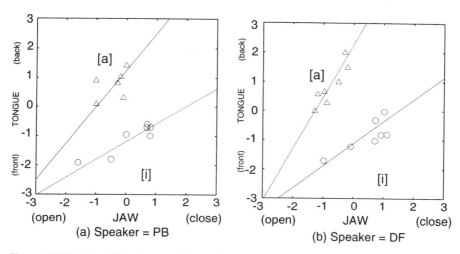

Figure 2.23 Relationships between "target" positions of the jaw opening and the tongue-dorsal position, for the two unrounded vowels [i] and [a]. (Reprinted with permission of Dr. Shinji Maeda.)

of the four vowels /i/, /u/, /a/, and /æ/ in /əpVp/ utterances in English. The space of this chart is defined by the normalized activities of two antagonistic pairs of major muscles, the posterior genioglossus (GGp)—hyoglossus (HG) and styloglossus (SG)—and anterior genioglossus (GGa), which are used in vowel production. The trajectory in this chart shows the contribution of these muscles. For example, HG plays the main role in articulating /a/ while the contraction of both GGp and GGa characterize the tongue shape for /i/. In addition, Honda argued that a similar shape (quadrilateral) of the vowel space can be obtained from the F1–F2 space (Figure 2.24(b)). With some supporting simulation study using an articulatory model for generating formant patterns based on the EMG data [55], he speculated the existence of a simple and robust system for mapping the muscle activities to the acoustic patterns.

2.4.3 Summary

Models of the speech production mechanisms have been presented. The modeling of speech production started from the basic kinematic mechanisms of individual articulators. However, recent modeling trials are investigating the underlying mechanisms in the control of articulatory movements performed by the central nervous system.

Recent progress in speech production modeling has enabled speech synthesis based on articulatory movement data [56, 57]. There

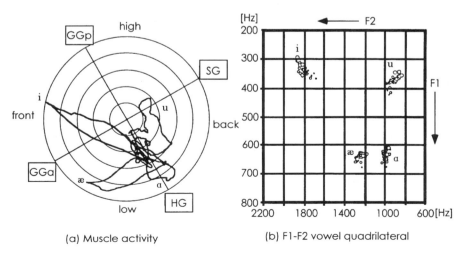

(a) Muscle activity   (b) F1-F2 vowel quadrilateral

Figure 2.24 (a) Muscle activity levels for the production of the vowels /i/, /u/, /a/, and /æ/. (b) The F1–F2 vowel quadrilateral. (Reprinted with permission of Dr. Kiyoshi Honda.)

is a possibility that speech production modeling can be applied not only to speech synthesis technology but also to speech recognition methodology. How can a prototype of the articulatory system be modeled as a speech production system? How can individual morphological differences in articulators be modeled? How can these modeling studies be combined? These questions will be main themes of discussion as further progress is made in speech modeling and its application to speech synthesis and recognition technology.

2.5 Conclusion

This chapter has dealt with the function of speech for transferring various kinds of information (linguistic, paralinguistic, and nonlinguistic information), the mechanism of speech production for implementing the information into segmental and suprasegmental features, and finally some examples of modeling for the speech production mechanisms. These topics should provide important information for understanding basic speech production. Originally, the objectives for the analysis of speech production mechanisms were to understand the physiological mechanisms. The findings obtained in such a basic study could be used for speech processing technology, such as the automatic speech recognition (ASR) of large-scale vocabulary and text-to-speech synthesis. However, in these technologies, findings from basic speech production studies have not been used effectively. Rather, enormous energy has been used for producing efficient parameters of the time-varying acoustic characteristics of the speech signal and for correlating acoustical features and linguistic information.

As described in Section 2.2, linguistic information is contained mainly in segmental features, and paralinguistic information and non-linguistic information are provided mainly by suprasegmental features. Therefore, the information contained in speech sound should be treated as inseparable materials. For example, if the information contained in suprasegmental features is properly processed for a speech under-standing system, linguistic information that might be influenced by the speaker's intention and emotional condition could be more effectively extracted. For text-to-speech synthesis technology, an analysis of speaker-specific characteristics due to individual differences in morphology and/or to differences in articulatory patterns can provide better quality speech sound with natural individual characteristics. As presented in Section 2.4, trials are being conducted to achieve text-to-speech synthesis by using the morphological and physiological characteristics of speech production organs, which were described in Section 2.3. It might be possible to apply such knowledge of speech production mechanisms to speech recognition or some other speech information processing in the near future.

Recently, the study of speech production has been recognized as a major field in multimodal (audiovisual) communication [1]. How does the morphological construction of the speech organs and face structure constrain the links between articulatory and facial movements during speech? How does the multimodal stimulus presentation (speech sound presented as the sound stimulus, and facial movements presented as a visual stimulus) affect the perception of linguistic, paralinguistic, and nonlinguistic information? Studies aimed at answering these questions are a recent trend, and their results should provide important information for studies on brain activity, which is currently recognized as one of the most important fields of scientific pursuit, and for future multimedia application.

Acknowledgment

I would like to thank Dr. Masahiko Wakumoto, Dr. Reiner Wilhelms-Tricarico, Dr. Shinji Maeda, and Dr. Kiyoshi Honda for providing me with the original versions of some of the figures used in this chapter. I would also like to thank Dr. Kikuo Maekawa, Dr. Peter Ladefoged, the Acoustical Society of Japan, American Speech-Language-Hearing Association, Blackwell Publishers, Springer-Verlag, and the Pierre Fauchard Academy for giving me their permission to use other illustrations that appear here. I would also like to thank the person who was the subject of the larynx image used in Figure 2.8 for her patience. And I would like to express my gratitude to Mr. Yasuhiro Shimada, Mr. Ichiro Fujimoto, and Mr. Yuji Nakamura at Takanohara Chuo Hospital for their technical support in taking the MRI images for vowel and consonant production (Figures 2.12 and 2.14). Lastly, I would like to offer special thanks to Dr. Kiyoshi Honda for his valuable assistance during the preparation of this chapter.

References

[1]   Rubin, P. and E. Vatikiotis-Bateson, Editorial, *Speech Communication*, Vol. 26, 1998, pp. 1–4. Special issue on auditory-visual speech processing.

[2]　Fujisaki, H., "Prosody, Models and Spontaneous Speech," in Y. Sagisaka, N. Campbell, and N. Higuchi, eds., *Computing Prosody*, New York: Springer, pp. 27–42.

[3]　Denes, P. B., and E. N. Pinson, *The Speech Chain: The Physics and Biology of Spoken Language*, New York: Anchor Press, 1973.

[4]　Fry, D. B., "Duration and Intensity as Physical Correlates of Linguistic Stress," *J. Acoust. Soc. Am.*, Vol. 27, 1955, pp. 65–768.

[5]　Hirose, K., H. Fujisaki, and M. Sugito, "Word Accent in Japanese and English: A Comparative Study of Acoustic Characteristics in Disyllabic Words." *Annual Bulletin of Research Institute of Logopedics and Phoniatrics*, No. 12, University of Tokyo, 1978, pp. 141–148.

[6]　Han, M. S., "Acoustic Manifestations of Mora Timing in Japanese," *J. Acoust. Soc. Am.*, Vol. 96, 1994, pp. 73–82.

[7]　Kubozono, H. "Mora and Syllable," in N. Tsujimura, ed., *The Handbook of Japanese Linguistics*, Oxford: Blackwell, forthcoming.

[8]　Maekawa, K., "ONSEI-GAKU" (Phonetics), in Y. Ohtsu, T. Gunji, Y. Takubo, M. Nagao, K. Hashida, T. Masuoka, and Y. Matsumoto, eds., *GENGO NO KAGAKU* (Linguistic Science), Tokyo: Iwanami Shoten, 1998, pp. 1–52 (in Japanese).

[9]　Kasuya, H., and C. Yang, "Voice Quality Associated with Voice Source," *J. Acoust. Soc. Jpn.*, Vol. 51, 1995, pp. 869–875.

[10]　Honda, K., "Trends of Articulatory Studies Based on the X-ray Microbeam System," *Journal of the Phonetic Society of Japan*, Vol. 2, 1998, pp. 8–18 (in Japanese).

[11]　Dang, J., and K. Honda, "Morphological and Acoustical Analysis of the Nasal and the Paranasal Cavities," *J. Acoust. Soc. Am.*, Vol. 96, 1984, pp. 2088–2100.

[12]　Fujimura, O. "Phonology and Phonetics—A Syllable-Based Model of Articulatory Organization," *J. Acoust. Soc. Jpn.* (E), Vol. 13, 1992, pp. 39–48.

[13]　Kawahara, H., "On Interactions Between Speech Production and Perception Using Transformed Auditory Feedback—Effects of Pitch Perturbation by Pseudo-Random Signal—," Research Report of the Committee of Hearing Researches, Acoustical Society of Japan, H-93–24, 1993 (in Japanese).

[14]　Kawahara, H., "Effects of Auditory Feedback Conditions on Fundamental Frequency Fluctuations," Technical Report of the Institute of Electronics, Information and Communication Engineers, EA94–48, 1994 (in Japanese).

[15]　Kawahara, H., "Effects of Natural Auditory Feedback on Fundamental Frequency Control," *Proc. Int. Conf. Spoken Language Processing*, Vol. 94, No. 3, 1994, pp. 1399–1402.

[16]　Fant, G., *Theory of Speech Production*, The Hague: Mouton, 1960.

[17]　Titze, I., *Principles of Voice Production*, Englewood Cliffs: N.J.: Prentice Hall, pp. 64–65.

[18]　Draper, M. H., P. Ladefoged, and D. Whitteridge, "Respiratory Muscles in Speech," *J. Speech and Hearing Res.*, Vol. 2, 1959, pp. 16–27.

[19]　Hixon, T., M. Goldman, and J. Mead, "Kinematics of the Chest Wall During Speech Production: Volume Displacements of the Rib Cage, Abdomen, and Lung," *J. Speech and Hearing Res.*, Vol. 16, 1973, pp. 78–115.

[20]   Hixon, T., J. Mead, and M. Goldman, "Dynamics of the Chest Wall During Speech Production: Function of the Thorax, Rib Cage, Diaphragm, and Abdomen," *J. Speech and Hearing Res.*, Vol. 19, 1976, pp. 279–356.

[21]   Hirose, H., "Investigating the Physiology of Laryngeal Structure," in W. J. Hardcastle, and J. Laver, eds., *The Handbook of Phonetic Sciences*, Oxford: Blackwell, 1997, pp. 116–136.

[22]   Hirano, M., W. Vennard, and J. Ohara, "Regulation of Register, Pitch and Intensity of Voice: An Electromyographic Investigation of Intrinsic Laryngeal Muscles," *Folia Phoniatrica*, Vol. 22, 1970, pp. 1–20.

[23]   Vatikiotis-Bateson, E., and D. J. Ostry, "An Analysis of the Dimensionality of Jaw Motion in Speech," *J. of Phonetics*, Vol. 23, 1995, pp. 101–117.

[24]   Wakumoto, M., et al., "Visualization of Dental Crown Shape in an MRI-Based Speech Production Study," *Int. J. of Oral and Maxillofacial Surgery*, Vol. 26, 1997, pp. 189–190.

[25]   Masaki, S., et al., "MRI-Based Speech Production Study Using a Synchronized Sampling Method," *J. Acoust. Soc. Jpn.* (E), Vol. 20, 1999, pp. 377–381.

[26]   Hardcastle, W. J. "The Use of Electropalatography in Phonetic Research," *Phonetica*, Vol. 25, 1972, pp. 197–215.

[27]   Wakumoto, M., and S. Masaki, "Three-Dimensional Visualization of Electropalatographic Data," *J. Acoust. Soc. Jpn.* (E), Vol. 20, 1999, pp. 137–141.

[28]   Wakumoto, M., S. Masaki, K. Honda, and T. Ohue, "A Pressure Sensitive Palatography: Application of New Pressure Sensitive Sheet for Measuring Tongue-Palatal Contact Pressure," *Proc. 5th Int. Conf. Spoken Language Processing*, Vol. 7, 1998, pp. 3151–3154.

[29]   Wakumoto, M., "Analysis of Articulation Using Tongue-Palatal Contact Pressure," *Journal of Pierre Fauchard Academy*, 1999, pp. 49–53 (in Japanese).

[30]   Wakumoto, M., S. Masaki, and K. Honda, "Pressure Sensitive Palatography for Measuring the Lingual-Palatal Contact Pressure Using a Pressure Sensitive Film," *Proc. Ann. Mtg. Logopedics Phoniatrics of Japan*, 1999, Vol. 154 (in Japanese).

[31]   Masaki, S., et al., "Comparison of Laryngeal/Velar Movements Between Voiced and Voiceless Consonants: Observation by MRI Movie," *Proc. Autumn Mtg. Acoust. Soc. Jpn.*, Vol. I, 1998, pp. 267–268 (in Japanese).

[32]   Flanagan, J. L., and L. Landgraf, "Self-Oscillating Source for Vocal Tract Synthesizers," *IEEE Trans. on Audio Electroacoustics*, Vol. AU-16, (1968), pp. 57–64.

[33]   Ishizaka, K., and J. L. Flanagan, "Synthesis of Voiced Sounds from a Two-Mass Model of the Vocal Cords," *Bell System Technical Journal*, Vol. 51, 1972, pp. 1233–1268.

[34]   Titze, I., "The Human Vocal Cords: A Mathematical Model," *Phonetica*, Vol. 28, 1973, pp. 129–170.

[35]   Fant, G., J. Liljencrants, and Q. Lin, "A Four-Parameter Model of Glottal Flow," Vol. STL-QPSR, 1985, Royal Institute of Technology, pp. 1–13.

[36]   Fant, G. "Frequency Domain Analysis of Glottal Flow: The LF-Model Revisited," in S. Kiritani, H. Hirose, and H. Fujisaki, eds., *Speech Production and Language*. Berlin: Mouton de Gruyter, 1997, pp. 77–110.

[37] Fujisaki, H., and K. Hirose, "Analysis of Voice Fundamental Frequency Contours for Declarative Sentences of Japanese," *J. Acoust. Soc. Jpn.* (E), Vol. 5, 1984, pp. 233–242.

[38] Fujisaki, H., S. Ohno, T. Ueno, and O. Tomita, "Analysis and Modeling of Fundamental Frequency Contours of English sentences," *Proc. Autumn Mtg. Acoust. Soc. Jpn.*, 1995, pp. 207–208 (in Japanese).

[39] Mixdorff, H., and H. Fujisaki, "Analysis of F0 Contours of German Utterances Using a Model of the Production Process," *Proc. Spring Mtg. Acoust. Soc. Jpn.*, 1994, pp. 287–288 (in Japanese).

[40] Hirose, K., H. Lei, and H. Fujisaki, "Analysis and Formulation of Prosodic Features of Speech in Standard Chinese Based on a Model of Generating Fundamental Frequency Contours," *J. Acoust. Soc. Jpn.*, Vol. 50, 1994, pp. 177–187 (in Japanese).

[41] Pierrehumbert, J., and M. Beckman, *Japanese Tone Structure*, Cambridge, Mass.: MIT Press, 1988.

[42] Hirai, H., K. Honda, I. Fujimoto, and Y. Shimada, "Analysis of Magnetic Resonance Images on the Physiological Mechanisms of Fundamental Frequency Control," *J. Acoust. Soc. Jpn.*, Vol. 50, 1994, pp. 296–304 (in Japanese).

[43] Honda, K., "Form and Function: Another View of Speech Production," ATR Technical Report, TR-H-235, 1997.

[44] Wilhelms-Tricarico, R., "Physiological Modeling of Speech Production: Methods for Modeling Soft-Tissue Articulators," *J. Acoust. Soc. Am.*, Vol. 97, 1995, pp. 3085–3098.

[45] Wilhelms-Tricarico, R., "A Biomechanical and Physiologically-Based Vocal Tract Model and Its Control," *Journal of Phonetics*, Vol. 24, 1996, pp. 23–38.

[46] Baer, T., P. Alfonso, and K. Honda, "Electromyography of the Tongue Muscles During Vowels in /pVp/ Environment," *Annual Bulletin of Research Institute of Logopedics and Phoniatrics*, University of Tokyo, Vol. 22, 1988, pp. 7–19.

[47] Guiard-Marigny, T., A. Adjoudani, and C. Benoît, "3D Models of the Lips and Jaw for Visual Speech Synthesis," in J. P. H. van Snatesn, R. Sproat, J. Olive, and J. Hirshberg, eds., *Process in Speech Synthesis*, New York: Springer, 1996, pp. 247–258.

[48] Benoît, C., and B. Le Goff, "Audio-Visual Speech Synthesis from French Text: Eight Years of Models, Designs and Evaluation at the ICP," *Speech Communication*, Vol. 26, 1998, pp. 117–129.

[49] Flanagan, J. R., D. J. Ostry, and A. G. Feldman, "Rotation and Translation of the Jaw during Speech," *J. of Speech and Hearing Research*, Vol. 33, 1990, pp. 550–452.

[50] Mermerstein, P., "Articulatory Model for the Study of Speech Production," *J. Acoust. Soc. Am.*, Vol. 53, 1972, pp. 1070–1082.

[51] Shirai, K., and M. Honda, "Estimation of Articulatory Motion," in M. Sawashima, and F. S. Cooper, eds., *Dynamic Aspects of Speech Production*, Tokyo: University of Tokyo Press, 1977, pp. 279–302.

[52] Maeda, S., "Compensatory Articulation During Speech: Evidence from the Analysis and Synthesis of Vocal-Tract Shapes Using an Articulatory Model," in W. J. Hardcastle, and A. Marchal, eds., *Speech Production and Speech Modelling*, Boston: Kluwer Academic Publishers, 1990, pp. 131–149.

[53] Yehia, H. C., "A Study on the Speech Acoustic-to-Articulatory Mapping Using Morphological Constraints," Ph.D. diss. (http://www.cpdee.ufmg.br/~hani/papers.html)

[54] Honda, K., "Organization of Tongue Articulation for Vowels," *J. of Phonetics*, Vol. 24, 1996, pp. 39–52.

[55] Maeda, S., and K. Honda, "From EMG to Formant Patterns of Vowels: The Implication of Vowel Systems and Space," *Phonetica*, Vol. 51, 1994, pp. 17–29.

[56] Dang, J., and K. Honda, "Speech Production of Vowel Sequences Using a Physiological Articulatory Model," *Proc. 5th Int. Conf. Spoken Language Processing*, Vol. 5, 1998, pp. 1767–1770.

[57] Kaburagi, T., M. Honda, and T. Tsumura, "A Study on the Synthesis of Speech Based on the Search of an Articulatory-Acoustic Codebook with Phoneme Labels," *J. Acoust. Soc. Jpn.*, Vol. 54, 1998, pp. 207–214 (in Japanese).

3

Speech Recognition

Shigeru Katagiri

3.1 Introduction

One aim of speech recognition is to encode acoustic speech signals into linguistic indexes such as words and sentences and also to interpret speech content by machine. Naturally, one of the most promising approaches to such speech recognition is to investigate the human process of speech perception and spoken language understanding and apply the resulting scientific findings to the development of a mathematical model of recognition or recognition machine (recognizer). However, this human-science-based approach has not yet achieved comprehensive analysis of the human mechanism. Though greatly supported by the scientific discipline of "learning the human mechanism," research into speech recognition has been strongly technologically-colored, i.e., it has long been conducted mainly in the research field of signal processing technology. The relationship between human recognition and machine recognition appears similar to that of "bird versus airplane." Actually, although the human mechanism has not yet been fully grasped, speech recognition technology is already providing many useful speech recognition systems and services in daily life. In this chapter, we shall discuss the basics of such speech recognition technologies.

Traditionally, speech recognizers have been developed using system structures other than artificial neural networks, such as a distance classifier using prototypes and a probabilistc classifier using hidden Markov models (HMMs). Accordingly, this chapter will incorporate these traditional system structures in its discussions. However, these classical structures actually have close relationships with neural networks, and the descriptions in this chapter will be useful for understanding the topics involving neural networks presented in later chapters.

3.2 Overview

3.2.1 Hearing and Machine Recognition

Usually, a speech signal to be recognized exists together with various types of acoustic signals, which inevitably interfere with the accurate recognition of the target speech. Let us imagine a lively party (see Figure 3.1). People are chatting and laughing. A band is playing music. Someone has pulled the cork out of a champagne bottle. In this loud environment, you can still listen to your conversational partner without difficulty. Humans live in a complex acoustic environment (auditory scene) in which various sounds coexist, but they can still easily distinguish a target sound from interference sounds. Recently, there has been a great deal of research interest in a challenging technology,

Figure 3.1 A party scene. Humans easily distinguish target sounds from interference sounds.

sometimes called auditory scene analysis, for recognizing speech sounds in complex realistic acoustic environments [1, 2]. However, there has been neither a thorough understanding of the human auditory mechanism nor full development of auditory scene analysis technology; in reality, most existing speech recognition systems (speech recognizers) assume that a speech signal with a high signal-to-noise ratio is extracted from its auditory scene before being given to a recognizer. In this chapter, we will concentrate on this simplified but practical task setting for recognition; we will not discuss the technologies for extracting or segregating a target speech signal from the background auditory scene. Figure 3.2 illustrates the assumed relationship between a realistic acoustic environment and the speech input for a recognition task.

The definition of "recognition" is not necessarily unique to speech recognition research. There are several ways of defining this decision operation. In this chapter, we define it as a process that maps an input speech pattern to one of several prescribed classes (such as phonemes, words, or word sequences) through three modules (sub-processes): (1) feature extraction, (2) segmentation, and (3) classification. Figure 3.3 illustrates this modular recognition process. In the figure, an analogy is also illustrated between the human auditory

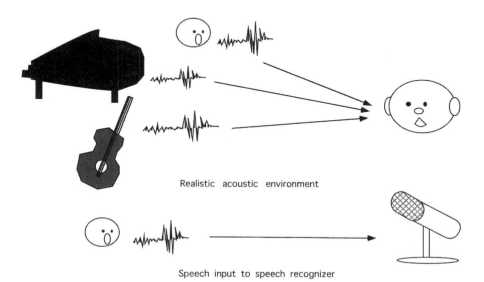

Realistic acoustic environment

Speech input to speech recognizer

Figure 3.2 Relationship between a realistic acoustic environment and a speech input assumed for recognition.

process and the recognition process. Encouraged by this analogy, the modular-process definition has been widely used in speech recognition, though obviously the modularity illustrated is not the only possible definition.

Figure 3.4 illustrates the basic procedures of speech recognition. Based on the definition cited in the above paragraph, let us assume that a speech wave signal, which is digitized with audio and A/D conversion equipment and extracted from its background acoustic signal, is provided to the recognizer. Recognition begins with feature extraction. This first-step module converts the input speech signal to a feature pattern that is suitable for recognition. The module corresponds to the human auditory (peripheral) system that transforms an acoustic input speech to a neural feature pattern by some feature extraction, which has not yet been fully clarified. Segmentation is an operation of dividing the feature pattern into segments (partial patterns), each corresponding to a linguistic unit such as phoneme or word. Then the final classification module maps a segment to one of the prescribed classes such as words and sentences. In this module, an acoustic model and a language model are assigned to each class in order to measure the degree to which the feature pattern belongs to one class. The acoustic model measures the acoustical class identity (or possibility) by which the feature pattern belongs to its corresponding class. The language model is a linguistic constraint or rule that enables one to predict and determine a possible sequence of linguistic units such as words.

In Figure 3.4, all of the modules are sequentially located, and the information is assumed to flow in a one-way, left-to-right manner. However, our experiences in speech perception and spoken language understanding suggest that there are significant interactions among the modules (e.g., [3]). Actually, we are not aware of the existence of such interactions in our daily communications using our mother tongue. However, we easily become aware of these interactions in the use of an unfamiliar foreign language: linguistic information such as vocabulary and syntax greatly helps identify (segment and classify) foreign words, which are often difficult to identify by only their acoustic information.

3.2.2 Recognition-Oriented Speech Feature Representation

3.2.2.1 Sound Spectrogram: Time-Frequency-Energy Representation

A speech wave is a one-dimensional signal having temporal structure. The signal can be considered a combination of different frequency sine-waves, and its acoustical characteristics are determined by the

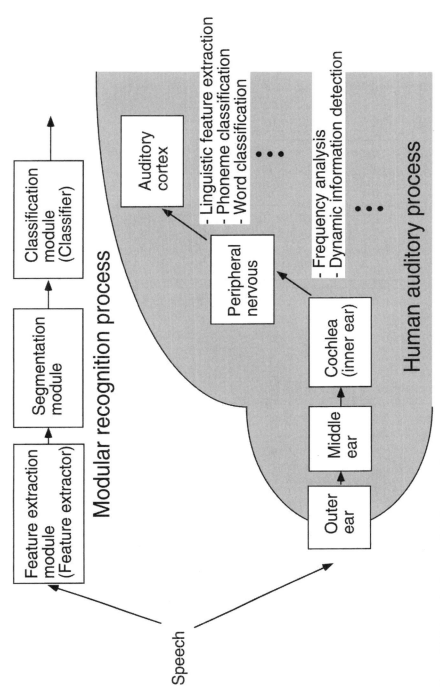

Figure 3.3 Schematic diagram of the human auditory process and a modular recognition process.

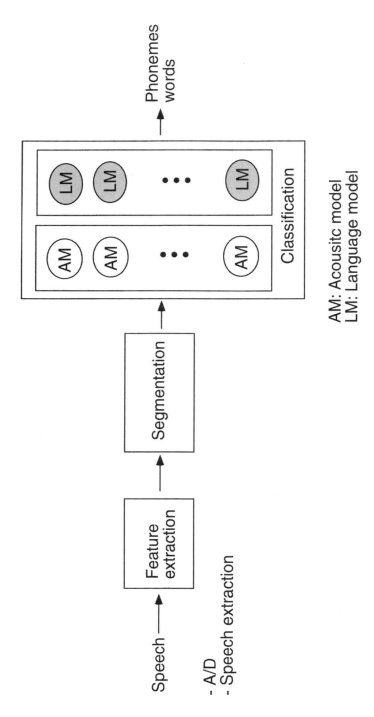

Figure 3.4 Basic procedures of speech recognition.

frequency, energy (amplitude), and phase of each component sine-wave. However, for speech recognition, a speech signal is usually converted to a three-dimensional, time-frequency-energy feature pattern that is similar to a sound spectrogram. The main reason for this selection of conversion is that (1) the human auditory system is not sensitive to the phase information of a speech signal, and (2) the three factors, i.e., time, frequency, and energy, determine the basic perceptual nature of phonemes.

Figure 3.5 is a sample sound spectrogram and its corresponding original speech wave for the word "Yokozuna."[1] In the sound spectrogram, the shade indicates the intensity of energy at every time-frequency point. The darker the shade, the higher the intensity is. The sample includes two segments of the identical vowel class $\langle\langle o \rangle\rangle$.[2] The time-fequency-energy patterns of these two segments look quite similar. As in this case, a speech segment is associated with the time-frequency-energy pattern specific to its corresponding linguistic class such as phoneme or syllable. Accordingly, a spoken word and sentence pattern is characterized by the concatenation of the time-frequency-energy patterns, each specific to the basic linguistic units such as phonemes.

Based on the change of speaking speed (rate), the speech signal of "yokozuna" varies in length. A slowly spoken $\langle\langle\text{yokozuna}\rangle\rangle$ is represented by a long sound spectrogram along the time axis. Temporal warping (extension and shrinking) is observed locally in short segments such as phonemes (e.g., $\langle\langle k \rangle\rangle$ and $\langle\langle o \rangle\rangle$) as well as in the entire (usually long) signal of word/sentence. Generally, the warping range is large for vowel classes, small for consonant classes. For example, for slowly spoken $\langle\langle\text{yokozuna}\rangle\rangle$, usually only vowel segments are lengthened while the length of consonant segments rarely changes. Such nonuniform change in temporal structure is called nonlinear time warping.

3.2.2.2 Acoustic Feature Vector

The gray-scaled sound spectrogram is originally a three-dimensional object with its height being the energy intensity. A section sliced at some time position in parrallel to the frequency axis can thus be observed as a two-dimensional frequency-energy pattern, i.e., short-time spectrum. The spectrum is illustrated in Figure 3.6. From this understanding, the sound spectrogram is treated as a sequence of short-time spectra.

---

1. Yokozuna is the rank of grand champion wrestler in Japanese sumo wrestling.
2. The expression $\langle\langle * \rangle\rangle$ denotes phoneme symbols (class names).

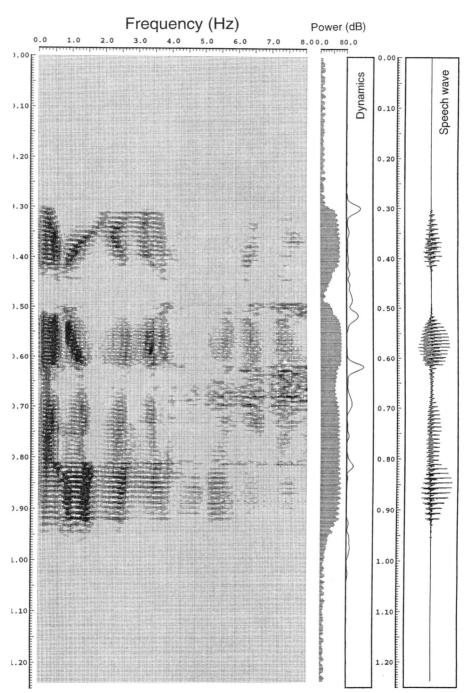

Figure 3.5 Sound spectrogram of the word "yokozuna."

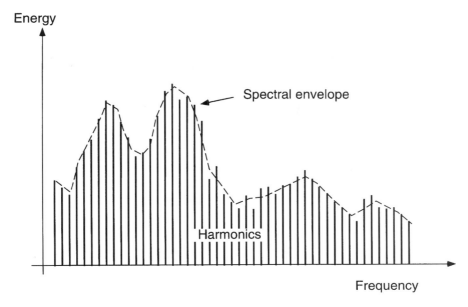

Figure 3.6 Short-time spectrum.

The short-time spectrum of Figure 3.6 shows two noticeable structures: spectral envelope and harmonics. The spectral envelope is an outline structure that smoothly varies and is mainly determined by the status (shape) of articulators such as lips and tongue. In Figure 3.6, the peaks of the spectral envelope correspond to acoustic resonance frequencies, which are determined by the articulator status and are referred to as *formants*. It is easy to observe that phoneme classes are determined by controlling the status of the articulators. Linguistic information, which is the result of articulator behavior (articulation) and thus important for speech recognition, is naturally included mainly in the spectral envelope. In particular, formants are considered a principal factor that determines the identity of vowel classes. In the sound spectrogram of Figure 3.5, one can observe that the formant structure varies as the phoneme class changes, or in other words, as the articulatory status changes. On the other hand, harmonic structure is caused by the vibration of the vocal cords; specifically, the spikes forming the harmonic structure appear at multiples of the fundamental frequency, which is the frequency of the vocal cords' vibration. This periodic structure is considered to mainly convey speaker identity and prosodic information.

Using the short-time spectra for speech recognition as they are would not be recommended because they each include both types of information, i.e., the one relevant to and the one irrelevant to speech recognition. Also, the direct digital representation of the short-time spectrum results in a vector of excessive dimensions,[3] and it is obviously inappropriate with regards to computation load for the later classification process. A way to efficiently represent the information relevant to speech recognition in low-dimensional features (feature vector) is obviously needed. Actually, in order to realize such feature vectors, various kinds of representation methods have already been investigated, e.g., the filter-bank method motivated by the human peripheral auditory system, linear predictive coding (LPC) to provide an efficient parametric model of the human speech production system, and another model of speech production, the cepstrum method. In the later pages of this chapter, the feature vectors derived from these representation methods will be collectively referred to as the acoustic feature vector.

3.2.2.3 Static Versus Dynamic Nature

As explained above, a single acoustic feature vector (or any feature set that is computed by further modifying the acoustic feature vector) can be considered a static feature in the sense that it represents a status (shape) of articulation at some time point. In contrast, a sequence of acoustic feature vectors or any set of features computed from the sequence are called dynamic features in the sense that they represent the dynamics (behavior) of articulators in the time period corresponding to the sequence. Generally, vowels such as $\langle\langle a \rangle\rangle$ and $\langle\langle i \rangle\rangle$, which are each produced by their own constant shape of the articulators, are characterized by static features. Such consonants as $\langle\langle k \rangle\rangle$ and $\langle\langle t \rangle\rangle$, which are each produced by their own rapid moving of the articulators (especially the lips and tongue) during a few tens msec, are characterized by dynamic features.

Usually, the dimensionality of an acoustic feature vector is fixed for every task and system design. Such a fixed-dimensional vector is often called a static pattern. In contrast, the duration of a speech signal or its corresponding sequence of acoustic feature vectors is essentially variable. Such a variable-durational pattern is called a dynamic pattern.

The definitions of "static" and "dynamic" in *static/dynamic feature* are different from those in *static/dynamic pattern*, and this

---

3. The sound spectrogram shown in Figure 3.5 was computed as a sequence of 256-dimensional short-time power spectra for the frequency range of 10 kHz.

inconsistency is incovenient and confusing. However, because this terminology has been commonly used in the speech recognition field, we shall continue to use it here.

3.2.3 Variety of Recognition Tasks

Even if a system input is assumed to be a speech signal, the system has to handle various types of speech inputs due to the variety of speaking styles. In daily-life speech communication, people rarely speak in the isolated word manner; rather, they speak connected words fluently and continuously and their speech often does not conform to proper grammar. On the other hand, from the viewpoint of system design, the isolated word recognition that attempts to recognize a word spoken in the isolated word-by-word mode is more productive than continous speech (or connected word) recognition, and also the recognition of read speech is more tractable than the recognition of conversational speech, which does not often conform to grammatical rules. It seems that achieving tractability for system design conflicts with acceptability/ comfort for system users (speakers). Figure 3.7 illustrates the relationship between speaking conditions (styles) and resulting characteristics of speech signals and also shows the taxonomy, with regard to this relationship, of speech recognizers. In general, the conditions in the left side are acceptable for system design but not for system users; the conditions in the right side are acceptable for system users but not for system design. The figure shows that there are many possibilities of system selection, ranging from limited-vocabulary isolated-word speaker-dependent recognition to open-vocabulary speaker-independent recognition.

3.2.4 Recognition Mechanism

3.2.4.1 Example Task Setting

Let us explain a recognition procedure by using the most fundamental example framework of recognizing words spoken in the isolated mode.

Figure 3.8 is a diagram of procedures employed in our example isolated-word recognizer. The recognizer is a modular system, as in Figure 3.3. An input speech is assumed to be a single word, such as "yokozuna," that is spoken as an answer to the question 'What do you call a grand champion sumo wrestler in Japanese?' Let us assume that our recognition task contains M word classes; C_j ($j = 1, \cdots, M$).

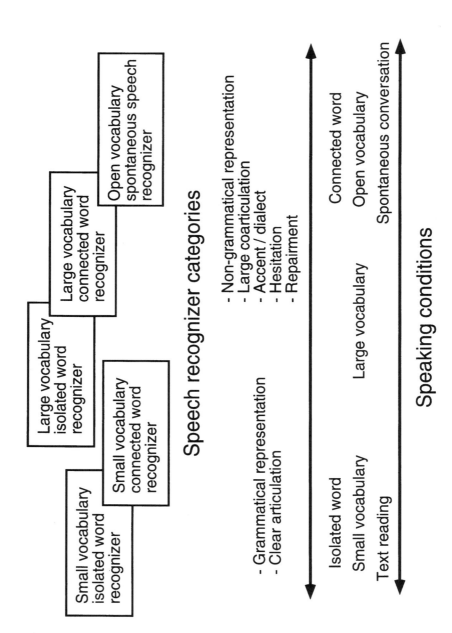

Figure 3.7 Relationship between speaking styles and speech sound characteristics.

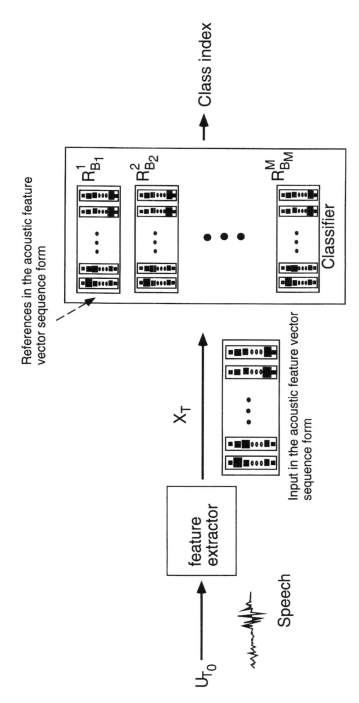

Figure 3.8 Diagram of isolated-word recognition.

Accordingly, the goal of recognizer design is to adjust (optimize) the trainable parameters of the recognizer, aiming to encode an input speech to a correct class among the M word classes. Let us denote an input word speech wave by U_{T_0}, where T_0 is the duration of U_{T_0}.

3.2.4.2 Distance-Based Recognition

There are various ways of implementing the feature extraction module. In this introductory section, we avoid details about possible implementations and simply assume that in the feature extraction module the utterance sample U_{T_0} is converted to acoustic feature vector sequence $X_T = (\mathbf{x}_1, \cdots, \mathbf{x}_\tau, \cdots, \mathbf{x}_T)$, where T is the duration of the vector sequence, $\mathbf{x}_\tau$ is the acoustic feature vector at time index τ in the sequence, and the acoustic feature vectors are of F-dimensions. Figure 3.9 illustrates the representation of the acoustic feature vector sequence.

A basic operation of the classification module is to compare an input feature pattern and class models designed beforehand, and it is traditionally implemented by measuring the distance between the input feature pattern and a model pattern that corresponds to one class and is defined in the same functional form as the input feature pattern. The smaller the distance for C_j, the larger the C_j's class identity of the input

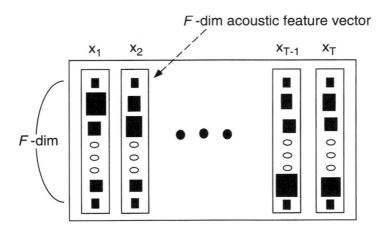

$$X_T = (x_1, x_2, \ldots x_{T-1}, x_T)$$

■ : Box size = Value of vector element

Figure 3.9 Schematic explanation of the acoustic feature vector sequence.

feature pattern, i.e., the degree to which the input belongs to C_j. According to the conventional approach, we consider here a distance-based recognizer that employs this distance-based classification strategy.

Let us define a formal representation of C_j's class model by acoustic feature vector sequence $R_{B_j}^j = (r_1^j, \cdots, r_\tau^j, \cdots, r_{B_j}^j)$, where B_j is the duration of the vector sequence, r_τ^j is acoustic feature vector at time index τ in the sequence, and r_τ^j's are all of F-dimensions. A class model defined as $R_{B_j}^j$ is traditionally referred to as a reference pattern in the speech recognition field because it works as a reference for comparison with the input. In practical designs, multiple references are often assigned to each class. However, for simplicity of discussion, we consider here the simplest case of using a single reference pattern for each class. A classification rule used by our recognizer then becomes

$$C(X_T) = C_i, \quad \text{iff } i = \arg\min_j g_j(X_T; R_{B_j}^j) \tag{3.1}$$

where $g_j(X_T; R_{B_j}^j)$ is the distance between X_T and $R_{B_j}^j$.[4]

3.2.4.3 Distance Computation Based on Dynamic Time Warping

Speech samples of a single word differ in duration from sample to sample. Also, as cited above, the durational warping rates differ from segment (such as phoneme) to segment. Thus, the distance computation between dynamic input/reference patterns is not so trivial. Recall that distance is usually defined between static patterns. Among many possible solutions to this difficulty, a dynamic-programming (DP)-based method has been traditionally employed in the speech recognition field.

Figure 3.10 illustrates the process of DP-based distance computation between input X_T and reference $R_{B_j}^j$. In the figure, each dot indicates where the distance between the two corresponding acoustic feature vectors, that of X_T and that of $R_{B_j}^j$, is computed. Such a dot is usually called a matching node, and the corresponding distance is called local distance. We then define path distance by accumulating the local distance values along some matching path that consists of the matching nodes. Because the temporal structure is essentially one-way (past-to-future), matching paths are expected to basically expand in the left-to-right manner. In Figure 3.10, an example of a matching path is also

---

4. The definition of this distance will be discussed in Section 3.2.4.3.

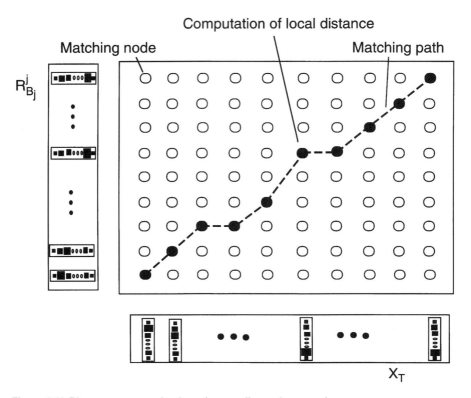

Figure 3.10 Distance computation based on nonlinear time warping.

illustrated by the broken line. Each path has its own path duration and corresponding distance. Consequently, path distance can be considered a duration-normalized distance measure between the dynamic patterns. Obviously, there are many possible ways of drawing the matching paths, and one can observe many different path distance values, one for each matching path. The most natural choice among the many possibilities is to use the minimum path distance, and this special selection is indeed a distance concept that has widely been used for speech recognition.

In the following paragraphs, we describe the procedure of computing the minimum path distance in more detail.

First, we define the distance measure between the acoustic feature vectors in a more concrete manner. As a natural candidate of distance in the fixed (F-dimensional) vector space, we can select a common distance measure such as Euclidean distance. By way of example, let us consider the squared Euclidean distance δ_{j,θ_τ} between $\mathbf{r}_i^j$, which is the

t-th component vector of $R_{B_j}^j$, and $\mathbf{x}_{\theta_\tau}$, which is the θ_τ-th component vector of X_T and is also selected with the matching node corresponding to $R_{B_j}^j$ along path θ. Because this distance is defined locally at every time index over the acoustic feature vector sequence, it is often called local distance. Accordingly, path distance, which is computed by accumulating the local distances along a path, is provided as

$$D_\theta(X_T, R_{B_j}^j) = \sum_{\tau=1}^{T} \delta_j^{\theta_\tau} \qquad (3.2)$$

Even though the temporal structure of an utterance sample can warp nonlinearly, such an extreme and unnatural matching as the one between the beginning feature vector of the input and the final feature vector of the reference pattern does not usually occur. Thus, we consider possible matching paths only in some reasonably preset restricted region when searching for the best path having the minimum

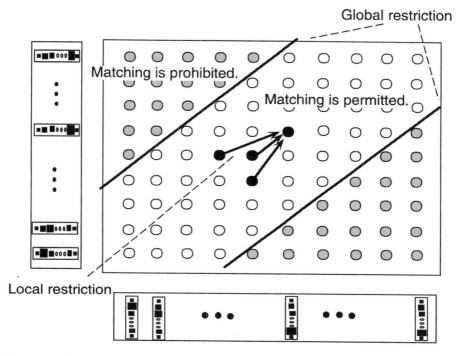

Figure 3.11 Restrictions to finding the minimum distance path: global restriction and local restriction.

Figure 3.12 Examples of implementing local restriction.

path distance.[5] Figure 3.11 illustrates the concept of restriction in finding the minimum distance path. There are mainly two types of restriction: global restriction and local restriction. Global restriction is applied to the entire input pattern, as shown by the hatched region in the figure. This is mainly used to avoid extreme warping from the global point of view in the matching space. On the other hand, local restriction is set to control the direction of path extension at every matching node. Let us consider an example case that x_{t_0} is matched to $r_{t_1}^j$. A question here is whether x_{t_0} can be matched to $r_{\hat{t}}^j$'s ($\hat{t} = 1, \cdots, t_1 - 1$). Based on the one-way (forward direction) nature of temporal signals such as speech, it may be reasonable to prohibit backward matching, such as the one between x_{t_0} and $r_{t_1-1}^j$. This local restriction is actually used to avoid the unreasonable extension of paths at every node. Several typical examples of local restriction are illustrated in Figure 3.12.

3.2.4.4   Remarks

Our example recognizer makes its classification decision according to rule (3.1). The decision rule using distance comparisons seems quite natural and conforms to our intuition. In fact, the rule is a typical implementation of the discriminant function (DF) approach to a fundamental classification decision rule, i.e., the Bayes decision rule. Details about the Bayes decision rule and the DF approach will be introduced in the next secion.

3.3   Bayes Decision Theory

3.3.1   Overview

In most cases, the present speech recognition technology is based on the Bayes decision theory, which is a fundamental principle of pattern

---

5. The best path is also called minimum distance path.

classification: the final stage of recognition, i.e., classification, is generally performed based on this classical theory. We shall focus in this chapter only on the fundamentals of the theory. A more mathematical formalism of the theory can be found in other sources (see [4, 5]).

A speech pattern is essentially dynamic, possessing nonlinearly-warping temporal structure. Accordingly, the pattern discussed in this chapter should be dynamic. However, handling dynamic patterns would require us to define new functions such as the probability of dynamic patterns and would make this introductory chapter unnecessarily complicated. Therefore, let us assume that our input pattern to the recognizer is static, as do most textbooks (e.g., [4]). Our example task here is thus to classify a static F-dimensional pattern $\mathbf{x} = ((x_1, \cdots, x_f, \cdots, x_F)^T)$ to one of M classes (C_j; $j = 1, \cdots, M$), where x_f is the f-th component of $\mathbf{x}$ and T is vector transpose. In addition, we assume that the input pattern $\mathbf{x}$ is directly put to the classification stage; which means that here we consider neither the feature extraction subprocess nor the segmentation subprocess. We also assume that $\mathbf{x}$ is a random vector that is characterized by unknown but fixed continuous distribution functions.

We assume that in order to accomplish the task, we are given (1) a classifier consisting of a set of designable (adjustable in the design stage) parameters Λ ($= \{\lambda_j\}_{j=1}^M$) and (2) a set of design samples $\mathcal{X}$ ($= \{\mathcal{X}_j\}_{j=1}^M = \{\mathbf{x}_n\}_{n=1}^N$) that is used for designing the classifier, where λ_j is the classifier parameters assigned to C_j, $\mathcal{X}_j$ is the set of design samples of C_j, N is the total number of design samples, and $\mathbf{x}_n$ is the nth design sample of $\mathcal{X}$ that is also assumed to belong to one of $\{\mathcal{X}_j\}$. Note here that λ_j can be either scalars, vectors, or a combination of scalars and vectors; in addition, the design samples are randomly extracted from the corresponding class sample distribution and they are mutually independent.

In the Bayes decision theory, the occurrence of a sample is treated as a probabilistic phenomenon. The probability with which the observation (observed sample) $\mathbf{x}$ occurs is denoted by $p(\mathbf{x})$; the probability with which an event of C_j occurs, i.e., the a priori class probability of C_j, is denoted by $P(C_j)$; the probability with which an event of C_j occurs conditioned by the observation of $\mathbf{x}$, i.e., the a posteriori probability, is denoted by $p(C_j|\mathbf{x})$. The classification that encodes $\mathbf{x}$ to a class is fundamentally performed by measuring the *a posteriori* probability $p(C_j|\mathbf{x})$. Basically, various types of classification decision rules are possible. Among the many, let us consider the most natural and simplest rule,

$$C(\mathbf{x}) = C_i, \quad \text{iff} \quad i = \arg\max_j p(C_j|\mathbf{x}) \tag{3.3}$$

where $C(\mathbf{x})$ is the classification operation. This rule requires $\mathbf{x}$ to be classified as the class having the largest a posteriori probability. This rule is called the Bayes decision rule, and a precise execution of the rule is shown to lead to the optimal, minimum classification error probability condition, where the probability with which an incorrect classification decision occurs is the smallest. In practice, however, we rarely use this rule directly because the direct measurement of $p(C_j|\mathbf{x})$ is quite difficult. The rule is usually used in the form

$$C(\mathbf{x}) = C_i, \quad \text{iff} \quad i = \arg\max_j p(\mathbf{x}|C_j)P(C_j) \tag{3.4}$$

which is derived according to the Bayes rule of probability

$$p(C_j|\mathbf{x}) = \frac{p(\mathbf{x}|C_j)P(C_j)}{p(\mathbf{x})} \tag{3.5}$$

where $p(\mathbf{x}|C_j)$ is the class-conditional probability.

Figure 3.13 illustrates a task of classifying two-class one-dimensional ($S = 1$) samples, where each class distribution is uni-modal. An optimal class boundary, which corresponds to the minimum classification error probability condition, exists at the crossing point of the two distribution curves. Moving the class boundary anywhere from the crossing point results in an increase in the error probability.

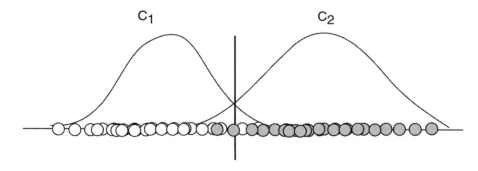

Figure 3.13 A two-class one-dimensional classification task.

As cited above, the precise execution of the Bayes decision rule leads to the optimal pattern classification. Thus, a central issue in classifier design becomes an accurate estimation of the probability functions included in the rule and also a faithful execution of the entire operation of the rule. Approaches to optimal classification include (1) maximum likelihood estimation approach (MLE approach), (2) Bayesian approach, and (3) discriminant function approach (DF approach). The design concepts underlying the MLE approach and the Bayesian approach are similar, and both aim at the accurate estimation of the probability functions. By contrast, the DF approach aims at a faithful execution of the decision operation by formalizing the decision rule as a loss (risk) minimization problem.

We shall explain the three approaches in more detail in the following sections.

3.3.2 Maximum Likelihood Estimation Approach

From (3.4), it is clear that the execution quality of the Bayes decision rule depends on the estimation accuracy of the a priori class probabilities and the class-conditional probabilities. The MLE approach then aims to achieve accurate estimation by computing the most likely estimates of the probabilities.

In principle, $P(C_j)$ is determined by some knowledge about the task that is independent of the observation of samples. Therefore, in the MLE approach, the a priori probabilities are assumed to be provided by a designer's knowledge and preferences, and the main design efforts are made to achieve the most likely (maximum likelihood) estimates of the class-conditional probabilities.

In order to estimate the class-conditional probabilities, the MLE approach employs two assumptions: (1) letting the functional form of the class-conditional probabilities be a function of the set of unknown but fixed (not random) variables (parameters) (for C_j, the class-conditional probability is expressed by using λ_j as $p(\mathbf{x}|C_j, \lambda_j)$), and (2) using the available design samples $\mathcal{X}_j$ in place of $\mathbf{x}$, which is unobservable in the design stage. Consequently, the design goal of the approach becomes finding the parameters that correspond to the maximum value of $p(\mathcal{X}_j|C_j, \lambda_j)$, i.e., the maximum likelihood estimate defined as

$$\hat{\lambda}_j \equiv \arg \max_{\lambda_j} p(\mathcal{X}_j|C_j, \lambda_j) \qquad (3.6)$$

Recall that the design samples are mutually independent. Thus (3.6) is rewritten as

$$\hat{\lambda}_j = \arg \max_{\lambda_j} \prod_{n=1}^{N} p(\mathbf{x}_n | C_j, \lambda_j) \mathbf{1}(\mathbf{x}_n \in C_j) \qquad (3.7)$$

where $\mathbf{1}(\cdot)$ is an indicator function that takes 1 for *true* and 0 for *false*. Intuition here indeed supports that $\hat{\lambda}_j$ is a reasonable solution to λ_j in the design strategy of using only available samples. The parameters λ_j determine the model function $p(\cdot | C_j, \lambda_j)$. In the MLE approach, we aim at making the value of $\max_{\lambda_j} p(\mathcal{X}_j | C_j, \lambda_j)$ as large as possible (as likely as possible) by repeating the selection of parameter type (type of λ_j) and the (3.7)-based optimal status search.

The computation of (3.7) is relatively easy. For example, when $p(\cdot | C_j, \lambda_j)$ is a Gaussian function, the maximum likelihood estimate of λ_j, i.e., $\hat{\lambda}_j$, can be computed with a simple gradient computation (in terms of λ_j) of $p(\cdot | C_j, \lambda_j)$. The principle of the expectation-maximization (EM) method [6], which is widely used for designing HHMs, also essentially relies on the MLE concept (the EM method will be further explained in Section 3.5). Indeed, based on the simplicity of the estimation, this approach has long been used as the most standard concept in a wide range of speech recognizer designs. However, in practice, it has also been pointed out that this simple MLE approach suffers from several serious problems. One basic problem is that a true functional form of the probability is usually more complex than the Gaussian form and is not accessible. To cope with this problem, a mixture Gaussian function, consisting of multiple Gaussian functions, is often used. However, because the true functional form is unknown, the problem of determining the structure of the mixture function, or in other words, the number of the Gaussian-form components, obviously remains unsolved. Another problem is that the number of available design samples is limited. In this limited resource environment, the estimation of the probability is almost always incomplete, even though the true functional form of the probability is known. Moreover, according to (3.7) the incompleteness is caused evenly by individual estimation errors, each computed at one design sample. That is, high accuracy in the estimation of the probability function, which corresponds to a large value of $p(\mathcal{X}_j | C_j, \lambda_j)$, does not necessarily mean high classification accuracy. There could even be an extreme case in which the estimated probability function is highly accurate in a region far from the class boundary and less than accurate near the class boundary, resulting in inaccurate classification decisions. The mismatch between the dotted curve and the hard curve in Figure 3.3 clearly illustrates this basic problem of the approach.

3.3.3 Bayesian Approach

In the Bayesian approach, we aim at computing the a posteriori probabilities used in (3.3) through the estimation of class-conditional probabilities. This design concept is apparently the same as that of the MLE approach. However, the Bayesian approach is clearly distinct from the MLE approach, because of the assumption that the classifier parameters λ_j's, each determining the functional form of the class-conditional probability, are treated as random variables. The Bayesian approach assumes that $\{\lambda_j\}$ possesses its own a priori probability function, and its design goal thus becomes computing the a posteriori probability for every class by estimating this a priori probability function, which is associated with the classifier parameter, and the class-conditional probability function through the observation of design samples.

The functional form of the class-conditional probability is assumed to be known, and it is expressed as $p(\mathbf{x}|C_j, \mathcal{X}_j)$ on the premise that all of the information used for estimating the functions are obtained through sample observation. Also, based on the assumption that λ_j is a random vector (variables), the approach introduces *a priori* probability $p(\lambda_j|C_j)$. Accordingly, the class-conditional probability $p(\mathbf{x}|C_j, \mathcal{X}_j)$ is rewritten as

$$p(\mathbf{x}|C_j, \mathcal{X}_j) = \int p(\mathbf{x}, \lambda_j | C_j, \mathcal{X}_j) d\lambda_j$$

$$= \int p(\mathbf{x}|\lambda_j) p(\lambda_j | C_j, \mathcal{X}_j) d\lambda_j \qquad (3.8)$$

where $p(\lambda_j|C_j, \mathcal{X}_j)$ is an estimate, which is obtained by observing the samples of $\mathcal{X}_j$, of the a priori probability $p(\lambda_j|C_j)$. Therefore, if $p(\lambda_j|C_j, \mathcal{X}_j)$ peaks sharply around $\hat{\lambda}_j$, and $\hat{\lambda}_j$ is a true value of λ, we can achieve an approximation of $p(\mathbf{x}|C_j, \mathcal{X}_j)$ as

$$p(\mathbf{x}|C_j, \mathcal{X}_j) \approx p(\mathbf{x}|\hat{\lambda}_j) \qquad (3.9)$$

The estimation method summarized above is quite reasonable, but it requires a large amount of computation for most functional forms of the a priori probability $p(\lambda_j | C_j, \mathcal{X}_j)$. Therefore, in practice, the Bayesian approach is usually applied only when the probability functions are assumed to belong to an exponential family, which is a special probability (and probability density) function group that enables us to reduce the computation load of the above estimation. Members of the exponential family include the normal (Gaussian) function, the Poisson function, and many other familiar distribution functions.

Similar to the MLE approach, the Bayesian approach has been used for speech pattern recognition. However, in reality, design results based on the Bayesian approach are often less than satisfactory: the true distributions of samples (probability functions) are usually more complicated than the exponential families, and it is unrealistic to collect enough design samples to accurately estimate the probability functions.

3.3.4 Discriminant Function Approach

The MLE and Bayesian approaches aim at computing the a posteriori probabilities, which are components of the decision rule (3.3). In contrast, the DF approach attempts to embed the entire rule in a design process. In this approach, in place of the *a posteriori* probability, a discriminant function is used to measure the degree to which an input sample belongs to some class, i.e., sample's class identity. As a discriminant function, one can use various familiar measures such as distance and similarity as well as probability. In the old days of insufficient computation power, this approach was actually a major design concept because of its simple computation. Embodiments of the approach include the linear discriminant function method and the perceptron learning method, which is a classic version of artificial neural network (ANN) learning.

For a long time, attempts to apply this classic approach to speech recognition were rare. However, after the reemergence of research interest in ANNs, the approach came to attract greater interest from speech researchers. The ANN-based recognition methods introduced in Part II of this handbook actually belong to the DF approach. Therefore, in this subsection, we shall elaborate on the fundamentals of the DF approach.

3.3.4.1 Example Task and Decision Rule

Let us consider the task of classifying M-class static samples by using the classic and most fundamental implementation of the DF approach, i.e., the linear discriminant function method. Given $\mathbf{x}$, a discriminant function $g_j(\mathbf{x}; \Lambda)$ is defined for C_j as

$$g_j(\mathbf{x}; \Lambda) = \sum_f \lambda_j[f] \cdot \mathbf{y}[f] \tag{3.10}$$

where $\mathbf{y} = (1, x_1, \cdots, x_F)$, $\mathbf{y}[f]$ is the f-th element of $\mathbf{y}$, and accordingly λ_j is an $(F + 1)$-dimensional weight vector and $\lambda_j[f]$ is the f-th element of λ_j. Note that the discriminant function is linear in terms of λ_j. As in the MLE and Baysian approaches, λ_j also forms a model of C_j in the DF approach. Let us refer to a classifier that is composed of the set $\{\lambda_j\}$ $(= \Lambda)$ as a linear discriminant function classifier. Figure 3.14 illustrates the computation of the linear discriminant function classifier, i.e., (3.10). A classification rule then becomes

$$C(\mathbf{x}) = C_i, \quad \text{iff} \quad i = \arg\max_j g_j(\mathbf{x}; \Lambda) \tag{3.11}$$

It should be noted here that classification rule (3.1) is essentially equivalent to rule (3.11), and distance-based classification decisions such as (3.1) accordingly belong to the DF approach.

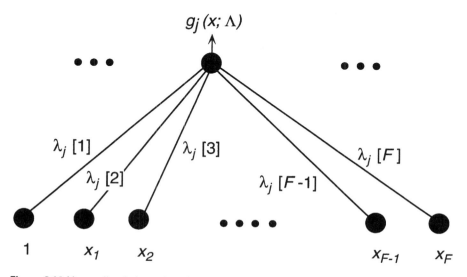

Figure 3.14 Linear discriminant function classifier.

3.3.4.2 Loss

In contrast to the likelihood maximization in the MLE approach, we aim at minimizing loss,[6] which corresponds to the evaluation of a classification result of each design sample. There are various implementations of DF-based design (e.g., see [4]), and each of them is characterized by the selection of loss function form and a minimization algorithm. Generally, the design of discriminant functions, which is based on loss minimization, is executed competitively in terms of classes. Thus, the design method of the DF approach is often referred to as discriminative learning and competitive learning. Also, the concept of using the classification results in the training (loss minimization), which is employed by the DF-based methods, can be considered a learning mechanism using supervision information. The DF-based classifier design methods are therefore sometimes referred to as supervised learning.[7]

Let us express a loss for $\mathbf{x}$ ($\in C_k$) as $\ell_k(\mathbf{x}; \Lambda)$. The classifier that we now try to design is expected to perform as correctly as possible for all future samples. An ideal way of design thus becomes the minimization of the expected loss that is defined over the entire set of possible samples as

$$L(\Lambda) = \sum_k P(C_k) \int \ell_k(\mathbf{x}; \Lambda) p(\mathbf{x}|C_k) d\mathbf{x} \qquad (3.12)$$

However, in realistic conditions where only a limited amount of design samples are available, this minimization is obviously impossible and thus we usually aim at the minimization of the empirical average loss that is defined over available design samples as

$$L_0(\Lambda) = \frac{1}{N} \sum_k \sum_n \ell_k(\mathbf{x}_n; \Lambda) \mathbf{1}(\mathbf{x}_n \in C_k) \qquad (3.13)$$

3.3.4.3 Optimization

Minimization of the empirical average loss can be considered a process of aiming at an *optimal* (corresponding to the minimum status

---

6. The loss is sometimes called risk, cost, or design objective. Typical loss functions include the perceptron loss, the squared error loss, and the classification error count loss (e.g., [4]).

7. In contrast, the design methods of the MLE and Bayesian approaches are categorized as *unsupervised learning* because they do not utilize (feedback) the evaluation results of system outputs (classification decisions) for training.

of the empirical average loss) status of Λ, i.e., *optimization*. Useful optimization methods here include the relaxation method, the gradient method, and simulated annealing (e.g., [7, 8]).

Originally, these are methods for optimizing optimizable parameters with a general class of objective functions (criteria) such as the empirical average loss. However, not all of the methods can necessarily achieve true optimization, i.e., global optimization (global minimization of objective function). Some of them result only in local optimization (local minimization of objective function); furthermore, some attempt to simply reduce the value of an objective function.

Generally, the relaxation methods lack a sufficient mathematical basis to guarantee the minimization of objective function, and its effectiveness is sometimes limited.

The gradient methods such as the steepest descent method are mathematically shown to lead to at least a local minimum of objective function by following the gradient directions of objective function. Actually, this method has been widely used because of its high practicality based on a simple computation mechanism.

Simulated annealing is a method that attracted recent research interest in the ANN field. An infinite repetition of adjustment has been demonstrated to lead to the minimum state of objective function in the probabilistic sense, though its execution needs high computational load.

The optimization methods can also be classified into *batch* optimization and *adaptive* (sequential) optimization. Batch optimization explicitly computes the empirical average loss over all of the design samples and adjusts Λ to achieve the minimum status of the empirical average loss. The steepest descent method, introduced above, is a typical example of batch optimization. On the other hand, adaptive optimization selects a limited number of design samples (usually one sample) randomly from a pool of design samples and adjusts Λ to reduce the individual loss (i.e., $\ell_k(\,\cdot\,; \Lambda)$) for the selected sample(s) in place of the explicit computation of the empirical average loss. Typical examples of adaptive optimization are methods based on stochastic approximation (e.g., [7]) and the probabilistic descent method [9, 10].

In a practical environment where only a limited number of design samples are available, the difference between batch optimization and adaptive optimization is small. However, adaptive optimization offers the potential of adapting a designed classifier to an unknown application environment, based on its sample(s)-by-sample(s) training mechanism.

3.3.4.4 Design of Linear Discriminant Function Classifier

From among the many possible selections of loss and optimization, we shall specifically select perceptron loss and batch optimization for designing our linear discriminant function classifier. It should be noted that the perceptron loss used here originated from perceptron learning, which played a historic role in neural network research, and also that the design method described in the following paragraphs is fundamentally equivalent to a neural network learning algorithm called the generalized error-correction rule (e.g., [11]).

First, let us initialize Λ to arbitrary random numbers. Obviously, in this initial stage, there is no guarantee that the classifier works accurately.

By using the initialized classifier, we next make a classification decision for $\mathbf{x}_n$ ($\in C_k$) that is randomly selected from design sample set $\mathcal{X}$. Obviously, it cannot be guaranteed that the classification in this initial stage is complete. The classifier works correctly in some cases and makes errors in other cases. Then we perform the adjustment (training) of the classifier parameters in a twofold fashion. That is, when $\mathbf{x}_n$ is correctly classified, we do not adjust Λ; when $\mathbf{x}_n$ is misclassified, in other words, at least one discriminant function of a class other than C_k reads a larger value than that of C_k (we refer to such a class as a confusion class), we adjust the parameters of both C_k and the confusion classes so as to increase the discriminant function of C_k and decrease the discriminant functions of the confusion classes.

The adjustment procedure described above is mathematically formulated in the following way. First, the loss for design sample $\mathbf{x}_n$ is expressed as

$$\ell_k(\mathbf{x}_n; \Lambda) = \frac{1}{N_n} \sum_{i \in \mathfrak{I}} (-g_k(\mathbf{x}_n; \Lambda) + g_i(\mathbf{x}_n; \Lambda)) \qquad (3.14)$$

where $\mathfrak{I}$ is the confusion classes defined as

$$\mathfrak{I} = \{C_j; g_j(\mathbf{x}_n; \Lambda) > g_k(\mathbf{x}_n; \Lambda)\} \qquad (3.15)$$

and N_n is the number of the classes included in $\mathfrak{I}$. Since the adjustment occurs only in the case of misclassification, the empirical average loss becomes

$$L_p(\Lambda) = \sum_n \sum_k \ell_k(\mathbf{x}_n; \Lambda) \mathbf{1}(\mathbf{x}_n \in C_k) \mathbf{1}(\mathfrak{I} \neq \emptyset) \qquad (3.16)$$

The adjustment based on the gradient optimization is then

$$\Lambda(t+1) = \Lambda(t) - \epsilon \nabla_\Lambda L_p(\Lambda) \qquad (3.17)$$

where t is the time index of adjustment repetition, ϵ is a small positive constant, and $\Lambda(t)$ represents the status of Λ at t. Consequently, the adjustment procedure for the weight vector of each class is given as

$$\lambda_i(t+1) = \lambda_i(t) - \epsilon \sum_n \sum_k \frac{\partial \ell_k(\mathbf{x}_n; \Lambda)}{\partial \lambda_i} \mathbf{1}(\mathbf{x}_n \in C_k)\mathbf{1}(\Im \neq \emptyset) \qquad (3.18)$$

where

$$\frac{\partial \ell_k(\mathbf{x}_n; \Lambda)}{\partial \lambda_i} = \begin{cases} -\mathbf{y} & (\text{for } i = k) \\ \dfrac{1}{N_n}\mathbf{y} & (\text{for } i \neq k) \\ 0 & (\text{otherwise}) \end{cases} \qquad (3.19)$$

3.3.4.5 Remarks

Compared with the MLE and Bayesian approaches that attempt to design class models separately (in terms of class), the DF-based design introduced above seems to be more direct in its classification decision rule: the loss minimization is a straight mechanism of feeding back classification decisions to the design stage, and also the definition of loss in (3.14) is a quite reasonable approximation of the search operation max in the decision rule.

However, it was shown that this linear discriminant function classifier led to accurate classification only in the simple task setting called the linear-separable case. Its performance under realistic task conditions is actually unsatisfactory. Moreover, the relationship between the minimum status of $L_p(\Lambda)$ of (3.16) and the desirable, minimum classification error status is unclear. The gradient computation for $L_p(\Lambda)$ is also mathematically inadequate: $L_p(\Lambda)$ is not differentiable in Λ due to $\mathbf{1}(\Im \neq \emptyset)$ of (3.16). Therefore, further improvement of the DF approach is clearly needed.

3.4 Acoustic Feature Extraction

3.4.1 Filter-Bank

Based on psychoacoustic findings, it is known that the human auditory system has a function of filter-bank-like frequency analysis of input

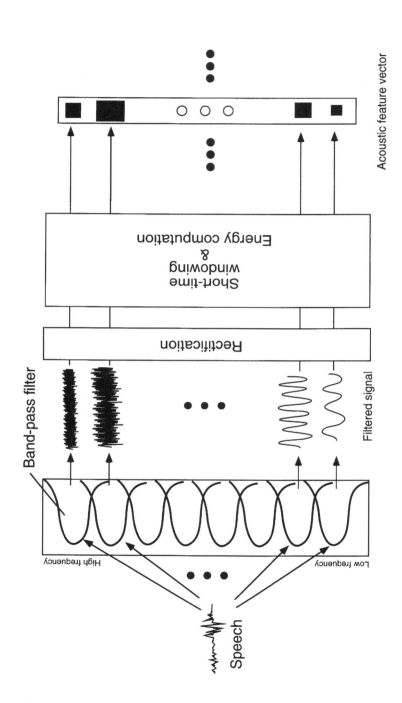

Figure 3.15 Diagram of the filter-bank feature extractor.

sound (e.g., [12, 13]). Thus, in a direct attempt to emulate this human system, various filter-bank feature extractors have been studied.

Filter-bank feature extraction is illustrated in Figure 3.15. A speech pattern U_{T_0} is passed through a bank of $\mathscr{I}$ bandpass filters. Generally, the individual filters overlap in frequency and are spaced according to a nonuniform frequency scale, such as the Mel scale or Bark scale, that are formed based on perceptual experiments. Each filter outputs a kind of short-time energy that can be observed through its corresponding frequency range.

Implementations of a filter-bank are mainly grouped into two classes: (1) perception-research-oriented attempts in which one emulates the human auditory (cochlea) system with real (infinite impulse response) filtering computation, and (2) engineering-oriented attempts in which one simply approximates the basic function of a filter-bank with Fourier transform.

3.4.1.1 Artificial Cochlea Filter

The human auditory peripheral system consists of the outer ear, the middle ear, the inner ear (cochlea), and the auditory nerve. Figure 3.16 illustrates a block diagram of the system.

The outer ear and the middle ear are basically considered a simple, linear and time-invariant bandpass filter. The cochlea works, in contrast, as a nonlinear, time-variant filter-bank that contains many narrow-band

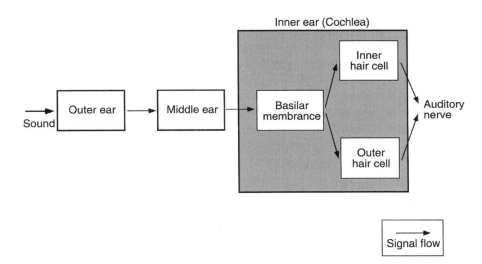

Figure 3.16 Block diagram of the human auditory peripheral system.

(high Q^8) bandpass filters. The main functions of the cochlea include (1) frequency scaling such as the Bark scale, (2) adaptive Q, which controls filter gain in response to the energy of signals through the filter channel, (3) frequency masking, and (4) rectification. Therefore, it would seem that emulating these functions of the cochlea would provide a direct route to effective modeling.

To date, there have actually been many implementation examples (e.g., [14, 15]). Among them, we briefly introduce in this chapter the artificial cochlea filter-bank that consists of adaptive Q cochlea filters [16]. Figure 3.17 illustrates a block diagram of the adaptive Q cochlea filter-bank. A major concern in this modeling is to simulate the function of the basilar membrane (in the cochlea) filtering, which changes its Q in response to the energy of the input signal. The filter-bank is formed by adding an adaptive Q circuit to a fixed Q cochlea filter that consists of a notch filter and a bandpass filter.

3.4.1.2 Fourier-Transform-Based Filter

The most standard technical approach to the frequency analysis of a temporal signal is to use the short-time Fourier transform. We can easily compute the power spectrum of a windowed, short input speech segment with the transform and can then compute energy in a form similar to bandpass filter output by accumulating spectral elements in a narrow frequency band. Figure 3.18 illustrates a block diagram of this Fourier-transform-based implementation.

3.4.2 Autoregressive Modeling

An alternative to the filter-bank feature extractor is speech analysis that uses autoregressive modeling [17, 18]. This statistical approach is often referred to as linear predictive coding (LPC), based on its prediction mechanism in the time domain.

A temporal signal in the natural world has dependency on the past. In other words, the present state of the signal is influenced at least to some extent by the preceding portion of the signal, and thus we can naturally expect the state of the signal at some time point to be predicted by using the signal segment before the time position. This is the modeling concept underlying autoregressive modeling, i.e., LPC.

For descriptive purposes, let us consider a discrete-sampled speech

---

8. Q comes from the quality factor that indicates the resonance sharpness. The larger Q is, the narrower is the bandwidth of its corresponding bandpass filter.

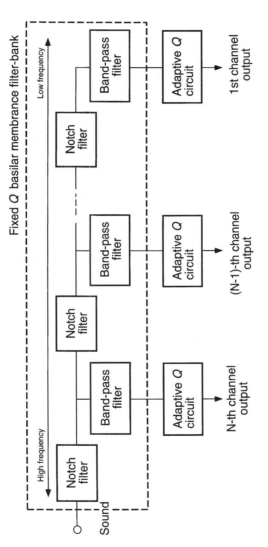

Figure 3.17 Block diagram of the adaptive Q Cochlea filter-bank.

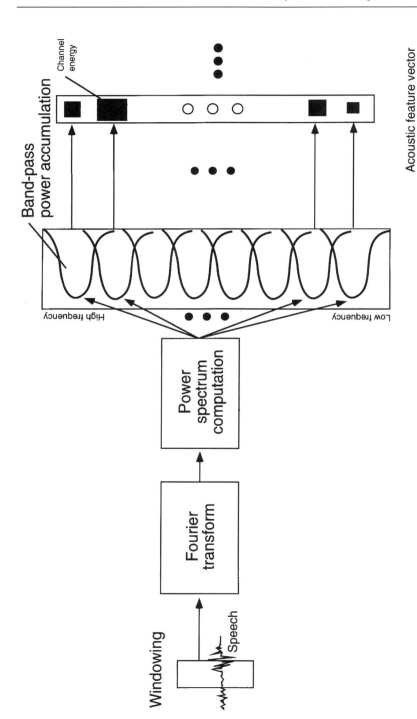

Figure 3.18 Block diagram of the filter-bank implementation using the Fourier transform.

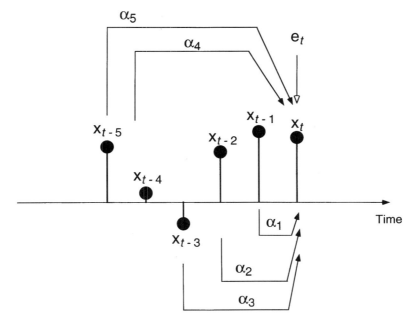

Figure 3.19 The mechanism of autoregressive modeling.

signal, which is a common representation in digital signal processing. The mechanism of the autoregressive modeling is illustrated in Figure 3.19. Discrete speech sample s_t at time point t is modeled as a linear combination of previous samples plus a nonpredictable (independent) additional noise e_t in the following way:

$$s_t = -\sum_{i=1}^{p} \alpha_i s_{t-1} + e_t \tag{3.20}$$

where p is the order of modeling and α_i's are called prediction coefficients. The goal of modeling is naturally to find the $\{\alpha_i\}$ that minimizes the mean-square error calculated over a windowed short speech segment, defined as

$$E_{rr} = \sum_{t=\tau_1}^{\tau_2} \{e_t\}^2 \tag{3.21}$$

where τ_1 and τ_2 are the left (past) and right (future) ends of the time window, respectively.

A simple computation converts (3.21), defined in the time domain, to equations in the correlation coefficient (in terms of the speech signal) domain. Two methods were developed to solve these correlation-domain equations: the correlation method and the covariance method. The correlation method, especially its efficient version, called the PARtial CORrelation (PARCOR) method or the Levinson-Durbin algorithm, has been extensively used due to its computation efficiency and the stability of the resulting model.

In addition to its natural modeling characteristics in the time domain, autoregressive modeling is characterized by its significant features in the frequency domain. In particular, this parametric modeling approximates a short-time spectrum of the windowed speech signal by using all-pole rational functional form. Naturally, the pole is assumed to correspond to one formant in the vowel sound case. This feature is observed by applying the z-transform to (3.20), as

$$S(z) = \frac{E(z)}{\sum\limits_{i=0}^{p} \alpha_i z^{-i}} = H(z)E(z) \tag{3.22}$$

where $S(z)$ and $E(z)$ are the z-transform of the speech signal and the noise, respectively, $H(z) = 1/\sum_{i=0}^{p} \alpha_i z^{-i}$, and $\alpha_0 = 1$. Importantly, the acoustical physics shows that the all-pole system function corresponds to a nonbranching acoustic tube, i.e., a nonbranching vocal tract model in speech. Therefore, from (3.22) we can conclude that the system function (spectrum) of speech signal $S(z)$ can be obtained as the product of the system function of the vocal tract $H(z)$ and the system function of the noise $E(z)$. Figure 3.20 illustrates a vocal tract based on autoregressive modeling. Consequently, noise can be interpreted here as a vocal source (excitation) signal.

From the above descriptions, it is obvious that the prediction coefficients convey the shape information of the vocal tract or, in other words, articulation information and the identity of linguistic class (such as phoneme). This is one of the reasons these coefficients and their derivatives have been widely used as acoustic feature vectors for speech recognition.

3.4.3 Cepstrum Modeling

As introduced in Section 3.2 and illustrated in Figure 3.20, a speech signal can be considered an output of a vocal tract filter ($H(z)$ in the case

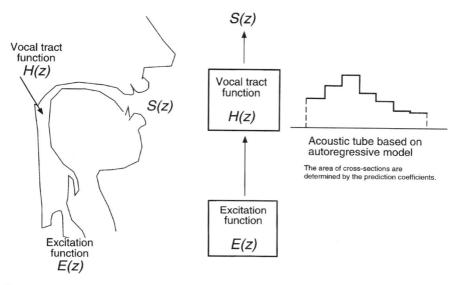

Figure 3.20 A vocal tract based on autoregressive modeling and an illustration of the human speech production system.

of autoregressive modeling) excited by an excitation function ($E(z)$ in the case of autoregressive modeling). For speech recognition, we need to extract only the information based on the vocal tract filter, which conveys linguistic class identity, from the speech signal. The acoustic feature representation/modeling using cepstrum has also been used extensively in the speech recognition field.

Cepstrum is an inverse Fourier transformation of a logarithmic power spectrum. A variable for cepstrum function is then defined as quefrency, which is effectively a time domain variable. Figure 3.21 illustrates the transformation relations. A speech signal is originally a scalar value function (discrete sample sequence after sampling) of time that is defined as the convolution of the excitation signal and the impulse response of a vocal tract filter. Through the Fourier transform, the speech signal is converted to its corresponding power spectrum, which is a function of frequency and is defined as the production of the vocal tract filter spectrum and the excitation spectrum. A simple logarithmic operation converts production to summation. In the log power spectrum domain, the speech spectrum becomes the summation of the two log spectra, that of the vocal tract filter and that of the excitation signal. Cepstrum is accordingly defined as the inverse Fourier transform of the summation of these two spectra.

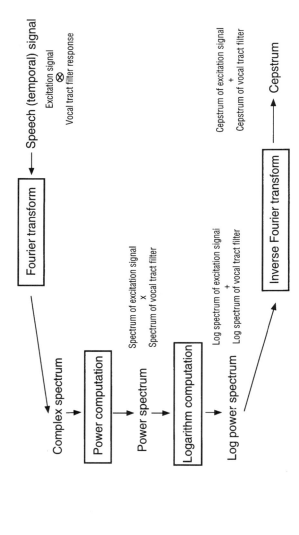

Figure 3.21 Relationships among a speech signal, a vocal tract filter, and a vocal source excitation signal.

As mentioned previously, the (log) spectrum of the vocal tract filter is smooth; the (log) spectrum of the excitation is flat for unvoiced speech segments and periodic for voiced segments. Thus the smooth spectrum of the vocal tract filter is transformed to cepstrum components in the low quefrency region, and the periodic spectrum of the (voiced) excitation signal is transformed to cepstrum components in the high quefrency region. The information localization in the quefrency domain is clearly useful for extracting the vocal tract filter information or, in other words, the speech-recognition-oriented linguistic class information.

Although we have defined cepstrum by using the Fourier transform in the above paragraphs, there are actually two types of cepstrum computation. The first type is based on the Fourier transform and the second type is derived from autoregressive modeling. Below, for reference purposes, we summarize the procedures of these two types of computation.

- **Cepstrum computation based on the Fourier transform**

$$
c(n) = \begin{cases} \dfrac{s(n)}{s(0)} - \dfrac{1}{n}\sum_{k=1}^{n-1} k \cdot c(k)\dfrac{s(n-k)}{s(0)} & (n \neq 0) \\[2ex] log\, s(0) & (n = 0) \end{cases} \qquad (3.23)
$$

where $s(n)$ is the n-th discrete-sampled speech signal, $c(n)$ is the n-th cepstrum coefficient, $x(n) = c(n) = 0$ for $n < 0$, and $x(0) \neq 0$.

- **Cepstrum computation based on autoregressive modeling**

$$
c(n) = -\alpha_n - \frac{1}{n}\sum_{i=1}^{n-1} ic(i)\alpha_{n-i} \quad n \geqslant 1 \qquad (3.24)
$$

where $\alpha_0 = 1$ and $\alpha_i = 0$ when $i > p$ (p is the order of autoregressive modeling).

3.4.4 Dynamic Feature Modeling

Originally, the Fourier transform and autoregressive modeling were methods for estimating the spectrum of a statistically stationary system (signal). However, speech signals are obviously non-stationary due to the dynamics of the speech production system, and this dynamics often plays a key role in the recognition of rapidly changing phonemes such as plosives and liquids. Therefore, as a practical compromise, these spectrum estimation methods are applied to a short speech segment that is assumed to be stationary and is extracted through a short time window (usually of about 20 msec). Consequently, in order to adequately represent the dynamics information for classification, a subsequence of the acoustic feature vectors is often used as an input pattern to the post-end classifier. The length of the subsequence is usually 30 msec–100 msec.

To date, two main methods of dynamics representation have been studied: (1) feature matrix and (2) delta-feature.

The first method simply uses a feature matrix that consists of successive acoustic feature vectors. Figure 3.22 shows an example of a feature matrix. An individual acoustic feature vector is a 16-dimensional vector of filter-bank outputs, and the matrix is composed of a sequence of 15 filter-bank output vectors. The size of the box indicates the normalized value of the filter-bank output (white for negative; black for positive). This type of representation has been widely used in ANN-based speech recognizers [19, 20].

The second method represents the direction of successive feature component change by linear regressive modeling. Figure 3.23 illustrates the concept and computation procedure of this representation. The figure illustrates an example sequence (in time window position) of the

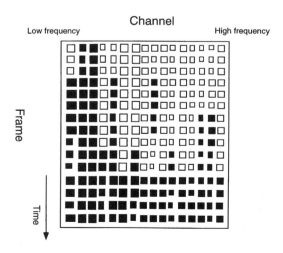

Figure 3.22 Example of a feature matrix used for representing the dynamics of a speech signal.

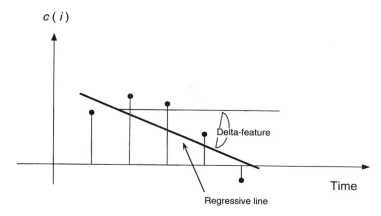

Figure 3.23 Concept of delta-feature used for representing the dynamics of a speech signal.

i-th cepstrum coefficients, $c(i)$. To represent the change direction (i.e., dynamics) in each feature component space, we attempt line fitting or, in other words, linear regressive modeling, over this coefficient sequence. The slant of this regressive function, called delta-feature, indicates the direction and amount of coefficient change. The delta-feature is computed for every original feature component, e.g., i-th delta-cepstrum $\delta(i)_\tau$ for i-th cepstrum $c(i)_\tau$ at time window position τ, and accordingly a new delta-feature vector having the same dimension as its corresponding original acoustic feature vector is computed. For recognition, a new high-dimensional feature vector consisting of the original acoustic feature vector and its corresponding delta-feature vector is used.

3.5 Probabilistic Acoustic Modeling Based on Hidden Markov Model

3.5.1 Principles of Hidden Markov Model

HMM is a system, or process, that probabilistically emits an observable output (observation) according to an unobservable state-transition mechanism. The mechanism is quite useful for modeling the temporal structure of speech signals, and the probabilistic formalization of HMM is also quite effective for modeling the acoustical variety of speech samples, due mainly to the variation in speaker and speaking style. Thus, HMM has become the standard technology in recent speech recognition research.

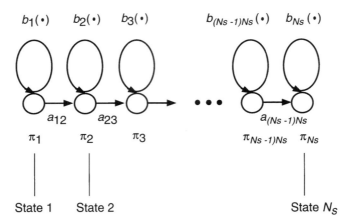

Figure 3.24 Example structure of left-to-right HMM.

Figure 3.24 illustrates a left-to-right HMM that is considered suitable for the modeling of speech signals. The model consists of N_s states, each indicated by a circle. It is assumed that the transition occurs in the probabilistic sense from one state to another or recursively within single state according to the initial state probabilities $\{\pi_i\}$ ($1 \le i \le N_s$; $\pi = \{\pi_i\}$) and transition probabilities $\{a_{ij}\}$ ($1 \le i, j \le N_s$; $A = \{a_{ij}\}$).

It should be emphasized here that this transition cannot be observed. The model is also supposed to emit an observation $\mathbf{o}_t$, which is a vector in the acoustic feature vector space at hand, at observation time index t. The emission of these observations is according to output probabilities $\{b_j(\cdot)\}$ ($b_j(\mathbf{o}_t)$ for observing $\mathbf{o}_t$ at state j; $B = \{b_j(\cdot)\}$). Therefore, an observation sequence O_T is defined as the sequence of model outputs that is probabilisticly produced through the state-transition and its corresponding observation emission; $O_T = \{\mathbf{o}_1\mathbf{o}_2 \cdots \mathbf{o}_t \cdots \mathbf{o}_T\}$. The probability for observing O_T is accordingly provided as

$$P(O_T|\lambda) = \sum_{\text{all } \mathbf{q}} P(O_T|\mathbf{q}, \lambda)P(\mathbf{q}|\lambda)$$

$$= \sum_{q_1, q_2, \cdots, q_T} \pi_{q_1} b_{q_1}(\mathbf{o}_1) a_{q_1 q_2} b_{q_2}(\mathbf{o}_2) \times \cdots$$

$$\cdots \times a_{q_{T-1} q_T} b_{q_T}(\mathbf{o}_T) \tag{3.25}$$

where $\mathbf{q}$ is a state sequence represented as $\mathbf{q} = (q_1 q_2 \cdots q_T)$, q_t is the state at t in $\mathbf{q}$, and $\lambda = (A, B, \pi)$.

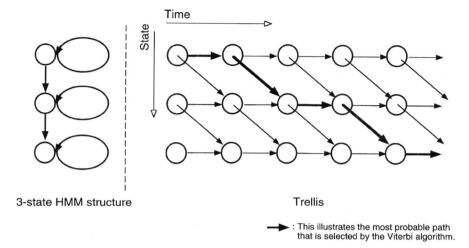

3-state HMM structure   Trellis

➤ : This illustrates the most probable path
  that is selected by the Viterbi algorithm.

Figure 3.25 Schematic explanation of state-transition: this diagram is usually called a trellis.

It should be noted here that the HMM probability (3.25) is the result of the probability accumulation over all possible state-transition paths. A diagram of these possible state-transition paths is referred to as a trellis and is illustrated in Figure 3.25.

Basically, for speech recognition a single HMM such as the model described above is assigned to every class. To classify an input acoustic feature vector sequence that belongs to some class, we then use the HMM probability, defined in (3.25), to measure the degree to which an input speech sample (its corresponding acoustic feature vector sequence) belongs to each class. A primary goal of classifier design is thus to find the desirable state of HMM parameters λ for every class.

In the MLE and Bayesian approaches, the HMM probability is used as the estimate of class-conditional probability. Accordingly, the design aims at the state of λ that leads to the maximum value of the HMM probability over design samples (of one class) in the class-by-class mode, by substituting each design sample to the observation sequence.[9] On the other hand, the design of the DF approach treats the HMM probability as the discriminant function and aims at the state of HMM parameters (of all classes) that corresponds to the minimum loss condition over the design samples of all prescribed classes.

---

9. The resulting HMM is the most likely model for observing the available design samples of its corresponding class.

The HMM formalization and its applications to speech recognition are described in detail in textbooks such as [21, 22].

3.5.2 Selection of Output Probability Function

Let us recall that an observed acoustic feature vector corresponds to the output of HMM. There are two major ways of probabilistic interpretation (definition) for the feature vectors: (1) interpreting the feature vector as an observation (output) from the corresponding discrete probability function, and (2) interpreting the feature vector as an observation from the corresponding continuous probability function. The HMM formalized in the former case is referred to as discrete HMM; the one formalized in the latter case is referred to as continuous HMM. We shall summarize these two major model classes in the following subsections.

3.5.2.1 Discrete Model

Discrete HMM is the Markov process that first encodes an observed acoustic feature vector into a code (symbol) and then models the sequence of the codes. Figure 3.26 illustrates the basic structure of discrete HMM. In the figure, the codebook is a set of codes used to encode an input speech pattern into a code sequence. Each code is associated with a static (fixed-dimensional) code vector, which is defined with the same form as the acoustic feature vector.

Generally, the code vectors are designed to minimize a distortion measure defined between the set of the code vectors and the design samples. The design is usually done by using the vector-quantization-based minimum distortion methods.

An input speech pattern is converted into its corresponding code sequence by encoding each acoustic feature vector to its nearest-neighbor code. When there are only two codes, i.e., $\{c_a, c_b\}$, both c_a and c_b are considered outputs of the binomial distribution. Similarly, in the case of using multiple codes (usually a few hundred codes for speech recognition), each code is considered an output of the multinomial distribution. The probability of the discrete HMM is accordingly determined by the output probabilities based on the multinomial distribution, the state-transition probabilities, and the initial probabilities.

3.5.2.2 Continuous Model

Continuous HMM directly models a speech pattern without transforming it into a code sequence. Figure 3.27 illustrates the basic structure of

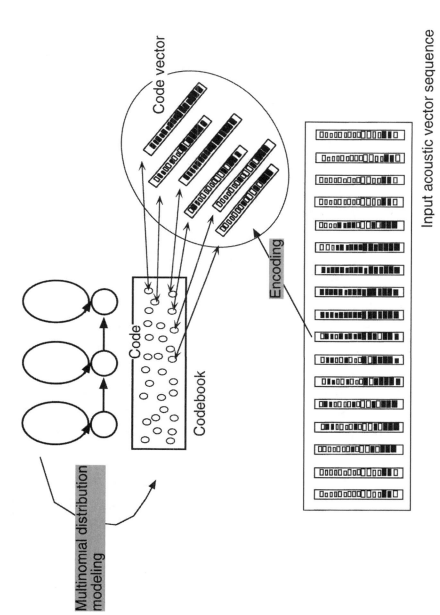

Figure 3.26 Structure of discrete HMM.

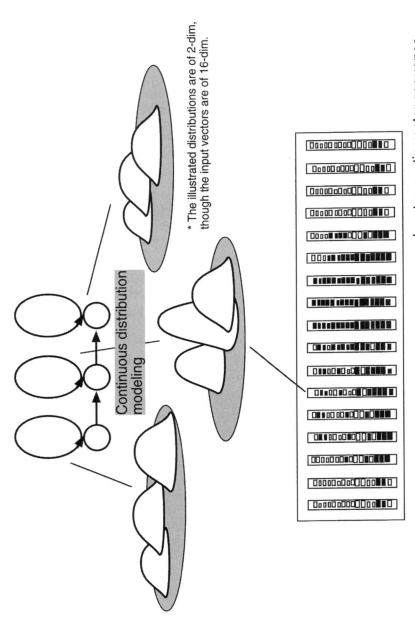

Input acoustic vector sequence

* The illustrated distributions are of 2-dim, though the input vectors are of 16-dim.

Continuous distribution modeling

Figure 3.27 Structure of continuous HMM.

continuous HMM. Each state is incorporated with a continuous output probability density function, such as the Gaussian probability function and the mixture Gaussian probability function. The acoustic feature vector of a speech pattern is treated as an output of the probability distribution function determined by the continuous probability density function. Therefore, continuous HMM consists of the continuous output probability functions, the state-transition probabilities, and the initial state probabilities.

When the output probability density function is the Gaussian function, the output probability is computed by the mean vector and covariance matrix of the function. A simple analysis in this particular case shows that the mean vector also works as a component acoustic feature vector of a dynamic (variable-length) reference pattern for distance-based classification such as (3.1); furthermore, it was shown that there is a close link between the Gaussian-based probability and the Euclidean distance. These results demonstrated that a classifier using the distance computation based on dynamic time warping, described in Section 3.2, is closely related to a continuous HMM classifier using the Viterbi algorithm (section 3.5.4).

3.5.3 MLE-Based Design Method

The design target of the MLE and Bayesian approaches is to achieve accurate estimates of the class-conditional probabilities and the a posteriori probabilities by computing the HMM probabilities, conditioned by observations of design samples. Most design efforts to date have actually been of the MLE approach and have attempted to gain the maximum likelihood (most probable) estimates of the HMM parameters by applying MLE optimization methods. The design policy based on the MLE concept is clear and reasonable. However, it is impossible to apply a standard and simple version of the MLE optimization, such as the one based on the gradient computation (section 3.3.2), because HMM has an unobservable part, i.e., the state-transition, which is intractable with simple gradient-based optimization. As a common solution to this problem, an advanced version of the MLE method, called the forward-backward method, has been used in a wide range of speech recognition systems.

3.5.3.1 Forward-Backward Method

The procedure that follows is a summary of the forward-backward method.

1. Define the current model as $\lambda = \{A, B, \pi\}$ and initialize it in any reasonable way.

2. Use λ to compute the right-hand sides of the following equations.

$\overline{\pi}_i$ = expected frequency (number of times) in state i at time $(t = 1)$

$$= \gamma_1(i) \tag{3.26}$$

$\overline{a}_{ij} = \dfrac{\text{expected number of transitions from state } i \text{ to state } j}{\text{expected number of transitions from state } i}$

$$= \frac{\sum\limits_{t=1}^{T-1} \zeta_t(i,j)}{\sum\limits_{t=1}^{T-1} \gamma_t(i)} \tag{3.27}$$

$\overline{b}_j(k) = \dfrac{\text{expected number of times in state } j \text{ and observing symbol } o_k}{\text{expected number of times in state } j}$

$$= \frac{\sum\limits_{t=1, s.t.o_t=o_k}^{T} \gamma_t(j)}{\sum\limits_{t=1}^{T} \gamma_t(j)} \tag{3.28}$$

where $\zeta_t(i,j)$ is the probability of being in state i at time t and state j at time $t+1$, given the model and the observation $(\zeta_t(i,j) = P(q_t = i, q_{t+1} = j | O_T, \lambda))$; $\gamma_t(i)$ is the a posteriori probability variable that is the probability of being in state i at time t, given the observation and the model $(\gamma_t(i) = P(q_t = i | O_T, \lambda))$.[10]

---

10. Here, $\zeta_t(i,j)$ is further rewritten in the iterative form as

$$\zeta_t(i,j) = \frac{P(q_t = i, q_{t+1} = j, O | \lambda)}{P(O|\lambda)}$$

$$= \frac{\alpha_t(i) a_{ij} b_j(o_{t+1}) \beta_{t+1}(j)}{P(O|\lambda)}$$

3. Define a reestimated model as $\overline{\lambda} = \{\overline{A}, \overline{B}, \overline{\pi}\}$ by using the left-hand sides of the above equations.

4. Repeat the reestimation of (2)–(3) by replacing λ with $\overline{\lambda}$ until the probability for observing the design samples exceeds a preset threshold.

The computation iteration of the forward-backward method was shown to increase the HMM probability monotonically at least up to its local maximum value unless the initial model is at a critical point of the probability function (e.g., [23, 24]). It is also known that this training optimality is a direct result of the EM method generally used for a wide range of problems of estimating system parameters under the missing-data condition, in which some part of the data is unobservable even though it is necessary for the estimation [6]. Consequently, the forward-backward method is often called the EM method. The method is also sometimes called the Baum-Welch method in honor of its inventors.

3.5.4 Trellis Algorithm and Viterbi Algorithm

In (3.25), the HMM probability was defined as the accumulation of the elemental probabilities such as the output probabilities over all state-transition cases. The forward-backward method is an efficient implementation of the way of computing the exact value of (3.25). Because all of the nodes on the trellis are used in the computation, this method is sometimes referred to as the trellis algorithm.

Even the efficient forward-backward method is time-consuming in such task conditions as large-vocabulary continuous speech recognition. To further reduce the computational load, the method of using only the most probable state-transition path has been widely used. The mechanism of searching for the most probable path is equivalent to

$$= \frac{\alpha_t(i)a_{ij}b_j(o_{t+1})\beta_{t+1}(j)}{\displaystyle\sum_{i=1}^{N}\sum_{j=1}^{N}\alpha_t(i)a_{ij}b_j(o_{t+1})\beta_{t+1}(j)}, \tag{3.30}$$

where $\alpha_t(i)$ is a forward variable defined as $\alpha_t(i) = P(o_1 o_2 \cdots o_t, q_t = i|\lambda)$, and $\beta_t(i)$ is a backward variable defined as $\beta_t(i) = P(o_{t+1}o_{t+2}\cdots o_T|q_t = i, \lambda)$. The name of the method originated from the mechanism of combining the forward and backward probability computations.

the Viterbi search and the dynamic programming search, and thus this simplified method is usually referred to as the Viterbi algorithm.

3.5.5   Discriminative Design Methods

As described in previous sections, the MLE and Bayesian approaches are often less than satisfactory for achieving high classification accuracy. The HMM classifier designed with the EM method also suffers from this problem. Thus, to increase classification accuracy, various DF-based design methods have been investigated.

The early methods, mainly studied in the 1980s, include maximization of mutual information [25] and the error-correction method [26]. Maximization of mutual information is discussed in more detail in Chapter 9, "Minimum Classification Error Networks." The design concept of the error-correction method is basically the same as the error-correction learning described in Section 3.3.4.

In the 1990s, the main research interest in discriminative design for HMM classifiers moved to the application of ANN-related methods. Although HMM can be considered a type of ANN [27–28], it is usually treated as a system structure different from neural networks. Therefore, ANN-based discriminative training methods, which are based on the minimization of squared error loss and sometimes on the maximization of mutual information, were applied to HMM-structure classifiers. Various examples of this type of ANN/HMM hybrid are elaborated in the literature [29–33].

Moreover, another type of ANN-related design method, i.e., learning vector quantization (e.g., [34]), was extensively applied to the design of HMM classifiers (e.g., [35, 36]).

The latest trend in discriminative design for HMM classifiers is based on the recent generalized probabilistic descent method [27, 37, 38]. Discriminative design, which has a close link to ANN-based recognition, will be discussed in more detail in several later chapters.

3.6   Language Modeling

3.6.1   Role of Language Modeling

There are several different levels of language information, e.g., syntax, semantics, and pragmatics. In most practical speech recognition designs, language modeling is treated as a process of controlling the syntactic information of language. In particular, it is used to estimate the a priori probabilities for the occurrence of an event of word class.

To understand the role of language modeling, let us consider an example task of classifying ten-figure telephone numbers, such as {0-1-2-9-8-7-6-5-4-3}. The vocabulary of the task consists of the following ten words: zero, one, two, three, four, five, six, seven, eight, nine. Our classifier is assumed to use an HMM-based acoustic word model for every digit word. A string input of ten figures is thus acoustically modeled by concatenating the HMM acoustic word models, and its corresponding class-conditional probability is computed through the accumulation of the individual HMM-based word probabilities.[11] Here, to achieve correct final decisions, we need accurate estimates of the a priori probabilities of all of the possible string classes.

In the framework of the above telephone number recognition task, let us assume three different example conditions in terms of the availability of knowledge about possible strings. The first condition is that no information is available. We have no knowledge about the telephone number system and the users' call histories in the task. The only reasonable attempt at modeling the a priori probabilities in this case would be to set an identical small value to each a priori probability. Obviously, however, this simple solution is inadequate. Such estimates of the a priori probabilities would not only be useless but in fact distort classification accuracy: some value of the a priori probability would be assigned to even a digit string class that corresponds to a nonexisting telephone number.

The second condition is that the information about the telephone number system is accessible but we have no information on the actual phone call histories. Knowledge is limited, but we can remove in this case the possibility of considering nonexisting phone number strings from the decision procedure of our classifier by setting the *a priori* probabilities of the corresponding nonexisting string classes to zero. The estimation quality of the a priori probabilities is still low, but the productive effect of removing the nonexisting classes from the list of possible class candidates would not be significant.

In the final case, we assume that in addition to the knowledge about the telephone number system, the call record corpus of telephone

11. HMMs are used in the concatenated form when classifying a long speech input such as concatenated words and sentences.

users is available. Clearly, this additional knowledge helps improve the estimation quality for the a priori probabilities, probably resulting in the accurate classification of the inputs. The above three example cases clearly demonstrate the role and effectiveness of incorporating language information into the classification decision.

3.6.2   *N*-Gram Language Modeling

In the recent literature, many types of language models have been investigated. Among them, probabilistic models have attracted much research attention. The probabilistic language models include the N-gram [39], the probabilistic LR parser [40], and the stochastic context-free grammars (e.g., [41]). In this subsection, we shall elaborate on N-gram language modeling, which is becoming a standard choice for modeling.

Let us consider a task of recognizing fixed-length connected word strings, where the vocabulary size is M and the string length is $\mathscr{L}_s$. No strict restriction of word string, such as the above phone-number restriction, is assumed. Thus the number of possible word string classes becomes $M^{\mathscr{L}_s}$. We denote here the j-th string class among the $M^{\mathscr{L}_s}$ classes as C_j. Therefore, the a priori probability $P(C_j)$ is expressed as

$$
\begin{aligned}
P(C_j) &= P(w_1, w_2, \cdots, w_{\mathscr{L}_s}) \\
&= P(w_{\mathscr{L}_s} | w_1, \cdots, w_{\mathscr{L}_s-1}) \\
&\quad \times P(w_{\mathscr{L}_s-1} | w_1, \cdots, w_{\mathscr{L}_s-2}) \times \cdots \\
&\quad \cdots \times P(w_2 | w_1) P(w_1) \\
&= \prod_{i=1}^{\mathscr{L}_s} P(w_i | w_1, w_2, \cdots, w_{i-1})
\end{aligned}
\tag{3.29}
$$

where w_i is one word class among the M vocabulary word classes, and $P(w_i | w_1, w_2, \cdots, w_{i-1})$ is the probability that w_i will be spoken given that word subsequence $w_1, w_2, \cdots, w_{i-1}$ was spoken previously. In principle, $P(w_i | w_1, w_2, \cdots, w_{i-1})$ is possible to estimate. However, we should note here that the accurate estimation of $P(w_i | w_1, w_2, \cdots, w_{i-1})$ requires us to observe (count) all of the M^i possible word strings. Under realistic task conditions, this computation is actually difficult to complete: most of the strings rarely occur and this insufficiency of observations makes the probability estimation inaccurate. Our daily-life experiences certainly show that we use only a small part of our vocabulary frequently and that we quite often and naturally face the difficulty of string class probability computation.

Even if some long word string rarely occurs, it often includes substrings that can be frequently observed. Therefore, it seems that approximating $P(w_i|w_1, w_2, \cdots, w_{i-1})$ by using a shorter, N-word-length history that is comparatively easy to observe is more feasible than pursuing the hopeless collection of rarer strings. This is the modeling concept of the N-gram language model. Naturally, the estimation of shorter versions of the model, where N is small, is easier than that for longer versions. Popular implementations of this model are the unigram ($N = 1$), the bigram ($N = 2$), and the trigram ($N = 3$).

N-gram modeling is a typical probabilistic kind of modeling based on the discrete multinomial distribution. Thus, we need as many observations as possible in order to increase the accuracy of the probability estimation. In reality, however, it is not easy to collect enough design strings to achieve the desired accuracy. Therefore, further improvement of the N-gram model remains one of the major research challenges in language modeling for speech recognition (e.g., [42]).

3.7 Concluding Remarks

3.7.1 Overview

We have explained the fundamental issues related to speech recognition by using simple example task settings of isolated-word recognition and static-pattern classification.[12] Unfortunately, however, these settings are primitive and limited. In this subsection, we shall give additional explanations of two important issues: (1) the selection of class model units and (2) open-vocabulary recognition.

3.7.2 Selection of Model Units

For most languages, the vocabulary size exceeds several tens of thousands, and even in our daily life we use more than several thousand words. The number of possible sentences that can be made

---

12. In Section 3.6, we introduced the task of recognizing word strings, but used it only as a task framework for describing language modeling.

by concatenating these words easily amounts to hundreds of thousands or millions. In the case of recognizing continuously spoken words or, in other words, sentences, it is obviously impractical to handle sentences such as {How long does it take to go from Kyoto to Tokyo?} and {How much does it cost to go from Kyoto to Tokyo?} as distinct classes and make class (sentence) models in a sentence-by-sentence way because of the huge number of possible sentences. Therefore, a short linguistic segment such as the phoneme is generally used as a class to recognize and as a model unit. Consequently, the class model of a long segment, such as a sentence, is formed by concatenating the short unit models.

In using short model units, recognition of words and sentences is executed by repeating two operations: one that segments an input utterance (word or sentence) to short segments, each corresponding to a model unit, and another that classifies the segment as the model's class. Generally, when the unit models are designed appropriately for the accurate classification of the corresponding segments, the classification of the entire input is accurate. Thus, in most cases, a basic recognizer design target is to increase the accuracy of the unit models. However, a simple concatenation of the short unit models that are separately designed does not necessarily result in the desired accuracy of word and sentence recognition; this shortfall is due to the coarticulation phenomenon, observed in the acoustic feature vector sequence, that is caused by the smooth behavior of speech production organs. The selection of model units is currently one of the most urgent research issues for speech recognizer design. Also, several design methods have been investigated with the aim of adjusting the short unit models to accurate word and sentence models (e.g., [43, 44]).

3.7.3 Open-Vocabulary Recognition

The recognition of conversational utterances is more exciting and promising in its potential applications than that of read-speech samples. However, it is hopelessly difficult to completely specify all of the possible acoustic and linguistic phenomena of conversational large-vocabulary speech signals. As one approach to alleviating this difficulty, the research on open-vocabulary recognition is becoming increasingly active.

The open-vocabulary condition means that a recognizer requires no restriction on vocabulary and thus its users can basically speak freely by using any word. Because it is almost impossible for the recognizer to model all of the possible word-string inputs, the recognizer attempts to spot and extract only the keywords that are useful for understanding the input word strings.[13] The system behavior would be similar to a person listening to a foreign language: because the foreign language listener's vocabulary size and listening ability for short segments such as phonemes are limited, he or she only tries to spot keywords and understand what the speaker intends.

The quality of open-vocabulary recognition mainly depends on the selection of keywords and the accuracy of keyword spotting. In particular, two types of techniques have been investigated for keyword spotting: (1) spotting based on threshold comparison and (2) spotting based on garbage modeling. Study on these techniques remains an important research issue.

3.7.4 Bibliographical Remarks

The main purpose of this chapter has been to provide fundamental information on speech recognition. However, the scope of this chapter is obviously insufficient for describing the details of the various recognition techniques.

Textbooks such as [21, 22, 45] are useful for readers to study the fundamentals of speech recognition in more detail. One recent textbook [31] invites readers to the new world of ANN/HMM hybrid speech recognition. Review papers [32, 46] provide a convenient means of grasping the present research environment. One handbook-style textbook [47] clearly explains the advanced topics in modern speech recognition.

To study the fundamentals of statistical pattern recognition, textbooks such as [4, 5, 48, 49] are recommended. Other textbooks [11, 50] provide ANN perspectives on pattern recognition.

References

[1] Bregman, A. S., *Auditory Scene Analysis*, Cambridge, Mass.: MIT Press, 1990.

[2] Cooke, M., A. Morris, and P. Green; "Missing Data Techniques for Robust Speech Recognition," *Proc. ICASSP97*, Vol. 2, 1997, pp. 863–866.

[3] Akahane-Yamada, R., Y. Tohkura, A. R. Bradlow, and D. B. Pisoni, "Does Training in Speech Perception Modify Speech Production?" *Proc. ICSLP96*, 1996, pp. 606–609.

---

13. Understanding here implies grasping the main points of recognizer input.

[4] Duda, R., and P. Hart, *Pattern Classification and Scene Analysis*, New York: Wiley, 1973.

[5] Fukunaga, K., *Introduction to Statistical Pattern Recognition*, New York: Academic Press, 1972.

[6] Dempster, A. P., N. M. Laird, and D. B. Rubin, "Maximum Likelihood from Incomplete Data via the EM Algorithm," *J. Roy. Stat. Soc.*, Vol. 39, No. 1, 1977, pp. 1–38.

[7] Fu, K., *Sequential Methods in Pattern Recognition and Machine Learning*, Academic Press, 1968.

[8] Geman, S., and D. Geman, "Stochastic Relaxation, Gibbs Distributions, and the Bayesian Restoration of Images," *IEEE Trans. on PAMI*, Vol. PAMI-6, No. 6, 1984, pp. 721–741.

[9] Amari, S., "A Theory of Adaptive Pattern Classifiers," *IEEE Trans. Electronic Computers*, Vol. EC-16, No. 3, 1967, pp. 299–307.

[10] Amari, S., "Information Theory II—Geometrical Theory of Information," Tokyo: Kyoritsu, 1968 (in Japanese).

[11] Nilsson, N., *The Mathematical Foundations of Learning Machines*, San Mateo, CA: Morgan Kaufmann Publishers, 1990.

[12] Bekesy, V. G., *Experiments in Hearing*, New York: McGraw-Hill, 1960.

[13] Zwicker, E., "Subdivision of the Auditory Frequency Range into Critical Bands," *J. Acoust. Soc. Am.*, Vol. 33, 1961, p. 248.

[14] Lyon, R. F., and C. Mead, "An Analog Electri Cochlea," *IEEE Trans. ASSP*, Vol. 36, 1988, pp. 1119–1134.

[15] Zwicker, E., and W. Peisl, "Cochlea Preprocessing in Analog Model, in Digital Model and in Human Inner Ear," *Hearing Research*, Vol. 44, 1990, pp. 209–216.

[16] Hirahara, T., "A Nonlinear Cochlea Filter with Adaptive Q Circuits," *J. Acost. Soc. Jpn.*, Vol. 47, 1991, pp. 327–335 (in Japanese).

[17] Atal, B. S., and S. L. Hanauer, "Speech Analysis and Synthesis by Linear Prediction of the Speech Wave," *J. Acoust. Soc. Am.*, Vol. 50, No. 2, 1971, pp. 637–655.

[18] Itakura, F., and S. Saito, "A Statistical Method for Estimation of Speech Spectral Density and Formant Frequencies," *Electronics and Communications in Japan*, Vol. 53-A, 1970, pp. 36–43.

[19] McDermott, E., and S. Katagiri, "LVQ-Based Shift-Tolerant Phoneme Recognition," *IEEE Trans. SP*, Vol. 39, No. 6, 1991, pp. 1398–1411.

[20] Waibel, A., et al., "Phoneme Recognition Using Time-Delay Neural Networks," *IEEE Trans. ASSP*, Vol. 37, No. 3, 1989, pp. 328–339.

[21] Huang, X., Y. Ariki, and M. Jack, "Hidden Markov Models for Speech Recognition," Edinburgh: Edinburgh University Press, 1990.

[22] Rabiner, L., and B.-H. Juang, *Fundamentals of Speech Recognition*, Englewood Cliffs, NJ: Prentice Hall, 1993.

[23] Baum, L. E., and T. Petrie, "Statistical Inference for Probabilistic Functions of Finite State Markov Chains," *Ann. Math. Stat.*, Vol. 37, 1966, pp. 1554–1563.

[24] Baum, L. E., and G. R. Sell, "Growth Functions for Transformations on Manifolds," *Pac. J. Math.*, Vol. 27, No. 2, 1968, pp. 211–227.

[25] Bahl, L., P. Brown, P. de Souza, and R. Mercer, "Maximum Mutual Information Estimation of Hidden Markov Model Parameters for Speech Recognition," *IEEE Proc. ICASSP86*, Vol. 1, 1986, pp. 49–52.

[26] Bahl, L., P. Brown, P. de Souza, and R. Mercer, "A New Algorithm for the Estimation of Hidden Markov Model Parameters," *IEEE Proc. ICASSP88*, Vol. 1, 1988, pp. 493–496.

[27] Katagiri, S., C.-H. Lee, and B.-H. Juang, "New Discriminative Training Algorithms Based on the Generalized Probabilistic Descent Method," *IEEE Neural Networks for Signal Processing*, 1991, pp. 299–308.

[28] Kung, S.-Y., "Digital Neural Networks," Prentice-Hall, 1993.

[29] Bengio, Y., R. De Mori, G. Flammia, and R. Kompe, "Global Optimization of a Neural Network-Hidden Markov Model Hybrid," *IEEE Trans. NN*, Vol. 3, No. 2, pp. 252–259.

[30] Bourlard, H., and C. Wellekens, "Links Between Markov Models and Multilayer Perceptrons," *IEEE Trans. PAMI*, Vol. 12, No. 12, 1990, pp. 1167–1178.

[31] Bourlard, H. and N. Morgan, *Connectionist Speech Recognition: A Hybrid Approach*, Norwell, MA: Kluwer Academic Publishers, 1994.

[32] Morgan, N. and H. A. Bourlard, "Neural Networks for Statistical Recognition of Continuous Speech," *Proc. of IEEE*, Vol. 83, No. 5, pp. 742–770.

[33] Niles. L., and H. Silverman, "Combining Hidden Markov Model and Neural Network Classifier," *IEEE Proc. ICASSP90*, Vol. 1, 1990, pp. 417–420.

[34] Kohonen, T., "The Self-Organizing Map," *Proc. IEEE*, Vol. 78, No. 9, 1990, pp. 1464–1480.

[35] Iwamida, H., S. Katagiri, E. McDermott, and Y. Tohkura, "A Hybrid Speech Recognition System Using HMMs with an LVQ-Trained Codebook," *J. Acoust. Soc. Jpn.* (E), Vol. 11, No. 5, 1990, pp. 277–286.

[36] Katagiri, S. and C.-H. Lee, "A New Hybrid Algorithm for Speech Recognition Based on HMM Segmentation and Learning Vector Quantization," *IEEE Trans. SAP*, Vol. 1, No. 4, 1993, pp. 421–430.

[37] Juang, B.-H., and S. Katagiri, "Discriminative Learning for Minimum Error Classification," *IEEE Trans. SP.*, Vol. 40, No. 12, 1992, pp. 3043–3054.

[38] Katagiri, S., B.-H. Juang, and C.-H. Lee, "Pattern Recognition Using a Family of Design Algorithms Based Upon the Generalized Probabilistic Descent Method," *Proc. IEEE*, Vol. 86, No. 11, 1998, pp. 2345–2373.

[39] Jelinek, F., "Self-Organized Language Modeling for Speech Recognition," IBM, T. J. Watson Research Center Report, 1985.

[40] Wright, J., "LR Parsing of Probabilistic Grammars with Input Uncertainty for Speech Recognition," *Computer Speech and Language*, Vol. 4, 1990, pp. 297–323.

[41] Lari, K., and S. Young, "The Estimation of Stochastic Context-Free Grammars Using the Inside-Outside Algorithm," *Computer Speech and Language*, Vol. 4, 1990, pp. 35–56.

[42] Nakamura, M., K. Maruyama, T. Kawabata, and K. Shikano, "Neural Network Approach to Word Category Prediction for English Texts," *Proc. COLING90*, pp. 213–218.

[43] Rabiner, L., J. Wilpon, and B.-H. Juang, "A Segmental k-Means Training Procedure for Connected Word Recognition," *AT&T Tech. Journal*, Vol. 65, No. 3, 1986, pp. 21–31.

[44] Chou, W., B.-H. Juang, and C.-H. Lee, "Segmental GPD Training of HMM Based Speech Recognition," *IEEE Proc. ICASSP92*, Vol. 1, 1992, pp. 473–476.

[45] Morgan, D. P., and C. L. Scofield, *Neural Networks and Speech Processing*, Boston: Kluwer Academic, 1991.

[46] Young, S., "A Review of Large-Vocabulary Continuous-Speech Recognition," *IEEE SP Magazine*, Vol. 13, No. 5, 1996, pp. 45–57.

[47] Lee, C.-H., F. K. Soong, and K. K. Paliwal, *Automatic Speech and Speaker Recognition*, Boston: Kluwer Academic, 1996.

[48] Iijima, T., *Pattern Recognition Theory*, Tokyo: Morikita, 1989 (in Japanese).

[49] Oja, E., *Subspace Methods of Pattern Recognition*, Letchworth: Research Studies Press, 1983.

[50] Pao, Y.-H., *Adaptive Pattern Recognition and Neural Networks*, Reading, MA: Addison-Wesley, 1989.

4

Speech Coding

Sridha Sridharan, John Leis, and Kuldip K. Paliwal

4.1 Introduction

Speech coding is a method whereby the amount of information needed to represent a speech signal is reduced. Speech coding has become an exciting and active area of research—particularly in the past decade. Due to the development of several fundamental and powerful ideas, the subject had a rebirth in the 1980s. Speech coding provides a solution to the handling of the huge and ever increasing volume of information that needs to be carried from one point to another, which often leads to existing telecommunications links operating at maximum capacity— even with the enormous channel capacities of fiber optic transmission systems. Furthermore, in the emerging era of large-scale wireless communication, the use of speech coding techniques is essential for the tetherless transmission of information. Not only communication but also voice storage and multimedia applications now require digital speech coding. The recent advances in programmable digital signal processing chips have enabled cost-effective speech coders to be designed for these applications.

This chapter focuses on key concepts and paradigms, with the intention of providing an introduction to research in the area of speech coding. It is not the intention of the authors to discuss the theoretical principles underpinning the various speech coding methods nor to provide a comprehensive or extensive review of speech coding research carried out to date. The fundamental aim is to complement other more technically detailed literature on speech coding by pointing out a number of useful and interesting techniques and referring the reader to additional sources. The chapter also considers aspects of the speech coder that need to be considered when a speech coder is incorporated into a complete system.

4.2 Attributes of Speech Coders

A large number of speech coding paradigms have been put forward in the literature. The particular choice for any given application scenario depends on the constraints for that application, and invariably a tradeoff must be made between two or more aspects of the coding method. This is not to imply that simultaneous improvements in several attributes of speech coders cannot be made—indeed, the balancing of seemingly conflicting requirements in the light of new approaches is likely to be a fruitful area of research for some time to come.

At first sight, it may seem that the primary goal of a speech coding algorithm is to minimize the bit rate. While this aspect is of major importance in many current applications, it is not the only attribute of importance and indeed other attributes may be more important in some cases. The main attributes of a speech coder are:

Bit rate This is the number of bits per second (bps) which is required to encode the speech into a data stream.

Subjective quality This is the perceived quality of the reconstructed speech at the receiver. It may not necessarily correlate to objective measures such as the signal-to-noise ratio. Subjective quality may be further subdivided into *intelligibility* and *naturalness*. The former refers to the ability of the spoken word to be understood; the latter refers to the "humanlike" rather than "robotic" or "metallic" characteristic of many current low-rate coders.

Complexity The computational complexity is still an issue despite the availability of ever increasing processing power. Invariably, coders that are able to reduce the bit rate require greater algorithmic complexity—often by several orders of magnitude.

Memory The memory storage requirements are also related to the algorithmic complexity. Template-based coders require large amounts of fast memory to store algorithm coefficients and waveform prototypes.

Delay Some processing delay is inevitable in a speech coder. This is due not only to the algorithmic complexity (and hence computation time), but also to the buffering requirements of the algorithm. For real-time speech coders, the coding delay must be minimized in order to achieve acceptable levels of performance.

Error sensitivity High-complexity coders, which are able to leverage more complex algorithms to achieve lower bit rates, often produce bit streams that are more susceptible to channel or storage errors. This may manifest itself in the form of noise bursts or other artifacts.

Bandwidth Refers to the frequency range that the coder is able to faithfully reproduce. Telephony applications are usually able to accept a lower bandwidth, with the possibility of compromising the speech intelligibility.

Some of these attributes are discussed in greater detail in Section 4.11 and in [1].

4.3 Basic Principles of Speech Coders

In essence, the fundamental aim of a speech coder is to characterize the waveform using as few bits as possible, while maintaining the perceived quality of the signal as much as possible. A waveform with a bandwidth of B Hz requires a sampling rate greater than $2B$ samples per second. Each sample in turn requires N bits in order to quantize it. For telephony, a bandwidth of 4 kHz and a quantization to 12 bits is usually required. Simplistic approaches merely use the nonlinear amplitude characteristics of the signal and the human perception of amplitude. The mathematical redundancy that exists between adjacent samples may also be exploited.

In order to achieve a truly low-rate coder, the characteristics of both the signal and the perception mechanism must be considered. A basic division often used to characterize a speech signal is that of *voiced* or *unvoiced* sounds, as illustrated in the upper panel of Figure 4.1. The voiced vowel evidently contains two or more periodic components, and one would expect that a simpler description of these components would suffice. The pseudostationary nature of the waveform means that such a parameterization would suffice over a small but finite time frame. An unvoiced sound as shown in the lower panel of Figure 4.1 appears to contain only random components. It might be expected that in order to code the unvoiced sound, a substantially larger number of bits would be

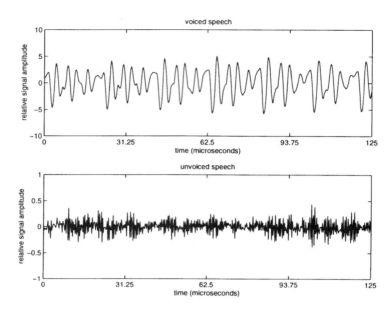

Figure 4.1 Time-domain waveforms for voiced (top) and unvoiced (lower) speech. The sampling rate is 8 kHz.

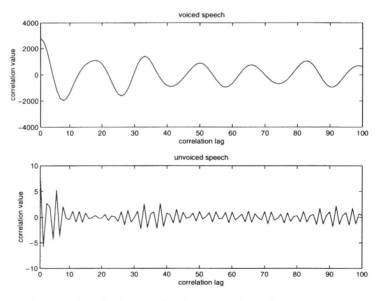

Figure 4.2 Autocorrelation of voiced (top) and unvoiced (lower) speech segments.

required. Although this is true in a mathematical sense, when the aural perceptual mechanism is taken into account the reverse is true.

One basic characterization of voiced sounds is that of the pitch. Figure 4.2 shows the autocorrelation function computed over a short time window for the time-domain waveforms previously shown. The pitch, which is due to the excitation of the vocal tract, is now quite evident for the voiced sound. Thus, the pitch is one parameter that gives the initial characterization of the sound.

In addition to providing the pitch for voiced sounds, the vocal tract and mouth modulate the speech during its production. Note that in Figure 4.3, the voiced sound contains a definite spectral envelope. The peaks of this envelope correspond to the *formants*, or vocal-tract resonances. A speech coder must be able to characterize these resonances. This is usually done through a short-term linear prediction (LP) technique (Section 4.5).

The unvoiced sound contains less obvious resonances. However, its power spectrum indicates a broader spread of energy across the spectrum. Transform techniques are able to exploit such a nonflat power spectrum by first transforming the time signal into transform-domain coefficients and then allocating bits in priority order of contribution to overall distortion or perceptual relevance. Noteworthy

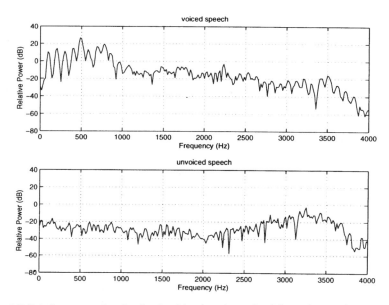

Figure 4.3 Relative power levels of voiced (top) and unvoiced (lower) speech segments.

here is that linear transform techniques have been the mainstay of such coding approaches—nonlinear techniques, although promising, have not been fully exploited.

4.4 Quantization

4.4.1 Scalar Quantization

Quantization is an essential component of speech coding systems. Scalar quantization is the process by which the signal samples are independently quantized. The process is based on the probability density function of the signal samples. An N-level scalar quantizer may be viewed as a one-dimensional mapping of the input range $\mathscr{R}$ onto an index in a mapping table (or codebook) $\mathbf{C}$. Thus

$$Q\colon \mathscr{R} \to \mathbf{C} \quad \mathbf{C} \subset \mathscr{R} \tag{4.1}$$

The receiver (decoder) uses this index to reconstruct an approximation to the input level. Optimal scalar quantizers are matched to the distribution of the source samples, which may or may not be known in advance. If the distribution is not known in advance, an empirical choice may be made (for example, a Gaussian or Laplacian distribution) for the purpose of designing the scalar quantizer [2].

4.4.2 Vector Quantization

Vector quantization is a process whereby the elements of a vector of k signal samples are *jointly* quantized. Vector quantization is more efficient than scalar quantization (in terms of error at a given bit rate) in that it accounts for the linear as well as nonlinear interdependencies of the signal samples [3].

The central component of a vector quantizer (VQ) is a codebook $\mathbf{C}$ of size $N \times k$, which maps the k-dimensional space $\mathscr{R}^k$ onto the reproduction vectors (also called *code vectors* or *code words*):

$$Q\colon \mathscr{R}^k \to \mathbf{C}, \quad \mathbf{C} = (\mathbf{y}_1\ \mathbf{y}_2\ \cdots\ \mathbf{y}_N)^T \quad \mathbf{y}_i \in \mathscr{R}^k \tag{4.2}$$

The codebook can be thought of as a finite list of vectors, $\mathbf{y}_i\colon i = 1, \ldots, N$. The codebook vectors are preselected through a clustering or training process to represent the training data. In the coding process of vector quantization, the input samples are handled in blocks of k samples, which form a vector $\mathbf{x}$. The VQ encoder scans the codebook for an entry $\mathbf{y}_i$ that serves best as an approximation for the current input vector $\mathbf{x}_t$ at time t. In the standard approach to VQ, the encoder minimizes the distortion $\mathscr{D}(\cdot)$ to give the optimal estimated vector $\hat{\mathbf{x}}_t$:

$$\hat{\mathbf{x}}_t = \min_{\mathbf{y}_i \in \mathbf{C}} \; \mathscr{D}(\mathbf{x}_t, \mathbf{y}_i) \qquad (4.3)$$

This is referred to as *nearest-neighbor encoding*. The particular index i thus derived constitutes the VQ representation of $\mathbf{x}$. This index, which is assigned to the selected code vector, is then transmitted to the receiver for reconstruction. Note that identical copies of the codebook $\mathbf{C}$ must be located in both the transmitter and the receiver. The receiver simply performs a table lookup to obtain a quantized copy of the input vector.

The *code rate* or simply the *rate* of a vector quantizer in bits per component is thus

$$r = \frac{\log_2 N}{k} \qquad (4.4)$$

This measures the number of bits per vector component used to represent the input vector and gives an indication of the accuracy or precision that is achievable with the vector quantizer if the codebook is well designed. Rearranging (4.4), it may be seen that $N = 2^{rk}$, and thus both the encoding search complexity and codebook storage size grow exponentially with dimension k and rate r.

Vector quantization training procedures require a rich combination of source material to produce codebooks that are sufficiently robust for quantization of data not represented in the training set. Examples of some of the conditions that might enrich the training set include varying microphones, acoustic background noise, different languages, and gender. In general, a large and diverse training set will provide a reasonably robust codebook but there is no guarantee that a new unseen application may not arise. A practical limitation is that codebook design algorithms, such as the Generalized Lloyd Algorithm (GLA) yield only locally optimized codebooks [4]. More recent methods, such as deterministic annealing [5] and genetic optimization [6] promise to overcome this drawback at the expense of greater computational requirements.

4.5 Linear Prediction

4.5.1 Linear Prediction Principles

Linear prediction is the most fundamental technique for removing redundancy in a signal. Linear prediction estimates the value of the current speech sample based on a linear combination of past speech samples. Let $s(n)$ be the sequence of speech samples and a_k be the kth predictor coefficient in a predictor of order p. The estimated speech sequence $\hat{s}(n)$ is given by

$$\hat{s}(n) = a_1 s(n-1) + a_2 s(n-2) + \cdots + a_p s(n-p)$$

$$= \sum_{k=1}^{p} a_k s(n-k) \tag{4.5}$$

The prediction error $e(n)$ is found from

$$e(n) = s(n) - \hat{s}(n) \tag{4.6}$$

By minimizing the mean square prediction error with respect to the filter coefficients we obtain the linear prediction coefficients (see, for example [7]). These coefficients form the *analysis filter*

$$A(z) = 1 - \sum_{j=1}^{p} a_j z^{-j}$$

$$= 1 - P_s(z) \tag{4.7}$$

The filter is sometimes known as a "whitening" filter due to the spectrum of the prediction error, which is (in the ideal case) flat. The process removes the short-term correlation from the signal. The linear prediction coefficients are an efficient way to represent the short-term spectrum of the speech signal.

For effective use of the linear prediction of speech, it is necessary to have a time-varying filter. This is usually effected by redesigning the filter once per frame in order to track the time-varying characteristics of the speech statistics due to the time-varying vocal tract shape associated with successive distinct sounds. Using this "quasi-stationary" assumption, typical speech coders use a frame size of the order of 20 ms (corresponding to 160 samples at an 8 kHz sampling rate).

4.5.2 Speech Coding Based on Linear Prediction

One way in which the linear prediction filter may be used in speech coders is as follows:

1. Subtract the predictable component $\tilde{s}(n)$ of a signal sample $s(n)$ forming the difference or error signal $e(n)$;
2. Quantize $e(n)$ to form $\hat{e}(n)$ and index i;
3. Digitally transmit i to the receiver;
4. At the receiver perform inverse quantization to recover $\hat{e}(n)$;
5. Add $\tilde{s}(n)$ to this quantized difference to form the final reproduction of $\hat{s}(n)$.

Note that the same prediction $\tilde{s}(n)$ has to be generated at the transmitter and the receiver. This is done by using the linear predictor operating on previous reconstructed speech samples $\hat{s}(n)$ to generate $\tilde{s}(n)$. Since $\hat{s}(n)$ is available both at the encoder and decoder, the same

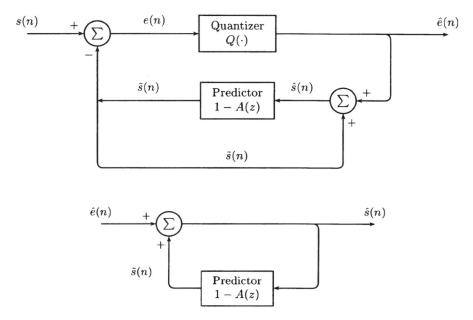

Figure 4.4 A DPCM coder. At the encoder, the prediction is based upon the *quantized* prediction error $\hat{e}(n)$ together with past predictions $\tilde{s}(n)$. At the decoder, the prediction (based on past quantized outputs) is added to the received error signal $\hat{e}(n)$ to generate each output $\hat{s}(n)$.

prediction is generated from either location. The distribution of the prediction error is normally such that scalar quantization may be applied using an appropriate quantizer [8]. The predictive quantization described here is the basis of the well-known differential pulse code modulation (DPCM) and adaptive differential (ADPCM)—an important standard for speech coding at rates of 24 to 48 kbps. The process is illustrated in Figure 4.4.

Speech coders such as ADPCM belong to the category of waveform coders that attempt to reproduce the original speech waveform as accurately as possible. Another class of coders, which is also based on linear prediction, is known as *parametric coders* or *vocoders*. These make no attempt to reproduce the speech waveform at the receiver. Instead, such coders aim to generate a signal that merely sounds similar to the original speech. The key idea is to excite a filter representing the vocal tract by a simple artificial signal that at least crudely mimics typical excitation signals generated by the human glottis. The excitation of the vocal tract is modeled as either a periodic pulse train for voiced speech, or a white random number sequence for unvoiced speech [8]. The speech signal is typically analyzed at a rate of 50 frames per second. For each frame, the following parameters are transmitted in quantized form:

1   The linear prediction coefficients;
2   The signal power;
3   The pitch period;
4   The voicing decision.

This process is shown diagrammatically in Figure 4.5.

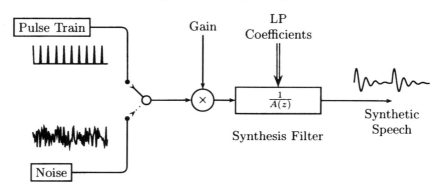

Figure 4.5 Linear predictive (LP) coder using a simple pulse train or noise excitation, corresponding to voiced or unvoiced speech. The LP coefficients and gain must be updated for each frame of speech.

The linear predictor $P_s(z)$ that specifies the analysis filter $A(z)$ is called the formant or short-term predictor. It is an all pole filter model for the vocal tract and models the short-term spectral envelope of the speech signal. The vocoder scheme can synthesize intelligible speech at the very low bit rate of 2,400 bps and has served as the underlying technology for secure voice communications. A version of the LP vocoder has been used for several years as the U.S. Government Federal Standard 1015 for secure voice communication (also known as LPC10 because it uses 10th-order linear prediction [8]). The bit allocation for this coder is summarized in Table 4.1.

The main weakness of the basic linear-prediction-based vocoder is the binary decision between voiced and unvoiced speech. Such binary voicing decisions result in low performance for speech segments where both periodic and aperiodic frequency bands are present. More recent work has resulted in the so-called mixed excitation linear prediction (MELP) coder, which has significantly increased the quality of the LPC coder. In this scheme the excitation signal is generated with different mixtures of pulses and noise in each of a number of frequency bands [10]. This scheme (with other innovations) has been selected as the new U.S. Government standard for 2,400 bps coding [11].

4.5.3 The Analysis-by-Synthesis Principle

An important concept in speech coding that has become central to most speech coders of commercial interest today is linear-prediction-based analysis-by-synthesis (LPAS) coding. In the LPC vocoder (as described in

Table 4.1
Summary of bit allocation of 2.4 kbps LPC-10 speech coder
(after [9])

Sample rate Frame size Frame rate	8 kHz 180 samples 44.44 frames/second
Pitch Spectrum (5, 5, 5, 5, 4, 4, 4, 4, 3, 2) Gain Spare	7 bits 41 bits 5 bits 1 bit
Total Bit rate	54 bits/frame $54 \times 44.44 = 2,400$ bits/sec

the previous section), the speech signal is represented by a combination of parameters (filter, gain, pitch coefficients). One method of quantizing each parameter is to compare its value to the stored values in a quantization table and to select the nearest quantized values. The index corresponding to this value is transmitted. The receiver uses this index to retrieve the quantized parameter values for synthesis. This quantization of the parameters is called *open-loop* quantization. An alternative is a process known as *closed-loop* quantization using analysis-by-synthesis. In this method the quantized parameters are used to resynthesize the original signal, and the quantized value that results in the most accurate reconstruction is selected. The analysis-by-synthesis process is most effective when it is performed simultaneously for a number of parameters.

A major reason for using the analysis-by-synthesis coding structure is that incorporating knowledge about perception is a relatively straightforward process. It can be achieved by incorporating a model of the human auditory system into the coder structure. It is well known that otherwise audible signals may become inaudible under the presence of a louder signal. This perceptual effect is called *masking* [12]. Analysis-by-synthesis coders commonly exploit a particular form of masking called *spectral masking*. Given that the original signal has a certain spectrum, the coder attempts to shape the spectrum of the quantization noise such that it is minimally audible under the presence of a louder signal. This means that most of the quantization noise energy is located in spectral regions where the original signal has most of its energy.

In the LPAS approach (Figure 4.6), the reconstructed speech is produced by filtering the signal produced by the excitation generator through both a long-term synthesis filter $1/P(z)$ and a short-term synthesis filter $1/A(z)$. The excitation signal is found by minimizing the weighted mean square error over several samples, with the error signal obtained by filtering the difference between the original and the reconstructed signals through a weighting filter $W(z)$. Both short-term and long-term predictors are adapted over time. The coder operates on a block-by-block basis. Using the analysis-by-synthesis paradigm, a large number of excitation configurations are tried for each block and the excitation configuration that results in the lowest distortion is selected for transmission. To achieve a low overall bit rate, each frame of excitation samples has to be represented such that the average number of bits per sample is small.

The multipulse excitation coder represents the excitation as a

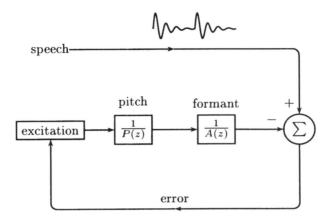

Figure 4.6 Analysis-by-synthesis minimization loop.

sequence of pulses located at nonuniformly spaced intervals [13]. The excitation analysis procedure has to determine both the amplitudes and positions of the pulses. Because finding these parameters all at once is a difficult problem, simpler procedures such as determining the locations and amplitudes one pulse at a time are used. For each pulse, the best position and amplitudes are determined, and the contribution of this pulse is subtracted before the next pulse is searched. The number of pulses required for acceptable speech quality varies between four to six pulses per 5 ms.

In the regular pulse excitation (RPE) coder [14], the excitation signal is represented by a set of uniformly spaced pulses (typically ten pulses per 5 ms). The position of the first pulse within a frame and the amplitudes of these pulses are determined during the encoding procedure. The bit allocation for the RPE coder as used in the GSM digital mobile telephony standard is shown in Table 4.2.

Code- or vector-excited coders (CELP) use another approach to reduce the number of bits per sample [16]. Here both the encoder and the decoder store a collection of N possible sequences of length k in the codebook, as illustrated in Figure 4.7. The excitation of each frame is described completely by the index to an appropriate vector in the codebook. The index is found by an exhaustive search over all possible codebook vectors and the selection of one that produces the smallest error between the original and the reconstructed signals. The bit allocation for CELP at 4,800 bps is summarized in Table 4.3.

The CELP coder exploits the fact that after removing the short- and long-term prediction from the speech signal, the residual signal has little

Table 4.2
Summary of bit allocation for GSM RPELTP 13 kbps speech coder
(after [8])

Sample rate **Frame size** **Frame rate** **Subframe size** **Pulse spacing**	**8 kHz** **160 samples (20 ms)** **50 frames/second** **40 samples (5 ms)** **3 (13 pulses/5 ms)**
Pitch Lag (40 to 120) (7, 7, 7, 7) Pitch Gain (0.1 to 1) (2, 2, 2, 2) Spectrum (6, 6, 5, 5, 4, 4, 3, 3) Excitation Pulse Position (2, 2, 2, 2) Subframe Gain (6, 6, 6, 6) Pulse Amplitudes (4 × 39)	28 bits 8 bits 36 bits 8 bits 24 bits 156 bits
Total Bit Rate	260 bits/frame 260 × 50 = 13,000 bits/sec

correlation with itself. A Gaussian process with slowly varying power spectrum can be used to represent the residual signal, and the speech waveform is generated by filtering a white Gaussian innovation sequence through time-varying long-term and short-term synthesis

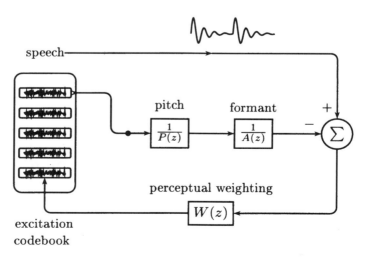

Figure 4.7 Analysis-by-synthesis using vector quantization of the excitation for the synthesis filter. This use of vector quantization is quite distinct from vector quantization of the short-term spectrum.

Table 4.3
Summary of bit allocation for FS1016 4.8 kbps CELP speech coder (after [8])

Sample rate Frame size Frame rate Subframe size	8 kHz 240 samples 33.33 frames/second 60 samples (4 subframes/frame)
Pitch lag (8, 6, 8, 6) Pitch gain (5, 5, 5, 5) Spectrum (3, 4, 4, 4, 4, 3, 3, 3, 3, 3) Excitation index (9, 9, 9, 9) Excitation gain (5, 5, 5, 5) Other (error protection etc)	28 bits 20 bits 34 bits 36 bits 20 bits 6 bits
Total Bit rate	144 bits/frame 144 × 33.33 = 4,800 bits/sec

filters. The optimum innovation sequence is selected from the codebook of random white Gaussian sequences by minimizing the subjectively weighted error between the original and the synthesized speech.

CELP can produce good quality speech at rates of 4.8 kbps at the expense of high computational demands due to the exhaustive search of a large excitation codebook (usually 512–1,024 entries) for determining the optimum innovation sequence. However, the complexity of the codebook search has been significantly reduced using structured codebooks. A thorough analysis and description of the above methods of LP-based coding may be found in [17] and [18].

4.5.4 Perceptual Filtering

One important factor in determining the performance of the LPAS family of algorithms at low rates is the modeling of the this system. By using the properties of this system, one can try to reduce the perceived amount of noise. Frequency masking experiments have shown that greater amounts of quantization noise are undetectable by the auditory system in frequency bands where the speech signal has more energy. To make use of this masking effect the quantization noise has to be properly distributed among different frequency bands. The spectral shaping is achieved by the perceptual filter $W(z)$ as shown in Figure 4.7. The filter is essentially a bandwidth expansion filter [18] of the form

$$W(z) = \frac{A(z)}{A(z/\gamma)} \tag{4.8}$$

with the bandwidth expansion controlled by the parameter γ. The effect of this is to broaden the LP spectral peaks by an amount Δf, which is related to the sampling frequency f_s by

$$\Delta f = -\frac{f_s}{\pi} \ln \gamma \quad Hz \tag{4.9}$$

Despite the error-weighting perceptual filter, it is not always possible to mask the noise in speech caused by the quantization of the excitation signal. By using a separate postprocessing filter after reconstruction by the decoder, the perceived noise can be further reduced. An adaptive postfilter of the form

$$H_{apf}(z) = \frac{\left(1 - \mu z^{-1}\right)\left(1 - \sum_{i=1}^{p} a_i \gamma_1^i z^{-i}\right)}{1 - \sum_{i=1}^{p} a_i \gamma_2^i z^{-i}} \tag{4.10}$$

may be incorporated [18]. The rationale here is to add a high-pass component controlled by μ to emphasize the formant peaks, and a pole-zero stage to "flatten" the spectral envelope. The degree of flattening is controlled by the relative values of γ_1 and γ_2, with large differences yielding a quieter but somewhat "deeper" voice [18].

4.5.5 Quantization of the Linear Prediction Coefficients

Many different representations of the linear prediction coefficients are possible. The line spectral frequency (LSF, also known as line spectrum pair, or LSP) transformation provides advantages in terms of quantizer design and channel robustness over other LP representations such as reflection coefficients, arc sine coefficients, or log area ratios.

To obtain the LSP pair representation of the LPC analysis filter, one must take the analysis filter $A(z)$ and its time reversed counterpart $A(z^{-1})$ to create a sum filter $P(z)$ and a difference filter $Q(z)$ as shown below

$$\left.\begin{array}{c} P(z) \\ Q(z) \end{array}\right\} = A(z) \pm z^{-(p+1)}A(z^{-1}) \qquad (4.11)$$

The analysis filter coefficients are recovered simply by

$$A(z) = \frac{P(z) + Q(z)}{2} \qquad (4.12)$$

The resulting line spectrum frequencies ω_k: $k = 1, \cdots, p$ are simply the alternating values of the roots of the sum and difference filters $P(z)$ and $Q(z)$ respectively. The roots are spaced around the unit circle and have a mirror image symmetry about the real axis (Figure 4.8). A number of properties of the LSF representation make it desirable in coding systems.

1. All roots of the polynomial $P(z)$ and $Q(z)$ are simple and are interlaced on the unit circle.

2. The minimum phase property of $A(z)$ can be preserved if property number (1) above is intact at the receiver. This minimizes the effect of transmission errors and ensures a stable filter for speech reconstruction at the receiver.

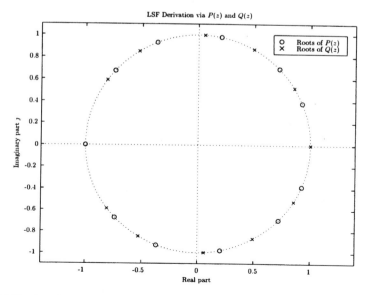

Figure 4.8 The interleaving of the roots of the polynomials $P(z)$ and $Q(z)$.

3. The LSF's exhibit frequency selective spectral sensitivity. An error in a single LSF will be confined to the region of the spectrum around that frequency.

Many quantization techniques have been developed for representing the LP coefficients with the smallest number of bits. A study of the quantization of linear prediction coefficients is reported in [19]. Scalar quantization techniques quantize the LP coefficients (or parameters derived from the LP coefficients) individually. Vector quantization quantizes a set of LP coefficients (or derived parameters) jointly as a single vector. Vector quantization exploits more efficiently the dependence of the components of the input vector, which cannot be captured in scalar quantization. Generally, vector quantization yields a smaller quantization error than scalar quantization at the same bit rate. It has been reported that it is possible to perform perceptually transparent quantization of LPC parameters using 32–34 bits per frame using scalar quantization techniques, and 24–26 bits per frame using vector quantization techniques [20]. For example, quantization of linear prediction coefficients in the FS1015 LPC10 vocoder uses 34 bit scalar quantization. In the new standard FS1017, 25 bit multistage vector quantization is used [11].

There are two major problems with LPC-VQ that need to be overcome. One is the high complexity associated with VQ algorithms, which has hindered their use in real-time applications. The second is that VQ usually lacks robustness when speakers outside the training sequence are tested.

The majority of vector quantizers used in spectrum quantization today fall within the realm of full, split, and multistage vector quantizers. The most basic is the full vector quantizer, in which the entire vector is considered as an entity for both codebook training and quantization. The two main drawbacks of full-vector VQ are the storage requirements and the computational complexity. Split and multistage VQ schemes (also known as "structured codebooks" or "product-code" schemes) have been introduced to reduce the complexity of VQ. Note that as techniques for more elaborate codebook structures are introduced, an increase in distortion level for the same bit rate is also observed when compared with exhaustive-search VQ. The main reason for different VQ structures is the necessity of lowering the complexity and storage requirements of speech coders.

Multistage VQ works by coarsely quantizing the input vector with a first-stage codebook, and in so doing creating an error vector

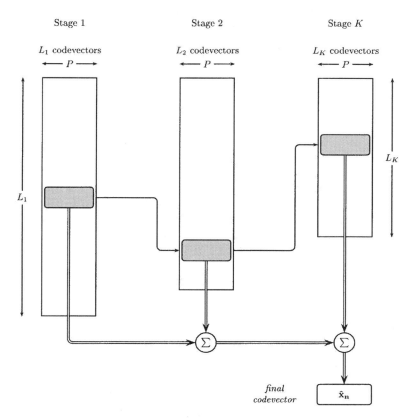

Figure 4.9 A multistage vector quantizer. Several separate codebooks are combined using addition to produce the final code vector. The number of vector elements P in each codebook is identical, and equal to the size of the final reconstructed code vector. Note that the number of entries in each codebook L_K need not be equal for each stage.

e_1 (Figure 4.9). The error vector is then finely quantized by a second stage, and if there are more than two stages, a second error vector e_2 is created by the quantization of e_1. This process continues until each stage of the codebook has been used.

The split VQ structure (Figure 4.10) divides the input vector into two or more subvectors that are independently quantized. For example, in [20], for a ten-dimensional LP coefficient vector, a two-way split partitions the vector between the fourth and fifth parameters. For a three-way split the partition could be between parameters three and four and six and seven, respectively. Splitting reduces search complexity by dividing the vector into a series of subvectors depending on how

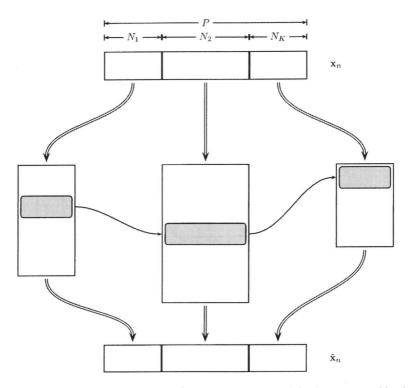

Figure 4.10 A split vector quantizer. Several separate codebooks are combined by concatenation of the subvectors to produce the final code vector.

many bits are used for transmission. For the same data, the two-way equal-split VQ requires half the computational complexity and half the storage capacity of the two-stage VQ. Comparisons of product code methods may be found in [21] and [22]. A generalized theory of product-code vector quantization may be found in [23].

Further bit rate reduction in the quantization of LP coefficients may be achieved by exploiting interframe correlation or time dependence between spectral parameters. Finite state vector quantization uses an adaptive codebook, which is dependent on previously selected code-words, in quantizing an input vector. Essentially these quantizers exploit the underlying content of speech to provide transmitter and receiver with mutual information that is not transmitted [24]. Variable frame rate VQ quantizes the LP coefficients only when the properties of speech signal have changed significantly. The LP coefficients of the nonquantized frames are regenerated through linear or some other interpolation of the parameters of other quantized frames [25]. These quantization techniques are useful for coding as low as 400 bps but are not sufficient for maintaining reasonable spectral distortion and speech intelligibility for data rates below about 300 bps [26].

Another approach using the time dependence of LPC is to perform VQ on multiple frame vectors. This method is referred to as matrix quantization or segment quantization. Segment quantization is an extension of vector quantization based on the principle that rate distortion performance is improved by using longer blocks for quantization. The time dependence of consecutive frames is included implicitly in the spectral segment, unlike the situation in variable frame rate VQ. This approach has the potential to achieve speech coding at very low rates (below 300 bps) despite the substantial computational burden.

Adaptive vector quantizers allow for low probability codevectors that have not been used in a specified time period to be eliminated from the codebook. These codevectors are replaced by higher-probability codevectors selected from the current input data but not adequately represented in the existing codebook. In effect, the current input data is added to the training set. These updated codevectors are then transmitted to the receiver during periods of silence or low speech activity. In the limit this, technique can approximate a vector quantizer that was trained on source material from the current user, yielding a decrease in overall distortion. The main drawback of this method is that the transmission of updated codevectors across the channel without errors is required [27].

4.6 Sinusoidal Coding

Instead of using LP coefficients, it is possible to synthesize speech with an entirely different paradigm—namely by generating a sum of sinusoids whose amplitudes, phases, and frequencies are varied with time [28]. This certainly seems to be a reasonable way to generate a periodic waveform. In fact it can also be applied to the synthesis of unvoiced speech. The synthesis can also be improved by generating a suitable mixture of random noise with a discrete set of sinusoids so that both unvoiced and voiced speech can be more effectively modeled. The synthesizer must seamlessly adjust the sinusoidal parameters to avoid discontinuities at the frame boundaries. Critical to this synthesis concept is effective analysis of the speech, which determines for each frame the needed sinusoidal frequencies, amplitudes, and phases. The key to this analysis is to examine and model the short-term spectrum of the input speech with a minimal and effective set of parameters.

The sinusoidal representation of the speech waveform using L sine waves, given an analysis frame of length N samples, is

$$\hat{s}(n) = \sum_{l=1}^{L} A_l \cos(n\omega_l + \phi_l) \qquad (4.13)$$

in which the sine waves are multiples of the fundamental frequency ω_l. The corresponding amplitudes A_l and phases ϕ_l are given by the harmonic samples of the short-time Fourier transform (STFT).

When the speech is not perfectly voiced, the STFT will have a multiplicity of peaks that is not necessarily harmonic. These peaks can be used to identify the underlying sine wave structure.

The types of parameters that are coded in sinusoidal coding differ significantly for different bit rates. At higher bit rates, the entire sparse spectrum (magnitudes, phases, and frequencies) and overall power are transmitted. At lower rates the phases are modeled and frequencies are constrained to be harmonics. Thus the fundamental frequency, signal power, a description of the sine wave amplitudes, and the parameters of the phase model are transmitted at low bit rates. The sine wave amplitudes can be modeled in a number of ways including all-pole modeling and differential quantization [28]. The phase information is transmitted only for bit rates above 9.6 kbps. Excellent quality output can be obtained with an analysis-synthesis system based on sinusoidal coding when the sparse magnitude and phase spectra are updated every 10 ms. At lower bit rates the phase spectrum of the reconstructed speech is obtained using a model. Different models are used for voiced and unvoiced speech. Sinusoidal coders are now viewed as a viable alternative to CELP, particularly at the rates of 2–4 kbps and below.

4.7 Waveform Interpolation Methods

A recently introduced paradigm is that of prototype waveform inter-polation (PWI), based upon a representative or characteristic waveform (CW). Several observations motivate this approach [29]. First, previous low-rate methods such as CELP are essentially waveform matching procedures that attempt to match the waveform on a frame-by-frame basis using a segmental signal-to-noise (SNR) criteria. As the bit rate is lowered, the quality of the match deteriorates—this is especially true for

CELP below about 4.8 kb/s. As pointed out in [29], the SNR is far from an ideal perceptual criteria. The waveform-matching criteria is therefore relaxed, and more emphasis is placed on the pitch waveform periodicity and pitch cycle dynamics (evolution over time). Thus, the *perceptual* quality improves, although the computed SNR decreases.

The operation of the PWI method is essentially as follows. Prototype waveforms are extracted at intervals of 20–30 ms. This extraction may be done in the LP residual domain, with the residual prototype quantized using standard analysis-by-synthesis techniques. Care must be taken to ensure continuity at the block boundaries, to guarantee the smoothness and periodicity of the resulting waveform. As pointed out in [30], the PWI method reduces to LP vocoding (a single pitch impulse) if only single impulses are used for the prototype excitation waveforms. The method is claimed to produce good speech at rates of 3–4 kbps, but is suitable only for the voiced sections of speech. More recent work has extended the method to unvoiced speech as well, decomposing the CW into a so-called rapidly evolving waveform (REW) and a slowly evolving waveform (SEW). Recent results have reported that a 2.85 kbps coder operating on this principle achieves perceptual quality comparable to the FS1016 CELP coder [31].

4.8 Subband Coding

In subband coding (SBC), the speech signal is filtered into a number of subbands and each subband is adaptively encoded. The number of bits used in the encoding process differs for each subband signal, with bits assigned to quantizers according to a perceptual criteria. By encoding each subband individually, the quantization noise is confined within its subband. The output bit streams from each encoder are multiplexed and transmitted. At the receiver demultiplexing is performed, followed by decoding of each subband data signal. The sampled subband signals are then combined to yield the recovered speech.

Note that downsampling of subband signals must occur at the output of the subband filters to avoid oversampling. The downsampling ratio is given by the ratio of original speech bandwidth to subband bandwidth. Conventional filters cannot be used for the production of subband signals because of the finite width of the bandpass transition bands. If the bandpass filters overlap in the frequency domain, subsampling causes aliasing, which destroys the harmonic structure of voiced sounds and results in unpleasant perceptual effects. If the bandpass filters don't overlap, the speech signal cannot be perfectly reconstructed because the gaps between the channels introduce an audible echo. Quadrature mirror filter (QMF) banks [32] overcome this problem and enable perfect reconstruction of the speech signal.

4.9 Variable-Rate Coding

4.9.1 Basics

The speech coders described above apply a specific unchanging coding technique to the continuously evolving speech signal without regard to the varying acoustic-phonetic character of the signal. Recently, new varieties of LPAS and sinusoidal coders have emerged which do not apply the same coding technique to each of the input frames. Instead, one of several distinct coding techniques is selected for each frame, each with its own bit allocation scheme for the parameters. The encoder in such a coding scheme selects one out of a predetermined set of techniques as the one best suited to the local character of the speech signal. The decoder, having this information, applies the corresponding decoding algorithm. This type of coding offers the opportunity to dynamically tailor the coding scheme to the widely varying local acoustic and phonetic properties of the speech signal. Variable bit rate (VBR) coding is an extension of the above coding techniques where the total allocation of bits for the frame is allowed to vary, adapting the rate to the local phonetic character and/or network conditions. VBR coders benefit from this additional degree of freedom by allocating each frame the minimum bits necessary for the decoder to adequately reproduce a frame. For example, in a typical telephone call, roughly 60% of the time, one of the speakers is silent and the signal contains only the background noise. In this situation a unimodal coder diligently encodes such nonspeech segments with the same resolution, the same algorithm, and the same bit allocation as it does for active speech segments. Clearly this is inefficient. In contrast, a VBR coder will use only a very minimal bit rate to encode the nonspeech segments. The overall quality will remain the same but the average bit rate will be much lower than a comparable unimodal fixed rate coder. VBR multimodal coders are particularly advantageous for voice storage systems, code division multiplex access (CDMA) wireless networks, and packetized communication systems.

4.9.2 Phonetic Segmentation

Further bit rate reduction can be achieved by adapting the coder to match phonetically distinct frames. In the work described in [33], speech is segmented into three major categories: voiced, unvoiced, and onset. The onset category is defined as the transition from an unvoiced to a voiced region. The voiced class is further subdivided into four categories through two layers of segmentation. Different coding strategies are used for different classes.

Another approach to very low bit rate coding based on variable rate is temporal decomposition (TD). In [34], the method is considered an efficient technique to reduce the amount of spectral information conveyed by spectral parameters through the orthogonalization of the matrix of spectral parameters. Specifically, a $p \times N$ dimensional vector of spectral parameters $\mathbf{Y}$ is approximated in the form

$$\hat{\mathbf{Y}} = \mathbf{A}\Phi \tag{4.14}$$

where Φ is an $m \times N$ matrix of *event functions*, $\mathbf{A}$ is a $p \times m$ matrix of weightings. It has been shown that almost all phonemes can be described by event functions (74% with only single event function) by using a suitable parameter set [34]. It is necessary to code both $\mathbf{A}$ and Φ for speech coding. In doing so, a considerable coding gain is achieved in coding the spectral parameters themselves. In [35] VQ was used in combination with TD to encode speech. Non-real-time speech coders at rates between 450 bps and 600 bps with naturally sounding speech were claimed. A new technique called hierarchical temporal decomposition (HTD) has recently been proposed [36, 37]. This technique reduces the computational complexity of temporal decomposition.

4.9.3 Variable Rate Coders for ATM Networks

The area in which much of the current variable rate speech coding research effort is being directed is asynchronous transfer mode (ATM) networks. ATM networks have been proposed in order to provide a common format for both high-speed data and real-time traffic such as speech and video. ATM uses short-length packets called *cells*, which are 53 bytes long [38]. Of these, 5 bytes are reserved for header information and 48 are for information. One way of using variable rate coding in ATM is to use a hierarchical "package" of bits for a given duration of speech. Through hierarchical packaging of bits (such as putting the more significant bits in one cell and the less significant bits in another cell), priorities can be attached to cells that can be used by the ATM network to control traffic flow. Low-priority cells can be dropped when the network becomes congested without severely reducing the speech quality. The technique by which an ATM network drops cells is known as *cell discarding*. Achieving higher compression through rate reduction is not straightforward in ATM systems. As the rate decreases, longer encoding frames must be used to fill fixed-length ATM packets. This has implications for both delay and recovery from cell loss due to corrupted headers or buffer overflow. There is a need to design more efficient speech coders for ATM's fixed-length packet technology.

4.9.4   Voice over IP

Another area of considerable interest at present is that of transmitting voice signals over Internet connections—known as Voice over IP (VoIP). This uses conventional IP infrastructure to transmit digitally encoded voice signals, promising a considerable reduction in costs to the end user. Because the TCP/IP protocols are based on packet store-and-forward technology with no provision for prioritized sending, traffic bursts may lead to a highly variable transmission time between the communication end points. This is the opposite of what is required for real-time voice communications. Problems such as variable network delay, packet errors, and missing packets combine to create a very difficult environment for real-time speech traffic. The bandwidth of the compressed speech is also of interest, as a lower bit rate requirement can go some way to reducing the delay and providing more space for buffering and error concealment. At present, a number of competing vendor implementations exist.

4.10   Wideband Coders

The 300–3400 Hz bandwidth requires a sampling frequency of 8 kHz and provides speech quality referred to as "toll quality." Even though this is sufficient for telephone communications and emerging applications such as teleconferencing, improved quality is necessary. By increasing the sampling frequency to 16 kHz, a wider bandwidth ranging from 5 Hz to 7,000 Hz can be accommodated. Extending the lower frequency range down to 50 Hz increases naturalness, presence, and comfort. At the other end of the spectrum, extending the higher frequency range to 7,000 Hz increases intelligibility and makes it easier to differentiate between sounds such as, for example, *s* and *f*. This results in a speech signal that is more natural. In speech transmission there is very little subjective improvement to be gained by further increase of the sample rate beyond 16 kHz.

In wideband coders the perceived quality of speech is very important and therefore the frequency masking properties of the human auditory system should be fully exploited to make the coding noise inaudible. One of the major challenges for wideband coder designs is to retain as much speech modeling as possible (to achieve high speech quality at low bit rates) while allowing other types of signals such as music to be encoded without significant degradation. The reason for this is that audio channels in applications such as tele-conferencing and high-definition television do not carry only a speech signal, even though speech is likely to constitute a large part of the information transmitted on these channels.

In 1988 the Consultative Committee for International Telegraph and Telephone (CCITT, now known as the International Telecom-munication Union, or ITU) recommended the G.722 standard for wideband speech and audio coding. This wideband coder provides high-quality speech at 64 kbps with a bandwidth of 50 to 7,000 Hz. Slightly reduced quality is achieved at 56 and 48 kbps. The G.722 coder is essentially a two-subband coder with ADPCM encoding of each subband [39]. New coding schemes for low rates at 16, 24, and 32 kbps are currently being studied for standardization. The G.722 standard will serve as a reference for the development of these alternative coding schemes.

4.11 Measuring Speech Coder Performance

The quality of speech output of a speech coder is a function of bit rate, complexity, delay, and bandwidth. It is important to consider all these attributes when assessing the performance of a speech coder. The four attributes are related. For example, a low bit rate coder will have higher computational complexity and higher delay compared with a higher bit rate coder.

The delay in a speech coding system consists of three major components [40]. Before speech can be coded it is usually necessary to buffer a frame of data. The delay due to this process, known as the *algorithmic delay*, cannot be avoided. The second is the *processing delay*, which is the time taken for the encoder to analyze the speech and the decoder to reconstruct the speech. This delay will depend on the hardware used to implement the speech coder/decoder. The *communications delay* is the third component, and is the time taken for all the data to be transmitted from the encoder to the decoder. The total of these three delays is known as the "one-way" system delay. If there are no echoes in the system then one-way delays of up to 400 ms can be tolerated. In the presence of echoes, the one-way delay can be only 25 ms.

Speech coders are usually implemented using digital signal processors. The complexity of the implementation can be measured by the computing speed requirement of the processor, together with the memory requirements. More complexity results in higher costs and greater power usage. For portable applications, greater power usage necessitates either reduced time between battery recharges or the use of larger batteries (which means more expense and weight).

Methods of assessment of speech quality have been important in the development of high-quality, low-bit-rate speech coders. Standardization activities over the past few years have resulted in an increasing need to develop and understand the methodologies used to subjectively assess new speech coding systems before they are introduced into the telephone network.

Speech quality has many perceptual dimensions but the most important are the *intelligibility* and *naturalness* of speech. These attributes are tightly coupled but not equivalent. For example, in speech synthesis, the output speech may sound artificial but could be highly intelligible. The perceived quality of speech in a system incorporating speech coders will depend on a number of factors. For example, in a cellular application, the performance depends on the transducers, speech coder, error correction, echo cancellation procedures, switches, transmitters, and receivers. Poor performance of any of these parts will affect the speech quality.

For coders that approximate the waveforms of the input signal, one can measure the difference between the input and the output to quantify the quality. Many objective measures have been studied. For example, the signal-to-noise ratio (SNR) may be computed over the complete duration of the signal. This approach has the drawback that the SNR value will be dominated by the regions that have high energy. Since speech is a nonstationary signal with many high- and low-energy sections that are perceptually relevant, a better approach is to compute the SNR for shorter segments and then compute the mean over the entire duration. The measure is referred to as segmental signal-to-noise ratio (SEGSNR). Further refinement of the SEGSNR may be obtained by clipping the maximum and minimum values and excluding silence segments.

Since SNR operates on the complete frequency band it will not give any information about the frequency distribution of the error signal. The frequency-dependent SNR can be computed by filtering the signals through a filter bank and computing the SNR for each frequency band.

Segmental SNR has been shown to correlate reasonably well with speech quality [12]. However, these methods are extremely sensitive to waveform misalignment and phase distortions that are not perceptually relevant. One approach that eliminates the effect of phase mismatches involves computing the differences between the power spectra. Since linear prediction techniques accurately model spectral peaks—which are perceptually relevant—the difference between the LP spectra of the original and coded speech may be used to measure the perceptual difference between the two signals. One of the more popular measures is based on the Euclidean distance between the cepstral coefficients. Another measure commonly used in the quantization of LPC coefficients is the *spectral distortion*, defined as

$$SD_n = \sqrt{\frac{1}{B} \int_R \left(10 \log P_n(\omega) - 10 \log \hat{P}_n(\omega)\right)^2 d\omega} \qquad \text{dB} \qquad (4.15)$$

where $P_n(\omega)$ is the power in the nth frame due to the short-term filter, and $\hat{P}_n(\omega)$ is the power in the quantized version.

As may be observed from (4.15), it is effectively an RMS measure of the difference in power between the quantized and unquantized spectra. For 8 kHz sampling, some authors use R as the band 125 Hz to 3,400 Hz, while others use a range of 0 to 3 kHz. This measure is useful in the design process for *objectively* determining spectral distortion—however, it still lacks a direct correlation with *subjective* measures. To this end, the bark spectral distortion (BSD) was proposed in [12], which is a step toward a fully objective metric that is useful in predicting the subjective quality of speech coders.

In a subjective test, speech is played to a group of listeners who are asked to rate the quality of the speech signal. In most tests the minimum number of listeners is 16 but could be as high as 64. The maximum number of listeners is usually limited by cost and time limitations. For most subjective tests, nonexpert listeners are used. The main reason for this is to better reflect the conditions under which the system will eventually be used. The mean opinion score (MOS) is an absolute category rating in which listeners are presented with samples of processed material and are asked to give ratings using a five-point scale— excellent (5), good (4), fair (3), poor (2), bad (1). The average of all votes obtained for a particular system represents the MOS. For some applications, the distribution of ratings is relevant. For example, in telecommunication applications the percentage responses that get a rating of "poor" or "bad" quality could identify future user nonacceptance.

Conducting subjective quality tests is an expensive and time-consuming procedure. It would be useful if one could predict the subjective performance through some computational process acting directly on the original and coded signals. One approach is to model the human auditory system and use both the unprocessed and processed speech as input to this model [41]. The output of the model is compared for both signals and the difference indicates the difference in quality. The remaining problem is how to correlate the differences in auditory model outputs to subjective scores. Both clustering and heuristic procedures have been used.

The quality of some speech coders is speaker dependent [42]. This is a direct result of some of the coding algorithms used (such as linear prediction or pitch prediction). It is therefore necessary to test the coder with a number of different speakers. A common number of speakers that has been used in many evaluations tests is at least four males, four females, and two children. For coders that are used in different countries, it is also important to assess if there is any dependency on the language [43].

Sometimes nonspeech signals such as music are presented to the coder. This situation can occur, for example, in telephone applications in which the caller is put on hold. During the hold time the caller commonly hears music, which is referred to as "music on hold." It is important to note that this situation is different from a background music signal, since now it is the primary signal. Although one cannot expect that the low-bit-rate speech coder would faithfully reproduce music signals, it is often required that no annoying effects be noticeable.

4.12 Speech Coding over Noisy Channels

The fading channels that are encountered on mobile radio systems often produce high error rates, and the speech coders used on these channels must employ techniques to combat the effects of channel errors. Due to the narrow spectral bandwidth assigned to these applications, the number of bits available for forward error detection and correction will necessarily be small. This has forced code designers to use different levels of error protection. Specifically, a given parameter being encoded with a B bit quantizer will have B_1 of these bits highly protected, an average level of protection will be given to the next B_2 bits and the remaining $B-B_1-B_2$ bits will be left unprotected. The selection of which bits should be placed in which class is usually done by subjectively evaluating the impact on the received speech quality of an error in a given bit position.

Unequal error protection can be exploited in the design of the quantizer (vector or scalar) to enhance the overall performance. In principle what is needed is to match the error protection to the error sensitivity of the different bit positions of the binary word representing a given parameter. This error sensitivity can be defined as the increase in distortion when that bit position is systematically altered by a channel error.

A tailoring of the bit error sensitivity profile can be accomplished in at least two ways [44]:

1. By a judicious assignment of the binary indices to the output levels in which errors in the more vulnerable bits will most likely cause the transmitted word to be received as one of the neighboring codewords;

2. By adjusting the quantizer design procedure so that the codevectors are more suitably clustered.

4.13 Speech Coding Standards

Speech coding standards are necessary for interoperability. For interoperability to be achieved, standards must be defined and implemented. All telecommunications applications clearly belong to this class, as do some storage applications such as compact discs. There are, however, speech coding applications in which interoperability is not an issue, so no standards are required. Examples of such applications are digital answering machines and voice mail storage—customized coders may be used for these applications. Many of the early speech coding standards were created by the U.S. Department of Defense (DOD) for secure speech coding applications. An important standards body is the

International Telecommunications Union (ITU). ITU defines standards for international telephone networks. Another important body is MPEG (Motion Picture Experts Group), which defines standards for compression released by the International Organization for Standardization/ International Electrotechnical Commission (ISO/IEC). Recently there has been a flurry of activity in developing standards for wireless cellular transmission.

A speech coding standard currently being prepared is the MPEG4 standard for compression of speech between 2 and 24 kbps. This standard uses code excited linear prediction (CELP) with harmonic vector excitation coding (HVXC) [45] over the range 2–4 kbps. A key feature of this standard is scalability [36].

4.14 Conclusions

This chapter introduced some of the issues involved in speech coding research and the design of practical speech coders. The fundamental model around which speech coders at low rates are based—the linear model approach—has remained the dominant research focus in recent times. The linear prediction approach—the heart of most current implementations—was introduced, with some examples to motivate its application. Vector quantization—which is essentially a template-based pattern-matching technique—was also introduced, and shown to be a method that can leverage substantial savings in bit rate at the expense of considerably greater complexity.

Significant advances in speech coding have been made by incorporating vocal tract models and aural perception models. The reason for this success is that these models have captured, however poorly, some basic properties of speech production and perception. Further improvement in speech coding systems will undoubtedly come from better understanding of the speech production and perception mechanisms.

An area of significant research activity currently is very low bit rate coding (below 1,000 bps). At these rates the authors believe that the use of language models will play a key role in improving coder performance. By studying the relationship between abstract elements of the language and how they manifest themselves in actual speech waveforms, a clear picture of the acoustic features of speech that need to be preserved in the coding process will be obtained.

To use language models in speech coding, the speech coder may incorporate the speech recognition process and the decoder may incorporate the speech synthesis process. Recent results in this area have been reported in [46, 47].

References

[1] Kleijn, W. B., and K. K. Paliwal, eds., *Speech Coding and Synthesis*, Amsterdam: Elsevier, 1995.

[2] Jayant, N. S., and P. Noll, *Digital Coding of Waveforms: Principles and Applications to Speech and Video*, Norwell, MA: Prentice-Hall, 1984.

[3] Gersho, A., and R. M. Gray, *Vector Quantization and Signal Compression*, Kluwer Academic Publishers, 1992.

[4] Linde, Y., A. Buzo, and R. M. Gray, "An Algorithm for Vector Quantizer Design." *IEEE Transactions on Communications*, Vol. **COM-28(1)**, January 1980, pp. 84–94.

[5] Nassar, C. R., and M. R. Soleymani, "Codebook Design for Trellis Quantization Using Simulated Annealing," *IEEE Transactions on Speech and Audio Processing*, Vol. 1, no. 3, October 1993, pp. 400–404.

[6] Choi, S., and W. K. Ng, "Vector Quantizer Design Using Genetic Algorithms," *Proc. IEEE Data Compression Conference*, Vol. DCC-96, 1996.

[7] Makhoul, J., "Linear Prediction: A Tutorial Review," *Proc. of the IEEE*, Vol. 63, no. 4, April 1975, pp. 561–575.

[8] Tremain, T. E., "The Government Standard Linear Predictive Coding Algorithm," *Speech Technology*, 1982, pp. 40–49.

[9] Parsons, T., *Voice and Speech Processing*, New York: McGraw-Hill, 1987.

[10] McCree, A. V., and T. P. Barnwell III. "A Mixed Excitation LPC Vocoder Model for Low Bitrate Speech Coding," *IEEE Transactions on Speech and Audio Processing*, Vol. 3, no. 4, July 1995, pp. 242–250.

[11] McCree, A. V., et al. "A 2.4 kbit/sec Coder Candidate for the New U.S. Federal Standard," *Proc. ICASSP'96*, 1996, pp. 200–203.

[12] Wang, S., A. Sekey, and A. Gersho. "An Objective Measure for Predicting Subjective Quality of Speech Coders," *IEEE Journal on Selected Areas in Communications*, Vol. 10, no. 5, June 1992, pp. 819–829.

[13] Atal, B. S., and J. R. Remede, "A New Model LPC Excitation for Producing Natural Sounding Speech at Low Bit Rates," *Proc. ICASSP'82*, 1982, pp. 614–617.

[14] Kroon, P., E. F. Deprettere, and R. J. Sluyter. "A Novel Approach to Efficient Multipulse Coding of Speech," *IEEE Transactions on Acoustics, Speech and Signal Processing*, Vol. 34, no. 5, October 1986, pp. 1054–1063.

[15] Kroon, P., and W. B. Kleijn, "Linear Predictive Analysis by Synthesis Coding," in *Modern Methods of Speech Processing*, Chapter 3. Norwell, MA: Kluwer Academic Publishers, 1995.

[16] Schroeder, M. R., and B. S. Atal, "Code Excited Linear Prediction (CELP): High Quality Speech at Low Bitrates," *Proc. ICASSP'85*, 1985, pp. 937–940.

[17] Salami, R., et al., "Speech Coding," in *Mobile Radio Communications*, Pentech, 1992.

[18] Kondoz, A. M., *Digital Speech: Coding for Low Bit Rate Communications Systems*, New York: Wiley, 1994.

[19] Paliwal, K. K., and W. B. Kleijn. "LPC Quantization," in *Speech Coding and Synthesis*, Amsterdam: Elsevier, 1995, pp. 433–466.

[20] Paliwal, K. K., and B. S. Atal. "Efficient Vector Quantization of LPC Parameters at 24 Bits/Frame," *IEEE Trans. Speech and Audio Processing*, Vol. 1, no. 1, January 1993, pp. 3–14.

[21] Wang, S., E. Paksoy, and A. Gersho. "Product Code Vector Quantization of LPC Parameters," in *Speech and Audio Coding for Wireless Network Applications*, Kluwer Academic Press, 1993, pp. 250–258.

[22] Collura, J. S., "Vector Quantization of Linear Predictor Coefficients," in *Modern Methods of Speech Processing*, Norwell, MA: Kluwer Academic Publishers, 1995, Chapter 2.

[23] Chan, W. Y., and A. Gersho, "Generalized Product Code Vector Quantization: A Family of Efficient Techniques for Signal Compression," *Digital Signal Processing*, Vol. 95, no. 4, 1994, pp. 95–126.

[24] Dunham, M. O., and R. M. Gray. "An Algorithm for the Design of Labeled-Transition Finite-State Vector Quantizers," *IEEE Transactions on Communications*, Vol. COM-33(1), January 1985, pp. 83–89.

[25] Chung, C.-J., and S.-H. Chen, "Variable Frame Rate Speech Coding Using Optimal Interpolation," *IEEE Transactions on Communications*, Vol. 42, no. 6, June 1994, pp. 2215–2218.

[26] Viswanathan, V., et al., "Variable Frame Rate Transmission: A Review of Methodology and Application to Narrowband LPC Speech Coding," *IEEE Transactions on Communications*, Vol. COM-30(4), April 1982, pp. 674–686.

[27] Paul, D. B., "An 800 bps Adaptive Vector Quantization Vocoder Using a Perceptual Distance Measure," *Proc. ICASSP-83*, 1983, pp. 67–71.

[28] McCaulay, R. J., and T. F. Quatieri, "Low-Rate Speech Coding Based on the Sinusoidal Model," in *Advances in Speech Signal Processing*, Marcel Dekker, 1992.

[29] Kleijn, W. B., "Encoding Speech Using Prototype Waveforms," *IEEE Transactions on Speech and Audio Processing*, Vol. 1, no. 3, October 1993, pp. 386–399.

[30] Kleijn, W. B., and W. Granzow. "Waveform Interpolation in Speech Coding," in *Speech and Audio Coding for Wireless Network Applications*, Norwell, MA: Kluwer Academic Press, 1993, pp. 111–118.

[31] Kleijn, W. B., and J. Haagen, "Transformation and Decomposition of the Speech Signal for Coding," *IEEE Signal Processing Letters*, Vol. 1, no. 9, September 1994, pp. 136–138.

[32] Esteban, D. and C. Galand, "Application of Quadrature Mirror Filters to Split Band Voice Coding Scheme," *Proc. ICASSP-77*, 1977, pp. 191–195.

[33] Wang. S., and A. Gersho, "Phonetic Segmentation for Low Rate Speech Coding," in *Advances in Speech Coding*, Norwell, MA: Kluwer Academic Publishers, 1991.

[34] Atal, B. S., "Efficient Coding of LPC Parameters by Temporal Decomposition," *Proc. ICASSP'83*, 1983, pp. 81–84.

[35] Cheng, Y. M., and D. O. O'Shaughnessy, "Short-Term Temporal Decomposition and Its Properties for Speech Compression," *IEEE Trans. on Signal Processing*, Vol. 39, no. 6, 1991, pp. 1281–1290.

[36] Edler, B., "Speech Coding in MPEG4," *International Journal on Speech Technology*, Vol. 2, 1999, pp. 289–303.

[37] Ghaemmaghami, S., M. Deriche, and S. Sridharan, "Hierarchical Temporal Decomposition: A Novel Approach to Efficient Compression of Speech," *Intl. Conf on Spoken Language Processing (ICLSLP-98)*, 1998, pp. 2567–2570.

[38] Fischer, W., et al., "Data Communications Using ATM: Architectures, Protocols and Resource Management," *IEEE Communications Magazine*, Vol. 32, no. 8, August 1994, pp. 24–33.

[39] Consultative Committee for International Telegraph and Telephone, "7 kHz Audio Coding at 64 kbits/sec," *Recommendation G.722, Fascile III.4*, Blue Book, 1988, pp. 269–341.

[40] Kleijn, W. B., "An Introduction to Speech Coding," in *Speech Coding and Synthesis*, Amsterdam: Elsevier, 1995, pp. 1–47.

[41] Ghitza, O., "Auditory Models and Human Performance Tasks Related to Speech Coding and Speech Recognition," *IEEE Trans. Speech and Audio Processing*, Vol. 2, no. 1, 1994, pp. 115–132.

[42] Kitawaki, N., "Quality Assessment of Coded Speech," in *Advances in Speech Signal Processing*, New York: Marcel Dekker, 1992.

[43] Montagna, R., "Selection Phase of the GSM Half Rate Channel," *Proc. IEEE Speech Coding Workshop*, 1993, pp. 95–96.

[44] de Marca, J. R. B., "On Noisy Channel Quantizer Design for Unequal Error Protection," in *Advances in Speech Signal Processing*, New York: Marcel Dekker, 1992.

[45] Nishiguchi, M., K. Iijima, and J. Matsumoto, "Harmonic Vector Excitation Coding at 2kbps," *Proc. IEEE Workshop on Speech Coding for Telecommunications*, 1997, pp. 39–40.

[46] Ribeiro, C. M., and I. M. Trancoso, "Phonetic Vocoding with Speaker Adaptation," *Proc. Eurospeech-97*, 1997, pp. 1291–1299.

[47] Lee, K-S., and V. Cox, "TTS Based Very Low Bit Rate Speech Coder," *Proc. ICASSP-99*, 1999.

Part II:
Current Issues in Speech Recognition

5

Discriminative Prototype-Based Methods for Speech Recognition

Erik McDermott

5.1 Introduction

The concept of using a prototype to represent a class of patterns is as old as Plato. A form, in Plato's theory, is the perfect, idealized essence of a range of possible (imperfect) physical instantiations of that form. Forms act as prototypes of particular categories of entities, and indeed define those categories. For instance, any given physical instantiation of a circle will vary in its precise size and shape, and will never, strictly speaking, be truly circular. However, the unchanging essence of all actual circles can be given by a mathematical description. For Plato, the latter level of description, unspoiled by the morass of variability and imperfection of the physical world, was in fact the ultimate reality.

Leaving aside the utility of the concept of prototypes in the field of metaphysics, it has clearly been a very useful and productive concept in the field of pattern recognition. Indeed, one of the simplest ways of designing a pattern classifier is to use a well-selected pattern sample as a representative of its class, and then to classify unknown patterns by matching them to all class representatives and selecting the class with the best match. Using as class representative an average of samples, rather than a single "well-selected" sample, is a refinement of this idea. This rough outline describes a large portion of research in speech recognition up to now, starting with the template-matching methods based on dynamic time warping that were prevalent in the 1970s and 1980s [1, 2], and including the hidden Markov model paradigm [3] that is dominant today.

The use of a class representative (or a few representatives), be it a sample or an average of samples, to classify new patterns, defines the prototype-based approach. A specific implementation within this approach has to address two principal issues: the particular representation of the prototypes and the nature of the matching between pattern vector and prototype. Given these structural choices, a central concern is the design or learning of prototypes that will yield good classification performance. In this chapter, several approaches to prototype design are discussed from the perspective of discriminative training, in particular as embodied in the framework of minimum classification error (MCE) [4, 5].

The organization of this chapter is as follows. After outlining the essentials of Bayes decision theory as they relate to prototype-based classification, classic approaches to example-based pattern recognition are reviewed. These methods provide a contrast that sheds light on the nature of prototype-based methods. The main section, on prototype-based approaches to speech recognition, begins with an overview of the MCE framework from a prototype-based perspective. Several prototype-based approaches to speech recognition, all of them discriminatively trained, are then described. These include learning vector quantization (LVQ), and different schemes for incorporating LVQ with hidden Markov models (HMMs); prototype-based methods that embed a dynamic programming (DP) procedure to match prototypes to input patterns; and hidden Markov models. For all of these methods, we will discuss how the MCE framework allows one to design prototypes in a manner directly aimed at minimizing classification error.

It should be noted that there are other approaches to discriminative training, such as maximum mutual information (MMI) [6]; however, these methods typically do not share the MCE framework's close link with the goal of attaining the optimal Bayes error, and furthermore, do not have the MCE framework's generality. It is not clear, for instance, how MMI could be applied to a prototype-based model that was not explicitly probabilistic.

5.2 Bayes Decision Theory

Bayes decision theory provides firm guidelines for attaining the goal of optimal classification performance. It is particularly important to be aware of these guidelines in the context of prototype-based methods. In this section, the fundamentals of Bayes decision theory will be given, followed by an outline of general aspects of prototype-based methods in the context of this theory. Specific implementations of prototype-based methods will be described in Sections 5.4–5.8.

5.2.1 The Bayes Decision Rule

The pattern recognition problem, at its most basic level, is to associate a category C with an unknown pattern vector $\mathbf{x} = (x_1, ..., x_d)$. The set of possible categories C_i, and the space of the pattern vectors $\mathbf{x}$, must of course be defined to make the problem tangible. Having done that, the essential task is then to design the mapping $\mathbf{x} \to C$ so as to optimize some measure of performance, typically the correct classification rate. Bayes decision theory, described in nearly all introductory texts, e.g., [7], provides a framework for this problem. In particular, the Bayes decision rule states that no matter how the mapping $\mathbf{x} \to C$ is implemented, it should respect the following decision rule:

$$\text{decide } C_j \text{ if } P(C_j|\mathbf{x}) > P(C_k|\mathbf{x}) \text{ for all } k \neq j \tag{5.1}$$

This rule minimizes the overall probability of classification error,

$$P(error) = \sum_{k=1}^{M} \int_{\mathcal{X}_k} P(\mathbf{x}, C_k) 1(\mathbf{x} \in C_k) d\mathbf{x} \tag{5.2}$$

where the integral is over the part $\mathcal{X}_k$ of the pattern space that will be incorrectly classified by the Bayes rule, i.e.,

$$\mathcal{X}_k = \{\mathbf{x} \in \mathcal{X} \mid P(C_k|\mathbf{x}) \neq \max_i P(C_i|\mathbf{x})\} \tag{5.3}$$

$P(error)$ is often referred to as the Bayes error.

5.2.2 Discriminant Functions

At first glance, if one takes the Bayes rule as the guiding principle of classifier design, it seems that one must explicitly represent the posterior probabilities $P(C_k|\mathbf{x})$ in order to obey the rule. However, it is easily seen that this is not so. First, the use of any quantity that is proportional or monotonically related to $P(C_k|\mathbf{x})$ is perfectly acceptable. For instance, if the prior probabilities $P(C_k)$ are all the same, the Bayes law $P(C_k|\mathbf{x})p(\mathbf{x}) = p(\mathbf{x}|C_k)P(C_k)$ readily shows that the use of $p(\mathbf{x}|C_k)$ in the decision rule yields exactly the same decisions, minimizing the misclassification rate. Similarly, $\log P(C_k|\mathbf{x})$ can of course also be used. Second, quantities that are not necessarily proportional or monotonically related to $P(C_k|\mathbf{x})$ can be used, as long as they yield the same decisions as would direct use of $P(C_k|\mathbf{x})$. A function that only crudely represents $P(C_k|\mathbf{x})$ can be used, as long as the category yielding the greatest value at any point $\mathbf{x}$ in the pattern space is the same as the category with the greatest $P(C_k|\mathbf{x})$.

Summarizing the above more formally, one can restate the Bayes rule in terms of a discriminant function $g_k(\mathbf{x})$ for each category. The discriminant functions will be used to make classification decisions:

$$\text{decide } C_j \text{ if } g_j(\mathbf{x}) > g_k(\mathbf{x}) \text{ for all } k \neq j \qquad (5.4)$$

One can see that the following choices of discriminant functions yield identical classification results:

$$g_k(\mathbf{x}) = P(C_k|\mathbf{x})$$

$$g_k(\mathbf{x}) = \log p(\mathbf{x}|C_k) + \log P(C_k)$$

$$g_k(\mathbf{x}) = \frac{p(\mathbf{x}|C_k)P(C_k)}{\sum\limits_{j=1}^{M} p(\mathbf{x}|C_j)P(C_j)}$$

$$g_k(\mathbf{x}) = p(\mathbf{x}|C_k)P(C_k) \sum\limits_{j=1}^{M} p(\mathbf{x}|C_j)P(C_j)$$

$$g_k(\mathbf{x}) = \log p(\mathbf{x}|C_k) + \log P(C_k) \qquad (5.5)$$

In these choices, $g_k(\mathbf{x})$ is explicitly related to $P(C_k|\mathbf{x})$, but it should be clear that any $g_k(\mathbf{x})$ such that

$$\arg\max_j P(C_j|\mathbf{x}) = \arg\max_j g_j(\mathbf{x}) \qquad (5.6)$$

will also yield identical classification results, and furthermore that this condition does *not* imply an explicit relation between $g_k(\mathbf{x})$ and $P(C_k|\mathbf{x})$, such as in the choices above. This is extremely important in the context of prototype-based approaches to pattern recognition. Often, the discriminant function $g_k(\mathbf{x})$ used in matching a pattern vector to a class prototype does not yield, and is not directly based on, the posterior probability $P(C_k|\mathbf{x})$. The lack of a probabilistic output might suggest that a prototype-based method cannot be related to the framework of Bayes decision theory, but this is not the case. If the condition in (5.6) holds over the whole pattern space, the classifier will in fact yield the optimal Bayes error rate.

Having argued why it is not strictly necessary to attempt to represent directly the posterior $P(C_k|\mathbf{x})$, it should of course be said that there is nothing wrong with trying to do so. Indeed, many methods use prototypes as kernels to estimate these posterior probabilities. However, being aware of the above considerations significantly broadens the range of applications of prototype-based approaches.

5.2.3 Discriminant Functions for Prototype-Based Methods

The framework described above is general enough to cover most pattern recognition methods, at least for static pattern recognition where the task is the recognition of single pattern vectors $\mathbf{x} = (x_1, ..., x_d)$. (The extension to dynamic or sequential patterns will be given later in this chapter.) The specific implementation of the discriminant function $g_k(\mathbf{x})$ is clearly the crucial point. Different methods will use different discriminant functions, which may or may not be explicitly related to the posterior probabilities used as the starting point in the preceding analysis. For prototype-based methods, the functions $g_k(\mathbf{x})$ will typically embody some type of distance between pattern sample and prototype. In the simplest case, the distance will be Euclidean distance. Or the distance could be a local pattern-to-template distance that is cumulated over a nonlinear matching of pattern and template according to a dynamic programming-based search over all possible matches. Many such "distances" are no longer, strictly speaking, distances, in that they fail to meet the elementary conditions on distances, such as symmetry and positiveness. In either case, classifier design should strive to represent a set of $g_k(\mathbf{x})$'s that respect the Bayes decision rule in the classifications that they make, as described in the preceding section.

Many distances used in the context of prototype/template-based approaches derive, either explicitly or implicitly, from a Gaussian probability density function (pdf),

$$N(\mathbf{x}, \mu, \Sigma) = \frac{1}{(2\pi)^{d/2}|\Sigma|^{1/2}} \exp(-\tfrac{1}{2}(\mathbf{x} - \mu)^t \Sigma^{-1}(\mathbf{x} - \mu)) \qquad (5.7)$$

Starting with the discriminant function

$$g_k(\mathbf{x}) = \log p(\mathbf{x}|C_k) + \log P(C_k) \qquad (5.8)$$

mentioned above, and using $N(\mathbf{x}, \mu_k, \Sigma_k)$ for $p(\mathbf{x}|C_k)$, we have

$$g_k(\mathbf{x}) = -\tfrac{1}{2}(\mathbf{x} - \mu_k)^t \Sigma_k^{-1}(\mathbf{x} - \mu_k) - \tfrac{1}{2}d \log 2\pi - \tfrac{1}{2}\log|\Sigma_k| + \log P(C_k)$$
$$(5.9)$$

Depending on one's assumptions about the nature of the classification problem, a number of simplifications can be made to this expression for the discriminant function [7]:

$$\Sigma_k = \sigma^2 I$$

When the features are independent, and the feature variance is σ^2 across all features and all categories, the covariance matrix Σ_k is the diagonal matrix $\sigma^2 I$ (as a result, $|\Sigma_k| = \sigma^{2d}$ and $\Sigma_k^{-1} = (1/\sigma^2)I$). The $|\Sigma_k|$ term now becomes an additive constant across all categories, as is the $\tfrac{1}{2}d \log 2\pi$ term; both of these can be removed without affecting classification decisions. This leaves

$$g_k(\mathbf{x}) = -\frac{(\mathbf{x} - \mu_k)^t(\mathbf{x} - \mu_k)}{2\sigma^2} + \log P(C_k) \qquad (5.10)$$

as the discriminant function. If the categories are assumed to be equally likely a priori, the $\log P(C_k)$ can also be removed. The $2\sigma^2$ term can then be dropped, and since $(\mathbf{x} - \mu_k)^t(\mathbf{x} - \mu_k)$ is just the square of the Euclidean distance between $\mathbf{x}$ and μ_k, $\|\mathbf{x} - \mu_k\|$, we see that using simple Euclidean distance for $g_k(\mathbf{x})$ and classifying $\mathbf{x}$ as the category C_k with the smallest $g_k(\mathbf{x})$ (i.e., by finding the nearest mean vector μ_k to $\mathbf{x}$) is equivalent to using the original expression, (5.9), under the assumptions of uniform variance and equal prior probabilities.

It should be noted that expanding the quadratic $(\mathbf{x} - \mu_k)^t(\mathbf{x} - \mu_k)$ and dropping another constant term (namely $\mathbf{x}^t\mathbf{x}$) allows one to rewrite

$g_k(\mathbf{x})$ as a linear discriminant function

$$\Sigma_k = \Sigma$$

A slightly weaker assumption is to allow the features to have different variances and to be dependent on one another, but to assume that the covariance matrix Σ_k is shared across all categories. The resulting $\Sigma_k = \Sigma$ is no longer diagonal. The $|\Sigma_k|$ and $\frac{1}{2}d\log 2\pi$ terms in (5.9) can again be dropped, resulting in the discriminant function

$$g_k(\mathbf{x}) = -\tfrac{1}{2}(\mathbf{x} - \mu_k)^t \Sigma^{-1}(\mathbf{x} - \mu_k) + \log P(C_k) \qquad (5.11)$$

Again, if we assume that the $P(C_k)$'s are equal, $\log P(C_k)$ can be dropped. This leads to the Mahalanobis distance:

$$D_m(\mathbf{x}, \mu) = \sqrt{(\mathbf{x} - \mu_k)^t \Sigma^{-1}(\mathbf{x} - \mu_k)} \qquad (5.12)$$

Clearly, using $g_k(\mathbf{x}) = D_m(\mathbf{x}, \mu_k)$ and choosing the category with the smallest such Mahalanobis distance is now equivalent to using the original expression, (5.9). As above, a linear discriminant function can be derived from this discriminant function, by expanding the quadratic $(\mathbf{x} - \mu_k)^t \Sigma^{-1}(\mathbf{x} - \mu_k)$ and dropping the constant term $\mathbf{x}^t \Sigma^{-1}\mathbf{x}$.

Whether or not the above assumptions are justified for a particular pattern recognition task, it is important to keep the resulting simplifications in mind, and remember that the use of, for example, Euclidean distance in a prototype-based classifier does have a link with a probability density function, namely the Gaussian pdf.

5.3 Example-Based Methods

Having loosely defined, above, a prototype-based pattern recognition method to be one in which a number of pattern class representatives are used to match unknown pattern vectors, one can ask what the alternatives are. One such alternative is the approach adopted in *example-based* methods. In a fundamental sense, these stand at the opposite end of the spectrum from the prototype-based approach. Rather than attempt to distill any general features from specific instantiations of a pattern, example-based methods simply collect all such instantiations, and represent each category by referring to the collection of data for that category. Contrasting prototype-based methods and example-based methods is instructive. Some classic example-based approaches are outlined below. This overview is based on that given in [7], to which the reader should refer for a more thorough exposition.

5.3.1 Density Estimation

The starting point for example-based modeling is to consider that a probability density $p(\mathbf{x})$ can be estimated by measuring the number k out of n sample data points $\mathbf{x}_1, ..., \mathbf{x}_n$ that fall inside a small region $\mathcal{R}$ of the pattern space:

$$p(\mathbf{x}) \approx \frac{k/n}{V} \tag{5.13}$$

where V is the volume of $\mathcal{R}$. This estimate comes from the following line of thought. The probability P that a pattern vector $\mathbf{x}$ falls inside $\mathcal{R}$ is

$$P = \int_{\mathcal{R}} p(\mathbf{x}')d\mathbf{x}' \tag{5.14}$$

The probability that k samples fall inside $\mathcal{R}$ is given by a binomial distribution sharply peaked around the expected number $E(k) = nP$ of samples falling inside this region, making the ratio k/n a good estimate of P. Now, assuming that $p(\mathbf{x})$ is continuous and that the region $\mathcal{R}$ is sufficiently small, we can make the approximation

$$P = \int_{\mathcal{R}} p(\mathbf{x}')d\mathbf{x}' \approx p(\mathbf{x})V \tag{5.15}$$

for a point $\mathbf{x}$ within $\mathcal{R}$. From this we arrive at the estimate in (5.13).

As more and more samples are collected, k_n/n converges to P, but if V remains fixed, the estimate (5.13) still only provides an averaged version of $p(\mathbf{x})$. In order to obtain $p(\mathbf{x})$ with greater resolution, V_n must be made to approach zero. If $p_n(\mathbf{x})$ is the n-th estimate of $p(\mathbf{x})$,

$$p_n(\mathbf{x}) = \frac{k_n/n}{V_n} \approx p(\mathbf{x}) \tag{5.16}$$

it can be shown that if $\lim_{n \to \infty} V_n = 0$, $\lim_{n \to \infty} k_n = \infty$, and $\lim_{n \to \infty} k_n/n = 0$, then $p_n(\mathbf{x})$ converges (in probability) to $p(\mathbf{x})$.

There are two practical approaches to designing a sequence of regions $\mathcal{R}_n$ with shrinking volumes V_n in a manner that satisfies the above conditions. In the k-nearest-neighbor approach, k_n is defined as a function of n, such as $k_n = \sqrt{n}$, and the volume is grown around any given $\mathbf{x}$ until k_n samples fall inside the region. In the Parzen window method, the volume is directly defined as a function of n, such as $V_n = 1/\sqrt{n}$. Both approaches yield estimates that converge to $p(\mathbf{x})$ as the number of samples goes to infinity. However, for the practical situation in which only a finite number of samples is available, the exact choice of k_n or V_n is up for grabs, and in fact has a large impact on the quality of the estimate. There are few guidelines for these choices. If the region is too large, the estimate is overly smooth and of low resolution. If the region is too small, the resolution is high, but the estimate is very noisy. The optimal region size lies in between these two extremes. We will later return to this central issue of the smoothness of the estimate and that of models in general.

5.3.2 Estimation of Posterior Probabilities

A slight modification of the above approach to density estimation allows the estimation of posterior probabilities $P(C_i|\mathbf{x})$ from a set of n samples with associated category labels. We consider a region $\mathcal{R}$ of volume V centered on $\mathbf{x}$ and containing k samples. We count the number k_i of samples from each category C_i contained in those k samples. The joint probability $p(\mathbf{x}, C_i)$ can be estimated as

$$p_n(\mathbf{x}, C_i) = \frac{k_i/n}{V} \tag{5.17}$$

$P(C_i|\mathbf{x})$ can then be estimated as

$$P(C_i|\mathbf{x}) = \frac{p_n(\mathbf{x}, C_i)}{\sum_{j=1}^{M} p_n(\mathbf{x}, C_j)} = \frac{k_i}{k} \tag{5.18}$$

Therefore, the posterior probability $P(C_i|\mathbf{x})$ is just the fraction of samples from C_i that are found among the k samples contained in $\mathcal{R}$. The Bayes decision rule can readily be implemented by selecting the category C_i with the greatest fraction k_i/k of samples.

5.3.3   The *k*-Nearest-Neighbor Method

The essentials of applying the k-nearest-neighbor method to the tasks of (*a*) estimating the density $p(\mathbf{x})$, (*b*) estimating the posterior probability $P(C_i|\mathbf{x})$, and (*c*) classifying $\mathbf{x}$ according to the Bayes decision rule, have now all been given. From (a), for any given pattern vector $\mathbf{x}$, one can grow a region of volume V around $\mathbf{x}$ until it captures k samples (out of a collection of n samples). To estimate $p(\mathbf{x})$ (5.16) can be used. From (b), determining the number k_i of samples from category C_i there are among the k nearest neighbors to $\mathbf{x}$ allows the estimate $P(C_i|\mathbf{x}) = k_i/k$. From (c), classifying $\mathbf{x}$ as the category with the greatest fraction k_i/k of the k nearest neighbors corresponds to the application of the Bayes decision rule using the estimate from (b).

5.3.3.1   The Nearest-Neighbor Rule

A considerable simplification of the k-nearest-neighbor rule for classification is to classify $\mathbf{x}$ according to the class label of the single nearest neighbor. As we shall see, in the context of prototype- or template-based methods for classification, such a nearest-neighbor classification procedure is common. In the context of example-based methods, however, using only a single neighbor appears somewhat risky. Interestingly, the single-nearest-neighbor rule yields respectable (though not optimal) performance that is never worse than twice the Bayes rate (and of course never better than the Bayes rate). These bounds on the error apply to the asymptotic case where the number of available samples approaches infinity. For the practical, finite sample case, few useful statements can be made on nearest-neighbor error bounds.

5.3.3.2   Error Bounds for *k*-Nearest-Neighbor Classification

In the limit as the number of available samples goes to infinity, and as k approaches infinity, the error rate P for k-nearest-neighbor classification is the optimal Bayes error rate P^*. This is to be expected, given the fact that the optimal decision rule chooses the category C_i to maximize the posterior probability $P(C_i|\mathbf{x})$, and given the preceding discussion of the convergence of the k-nearest-neighbor estimate k_i/k to the posterior probabilities $P(C_i|\mathbf{x})$ as the number of samples, and k, go to infinity. Again, little can be said about error bounds for the finite sample case.

5.3.4   Parzen Windows

This section on example-based methods began with a discussion of density estimation based on counting the number k of samples (out of a total of n samples) that fall inside a small region $\mathcal{R}$ of volume V:

$$p(\mathbf{x}) \approx \frac{k/n}{V} \qquad (5.19)$$

While the nearest-neighbor method chooses a particular value of k and then grows the volume of the region until it encompasses k samples, the starting point for Parzen window estimation is to choose a region volume V and count the number k of samples falling within that region. The window function used to count samples is then generalized to a broader kind of window function than can be used to interpolate the density $p(\mathbf{x})$ from the training samples. The window function is often referred to as a "kernel function."

A straightforward window function that can be used to count samples falling in a region is the unit hypercube window:

$$\phi(\mathbf{u}) = \begin{cases} 1 & \text{if } |u_j| \leq 1/2, \quad j = 1, ..., d \\ 0 & \text{otherwise} \end{cases} \qquad (5.20)$$

If a d-dimensional hypercube has edge length h, its volume is $V = h^d$. Therefore, $\phi((\mathbf{x} - \mathbf{x}_i)/h)$ will be one if $\mathbf{x}$ falls within the hypercube of volume V centered at $\mathbf{x}_i$, and zero otherwise. We can now express the total number of samples falling inside this hypercube as:

$$k = \sum_{i=1}^{n} \phi\left(\frac{\mathbf{x} - \mathbf{x}_i}{h^d}\right) \qquad (5.21)$$

Using this expression for k in (5.13) yields the Parzen window estimate for $p(\mathbf{x})$:

$$p_n(\mathbf{x}) = \frac{1}{n} \sum_{i=1}^{n} \frac{1}{h^d} \phi\left(\frac{\mathbf{x} - \mathbf{x}_i}{h^d}\right) \qquad (5.22)$$

The central insight of Parzen window estimation is to consider that the window function need not be limited to the preceding sample counting type, and to consider more general types of window functions that still allow the estimation of the density $p(\mathbf{x})$ as a (properly normalized) sum of window functions centered on the samples. This can be seen as a kind of interpolation of $p(\mathbf{x})$ from the samples. Perhaps the easiest way of assuring that the estimate $p_n(\mathbf{x})$ is a well-behaved density is to require that the window functions $\phi(\mathbf{u})$ themselves be probability densities, subject to the usual conditions

$$\phi(\mathbf{u}) \geqslant 0 \qquad (5.23)$$

$$\int \phi(\mathbf{u})d\mathbf{u} = 1. \qquad (5.24)$$

Substituting $(\mathbf{x} - \mathbf{x}_i)/b^d$ for $\mathbf{u}$, one sees that using $1/b^d\phi((\mathbf{x} - \mathbf{x}_i)/b^d)$ as window function satisfies these conditions if $\phi(\mathbf{u})$ does. As a result, $p_n(\mathbf{x})$ satisfies these conditions too and is a legitimate density.

As discussed above, the volume V will affect the reliability or stability of the estimate, as well as its resolution or sharpness. As the number of samples increases, V should decrease, but for a finite sample scenario, the choice of V is a compromise between an estimate that has high resolution but poor stability and one that is stable put has poor resolution. This behavior can be seen in the effect of the window width b on the Parzen window estimate. If b is large, $1/b^d\phi((\mathbf{x} - \mathbf{x}_i)/b^d)$ will have a small, slowly varying amplitude over a large region. If b is small, the window function will be narrowly peaked over a small region centered on $\mathbf{x}_i$. The estimate $p_n(\mathbf{x})$, being a sum of window functions, will reflect these characteristics by being very smooth for a large b, and sharper but more jagged (in a way that reflects the particular nature of the available samples) for a small b. The good news is that as the number of samples approaches infinity, if one decreases b (hence V) in the right way, $p_n(\mathbf{x})$ converges to the true density $p(\mathbf{x})$. The bad news is that, again, we can say very little about the proper choice of b for a given fixed number of training samples.

5.3.4.1   An Example

An example helps to illustrate the nature of Parzen window estimation. Data samples were generated according to the probability density

$$p(\mathbf{x}) = \tfrac{1}{8}U(\mathbf{x}, 0, 3) + \tfrac{3}{8}N(\mathbf{x}, 5, 1) + \tfrac{1}{2}U(\mathbf{x}, 6, 10) \qquad (5.25)$$

where $U(\mathbf{x}, l, r)$ denotes a uniform distribution between l and r, and $N(\mathbf{x}, \mu, \sigma)$ denotes a Gaussian distribution, i.e.,

$$N(\mathbf{x}, \mu, \sigma) = \frac{1}{\sqrt{2\pi}\sigma}\exp\left(-\frac{1}{2}\left(\frac{\mathbf{x} - \mu}{\sigma}\right)^2\right) \qquad (5.26)$$

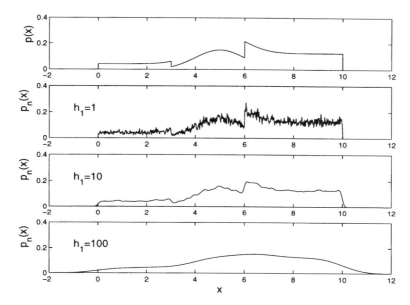

Figure 5.1 True density $p(\mathbf{x})$ (top), and three Parzen window estimates, for $h_1 = 1$, 10, and 100. Window size is $h_n = h_1/\sqrt{n}$.

A total of 20,000 samples were generated according to $p(\mathbf{x})$, and used for Parzen window estimation. The window width h is made to depend on the number of samples and an initial window width h_1, according to $h_n = h_1/\sqrt{n}$. Three choices for h_1 were considered: $h_1 = 1$, $h_1 = 10$, $h_1 = 100$. The resulting Parzen estimates $p_n(\mathbf{x})$ are plotted in Figure 5.1, with the true density shown as a dotted line. The choices for h_1 result in estimates that are too jagged ($h_1 = 1$), too smooth ($h_1 = 100$), and perhaps just about right ($h_1 = 10$). The latter judgment, of course, can only be made here because we know the true density $p(\mathbf{x})$.

5.3.5   Advantages and Limitations of Example-Based Methods

The main advantage of the example-based methods described here, and of these methods in general, is that they do not require that we make strong assumptions concerning the form of the probability distribution that is the target of estimation (or that underlies the target classification problem). Another advantage is that there is no training phase to speak of. The training data is memorized in one shot; no time is spent designing models, since there is no model.

These advantages come at a cost. The primary limitation of

example-based methods is their heavy storage requirement. Both the k-nearest-neighbor method and the Parzen window method require that the entire set of training samples be stored in memory. There is an associated computational cost, since all the samples must be used to calculate $p(\mathbf{x})$ for any given $\mathbf{x}$. Finally, as we saw, the lack of assumptions about the nature of the underlying probability distribution means that for a finite sample set, the estimate must be smoothed, yet the appropriate degree of smoothing is task-dependent and unknown. (All that one can say is that the smaller the training set, the greater the need for smoothing.)

In these days of cheap computer memory, the large storage requirement may seem less important than the computational burden. Tree-based search techniques that speed up the search for the nearest neighbors [8] can be used to make the computational load quite small, in which case an example-based method may seem quite appealing. Furthermore, there are approximations to methods like Parzen estimation that considerably reduce the memory and computational requirements. For instance, the window function can be approximated by a polynomial expansion, such that $p_n(\mathbf{x})$ itself becomes a polynomial, whose coefficients condense the information in the training samples [7].

The problem concerning the appropriate amount of smoothing can in many cases be addressed by evaluating different smoothing factors on a cross-validation data set [9]. If the task is classification, the smoothness parameter can be determined by finding the value of that parameter which yields the best performance on the cross-validation set.

5.3.6   Smoothing

It should be pointed out that the choice of smoothing factor is not an issue unique to example-based methods. Since few real-world problems come with known forms of probability distribution, the prototype- or model-based methods that nominally assume a particular type of distribution in practice must allow some aspect of the model structure to vary so as to best represent the unknown distribution underlying the task at hand. Usually this is the number of parameters, for example the number of Gaussians in a Gaussian mixture model. In a manner directly analogous to the example-based scenario, the choice of the number of parameters is a tradeoff between being able to represent a distribution with sufficient resolution, and obtaining stable estimates that do not vary too much from training set to training set. Hence, one can view the problem of choosing the right number of parameters in a model-based approach as another manifestation of the need for smoothing. We will return to this topic frequently in the remainder of this chapter.

5.3.7 Applications to Speech Recognition

Though there have been applications of k-NN and Parzen window estimation to problems in speech recognition (e.g., [10]), these remain relatively infrequent. The primary reason for this is the large computational requirement of the example-based methods. For a database of a few hours of speech, analyzed using speech frames calculated every 10 ms, the number of frames in the training data quickly approaches a million. This is the number of Parzen window functions (or potential nearest neighbors) that would have to be calculated for every frame in the unknown test data. Such a computational load is somewhat prohibitive.

Since speech recognition involves the recognition of a sequence of pattern vectors (i.e., the speech frames in an utterance), and not just the recognition of individual, static frames, any application of k-NN or Parzen window estimation would somehow have to extend the basic framework described here to handle the dynamic nature of speech. One idea for doing so is simply to incorporate k-NN or Parzen window estimation into a hidden Markov model framework. Instead of the usual Gaussian mixtures used to model state-based observation probabilities, the latter could be estimated using k-NN or Parzen window estimation. The Viterbi algorithm could then be used to find the best state alignment, and hence the recognition output, for any given unknown utterance. This proposal glosses over the issue of assigning speech frames in the training data to HMM states. The manually generated phoneme labels that are often available could be spliced into state labels; or perhaps an iterative scheme could be used in which a given state segmentation is used to estimate the observation probabilities, which are in turn used to generate a new state segmentation.

5.4 Prototype-Based Methods for Speech Recognition

A natural extension of the example-based methods described in the previous section is to overcome the large computational requirement by using only a subset of the examples, merging examples together, or some other way of condensing the relevant information in the samples into a small number of parameters, while at the same time preserving the generality of the example-based methods.

One approach to approximating the Parzen window function by using a polynomial expansion has already been described. But if one departs further from the example-based methods covered here and asks, more generally, how the relevant information in the training samples can be condensed into a small number of parameters, a very large number of methods can come into play, indeed covering nearly the entire field of statistical pattern recognition. For instance, a multi-layer perceptron (MLP) trained to generate posterior probabilities can be seen as an extremely compact representation of the aspects of the training data that are useful for density estimation or classification. Another example, more squarely in the model-based tradition, is to use a mixture of Gaussians as the model structure, and estimate the mean and variance parameters of the mixture using maximum like-lihood estimation (MLE) with the expectation-maximization (EM) algorithm [11]. It should be kept in mind that the gain in compactness is nearly always offset by a loss of generality, in that stronger assump-tions must be made about the probability distribution underlying the problem. These assumptions are sometimes not as explicit as assuming that, for example, the distribution in question is a Gaussian mixture, but nonetheless apply, in that the small number of parameters used (compared with the example-based methods) is not sufficient to represent arbitrarily complex distributions.

Prototype-based (often referred to as "template-based") methods are a subset of these many alternatives to the example-based approach and have been extensively studied. They are also the most closely linked to the example-based point of departure assumed here. This link can be seen in the nature of the prototype representation, which is usually identical to that of individual examples. Though the prototypes may be averages of large numbers of examples, or may be the result of a complex learning algorithm, they usually exist in the same space as the pattern vectors themselves.

There are several ways in which the prototypes can be used. The classic practice is to use the nearest-neighbor rule, where a simple distance metric is used to find the nearest prototype to an unknown pattern sample. The class label of that prototype is then given as the categorization of the pattern sample. This approach, clearly reminiscent of the k-NN rule, is the basis for the LVQ method that will be described in Section 5.6. Another usage, this time paralleling the Parzen window approach, is to define a kernel function around each prototype, and model an unknown function as a sum of kernel functions. This is the approach used in radial basis functions [9] and Gaussian mixture estimation. The different usages of the prototypes entail different learning rules to design the prototypes appropriately. This is an aspect that is not present in the example-based methods, in which the examples are fixed, regardless of how they are used. The computational savings of using a small number of prototypes are in a sense displaced to the sometimes large computational cost that goes into prototype design. The type of design that is of most interest to us here is discriminative training for classification tasks. Discriminative training will attempt to design prototypes that optimally divide the pattern space into categorical regions, in accordance, ideally, with the Bayes decision rule.

In the following sections, we describe a general framework for discriminative design of prototype-based classifiers, and then review recent work with prototype-based approaches to pattern recognition, in particular speech recognition.

5.5 Prototype-Based Classifier Design Using Minimum Classification Error

The Bayes decision rule, the guiding principle of classifier design, was described in the introductory section. The minimum classification error (MCE) framework for discriminative training provides a direct link between classifier design and Bayes decision theory. Applied to prototype-based classifiers, MCE allows one to design prototypes in a manner directly aimed at achieving minimum error classification.

The MCE formalization of classifier design [4, 5] is based on a three-step definition of a loss function that reflects classification performance for a single training token. This local loss function is a continuous, differentiable function of the input values and the classifier parameters, which is close to one for incorrect classifications and close to zero for correct classifications. An overall loss function can be defined by summing the local loss over all training tokens; the overall loss is then a continuous differentiable function of the classifier parameters and all input values, and approximates the performance of the classifier over all training tokens.

5.5.1 Definition of Discriminant Function

The first step of the definition of local loss has already been touched on. This is simply the definition of the discriminant function for each category, $g_k(\mathbf{x})$. To emphasize the dependence of this function on the system parameters, we can use Λ to denote the system parameters that are subject to adaptation, and rewrite the discriminant function to be $g_k(\mathbf{x}, \Lambda)$. In the MCE framework, it is vital that this function be a (mostly) continuous, first-order differentiable function of Λ. The reason for this is that MCE is typically used with gradient-based methods to adapt the system parameters. As discussed in the context of Bayes decision theory, the discriminant function is the heart of the classifier; classification decisions for a given input pattern token are made by choosing the category with the best discriminant function value.

5.5.2 Definition of Misclassification Measure

The next step in the MCE formalism is to define a misclassification measure that compares the discriminant function for the correct category C_k with the discriminant functions from all other categories, with an (adjustable) emphasis on the categories that are most competitive with the correct category. If the discriminant function is a type of distance (as it usually is with prototype-based methods), the misclassification measure has the form:

$$d_k(\mathbf{x}, \Lambda) = g_k(\mathbf{x}, \Lambda) - \left[\frac{1}{M-1} \sum_{j \neq k} g_j(\mathbf{x}, \Lambda)^{-\psi} \right]^{-1/\psi} \quad (5.27)$$

where M is the number of categories, and ψ is a smoothing factor that controls the degree to which the top incorrect category dominates the bracketed expression. (For a very large ψ, the expression is approximately the discriminant function value of the best incorrect category.) One can see that when $g_k(\mathbf{x}, \Lambda)$ is smaller (i.e., better) than the discriminant functions for the other, incorrect, categories, $d_k(\mathbf{x}, \Lambda)$ is negative. This situation corresponds to a correct classification: C_k would be chosen as the best category. Conversely, when a classification error occurs, $d_k(\mathbf{x}, \Lambda)$ is positive. Thus, the sign of the misclassification measure reflects the correctness or incorrectness of the classification. Note that the misclassification measure is (mostly) continuous and differentiable. (The case $g_j(\mathbf{x}, \Lambda) = 0$ has to be handled separately.)

5.5.3 Definition of Local Loss Function

Given this definition of misclassification measure, the next step in the MCE formalism is to define a loss function that maps the misclassification measure to a loss function between zero and one. This can be done with a number of definitions, for example the sigmoidal loss function:

$$\ell(d_k(\mathbf{x}, \Lambda)) = \frac{1}{1 + e^{-\alpha d_k(\mathbf{X}, \Lambda)}} \tag{5.28}$$

This is the local MCE loss function. Again, it is a continuous, differentiable function of the classifier parameters Λ.

5.5.4 Overall Loss Function and Optimization

Using the preceding definition of loss, an overall loss function can be defined. Ideally, the overall loss would be found by integrating the local loss over all M categories and their probability densities:

$$L(\Lambda) = E[\ell(\mathbf{x}, \Lambda)] = \sum_k^M P(C_k) \int \ell_k(\mathbf{x}, \Lambda) p(\mathbf{x}|C_k) d\mathbf{x} \tag{5.29}$$

Though we do not, in general, know the true probability densities that this expression is based on, the probabilistic descent theorem [12] provides one way of minimizing this definition of overall loss. In particular, this theorem shows that for an infinite sequence of random samples $\mathbf{x}_t$ (where t denotes a discrete time index in a training sequence), and step size sequence ϵ_t that satisfies the conditions

$$\sum_{t=1}^{\infty} \epsilon_t \rightarrow \infty \tag{5.30}$$

and

$$\sum_{t=1}^{\infty} \epsilon_t^2 < \infty \tag{5.31}$$

adapting the system parameters according to

$$\Lambda_{t+1} = \Lambda_t - \epsilon_t U \nabla \ell_k(\mathbf{x}_t, \Lambda_t) \tag{5.32}$$

where Λ_t denotes the state of Λ at t, converges with a probability of one to a local minimum of $L(\Lambda)$. (U is a positive definite matrix that allows one to scale the learning rate differently for different model parameters.

In the simplest case, this matrix is the identity matrix.) An example that satisfies the constraints on the step size is $\epsilon_t = 1/t$. This is the optimization approach adopted in the generalized probabilistic descent (GPD) method [4–5]. Through GPD, classifier parameters Λ can be adapted to minimize the MCE loss function.

A more practical definition of overall loss is the empirical loss function obtained by summing the local loss function over the available training samples:

$$L_1(\Lambda) = \frac{1}{N} \sum_{k}^{M} \sum_{i=1}^{N_k} \ell_k(\mathbf{x}_{ik}, \Lambda) \qquad (5.33)$$

This is amenable to optimization using many different methods. The token-by-token GPD training scheme can be applied by repeatedly cycling over the training data, while diminishing the step size appropriately. (In practice, training can be stopped after a fixed number of iterations.) Conjugate gradient methods, as well as second-order techniques such as modified Newton's methods, can also be used. The reader is referred to [13] for a good review of different optimization methods. Many of these methods can be used in "batch" mode, where the classifier parameters are only updated after a presentation of the entire training set. This makes it possible to parallelize the optimization phase by distributing training data over several machines. This is not so easy to implement in the token-by-token GPD adaptation.

5.5.5 Modified Newton's Method: The Quickprop Algorithm

We here describe one practical approach to second-order optimization, the quickprop algorithm [14]. This algorithm is easy to implement and parallelize, and illustrates the basic principles of Newton's method. It has been used to train both MLPs and HMMs [15], and may be an attractive method for prototype-based classifier design.

Quickprop can be seen as a rough approximation of the classic Newton's method. The central idea in Newton's method is to build a *model* $M(\ )$ of the function of interest $F(\ )$ using the first three terms of the Taylor series expansion of the function, around the current point Λ, for a given step s:

$$F(\Lambda + s) \approx M(\Lambda + s) = F(\Lambda) + \nabla F(\Lambda)'s + \tfrac{1}{2}s'\nabla^2 F(\Lambda)s \qquad (5.34)$$

Newton's method is to solve for the step s^N that leads to a point where the gradient of the model is zero, i.e., $\nabla M(\Lambda + s^N) = 0$. This corresponds to the linear system

$$\nabla^2 F(\Lambda) s^N = -\nabla F(\Lambda) \qquad (5.35)$$

The solution s^N is the Newton step, which can be written:

$$s^N = -(\nabla^2 F(\Lambda))^{-1} \nabla F(\Lambda) \qquad (5.36)$$

This step is then used in the optimization procedure to update the parameter vector Λ, i.e., $\Lambda_t = \Lambda_{t-1} + s^N$. In the ideal situation, if the quadratic model of $F(\ )$ is correct, and the Hessian matrix $\nabla^2 F(\ )$ is positive definite, the minimum will be reached in one iteration. In general, if the Hessian matrix is positive definite, and the initial value of Λ is sufficiently close to the optimum, Newton's method converges "q-quadratically" [13].

The first difficulty in using Newton's method for practical optimization problems is that calculating the Hessian matrix (whose size goes up as the square of the size of the parameter vector Λ) may be prohibitive. Secondly, if the Hessian is not positive definite, there will be directions of negative curvature (i.e., directions along which the model always decreases) that will lead to infinitely large steps s^N. Furthermore, even if the Hessian is positive definite, it may be ill-conditioned, a source of numerical problems for computer implementations.

Many (modified Newton's methods) address the problems of positive definiteness and conditioning of the Hessian by considering suitable modifications to the Hessian that ensure its positive definiteness and proper conditioning. One approach is to add to the Hessian a simple diagonal matrix of the form μI (i.e., the identity matrix times a scalar) so that $[\nabla^2 F(\ ) + \mu I]$ is positive definite and well-conditioned. The modified Cholesky factorization can be used to find a proper value for μ [13]. It is interesting to note that adding the diagonal matrix μI to the Hessian results in an optimization procedure that combines simple gradient descent and Newton's method. The step s^N becomes:

$$s^N = -[\nabla^2 F(\Lambda) + \mu I]^{-1} \nabla F(\Lambda) \qquad (5.37)$$

When the original Hessian is sufficiently positive definite and well-conditioned, μ will not have to be large, and the step will be close to the original Newton's step. When the Hessian is far from positive definite, and very poorly conditioned, a large μ will be required, and the addition μI will tend to dominate. In this case, the step is proportional to the negative of the simple gradient:

$$s^N = -[\nabla^2 F(\Lambda) + \mu I]^{-1} \nabla F(\Lambda) \approx -\frac{1}{\mu} \nabla F(\Lambda) \qquad (5.38)$$

Quickprop is closely related to this compromise between Newton's method and simple gradient descent. The quickprop algorithm uses an approximation diagonal to the Hessian matrix. For each component of the parameter vector, the previous and current values of the gradient are divided by the previous step size to yield a numerical approximation of the diagonal components of the Hessian. All other components are ignored:

$$\frac{\partial^2 F(\Lambda_t)}{\partial \lambda_i^2} \approx \frac{\dfrac{\partial F(\Lambda_t)}{\partial \lambda_i} - \dfrac{\partial F(\Lambda_{t-1})}{\partial \lambda_i}}{\Delta \lambda_{i,t-1}} \qquad (5.39)$$

Using this approximation of the Hessian, the quickprop algorithm then uses a heuristic to determine whether or not the Hessian model is sufficiently positive definite with respect to each dimension of the parameter vector. If the Hessian is positive definite quickprop compares the slope of the error (for a particular dimension) at times t and $t - 1$. If the current slope is opposite in sign from the previous slope, the Hessian is considered strongly positive definite, and is used as is for the update of Λ. Otherwise, if the current slope is in the same direction as the previous slope, the Hessian is considered insufficiently positive definite along that dimension. (In the worst case, if the current slope is in the same direction as the previous slope, and equal in magnitude, it would appear that the minimum lies at infinity.) In this case, the simple gradient, multiplied by a scalar, is added to the diagonal approximation of the Hessian. If the Hessian model is not positive definite, it is ignored, and the simple gradient is used alone.

This heuristic, dimension-by-dimension modification of the (approximated) Hessian closely resembles the modified Newton's method described above. However, there is still no guarantee that the resulting approximated Hessian is in fact positive definite and well-conditioned. As a further measure, the quickprop algorithm limits the step size that results from using the approximated Hessian. The step is first calculated according to (5.36), but its magnitude is then limited to μ

times the previous step magnitude for that parameter component. Furthermore, if the step magnitude is larger than an absolute (task-dependent) limit, or if (in spite of the above precautions) the step is in the opposite direction from the current gradient (i.e., the step points uphill on the current slope), μ times the previous step is used instead.

Though quickprop is a heuristic algorithm, it has enjoyed widespread use. It can be used in both batch and online modes. As mentioned, the batch mode is attractive if one wants to parallelize the learning procedure over the training data. It is also possible to consider hybrid batch/online modes, where the parameter update is calculated every N training tokens, for an N that is less than the total number of training tokens. Updating frequently shares the advantages of stochastic gradient descent in that it allows the optimization procedure to exploit redundancies in the data [16].

Applying quickprop to MCE-based optimization is straightforward. The gradient of the local MCE loss is summed over all training tokens in one update period, and used as $\nabla F(\Lambda)$ above.

Figure 5.2 illustrates the differences between (first-order) optimization based on GPD, and the kind of second-order optimization performed in quickprop.

5.5.6 Relation of MCE Loss to the Bayes Error

The MCE link to the Bayes error is discussed in depth in Chapter 9, "Minimum Classification Error Networks," to which the reader is referred. The essential result is that minimizing the MCE cost function

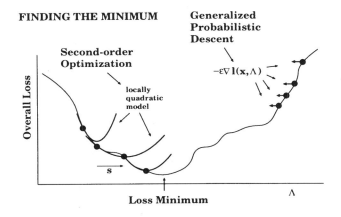

Figure 5.2 Second-order optimization versus probabilistic descent.

(5.29) (in an ideal scenario in which sufficient parameters are used, sufficient data is available, and the global minimum of the cost function can be attained) yields the optimal Bayes error rate. The key property of the MCE loss function $\ell(d_k(\mathbf{x}, \Lambda))$ is that it approximates the actual classification performance of the system. Minimizing an overall cost based on this loss function directly minimizes the overall misclassification rate. MCE-based optimization is thus directly aimed at achieving the lowest possible misclassification rate, i.e., the Bayes error rate.

5.5.7 Choice of Smoothing Parameters for MCE-Based Optimization

The parameters ψ and α control the extent to which the MCE loss function (5.28) approximates the classification correctness or incorrectness for a given input token $\mathbf{x}$. As both are made to approach infinity, the MCE loss approaches the binary, all or nothing, "ideal" loss function. However, the choice of ψ and α, in particular α, depends on the amount of training data available. In a situation very similar to the k-NN or Parzen window scenarios, the larger the training set, the lesser the need for smoothing; the smaller the training set, the greater the need. Specifically, when the training set is large, α may be large, resulting in an MCE loss close to the ideal loss; but for smaller training sets, α will have to be smaller.

Another important aspect of the loss function smoothness is its effect on the optimization procedure. A smooth error surface will have fewer local minima and will in general be more tractable to gradient-based optimization. An example of this will be provided below.

5.6 Learning Vector Quantization

Learning vector quantization (LVQ) [17, 18] is closely related to the learning subspace method (LSM) and the self-organizing map (SOM) [18, 19]. The SOMs offer a way of modeling data using a set of prototypes (reference vectors of fixed dimensionality, corresponding to that of the pattern space) organized in a low-dimensional grid. The training procedure is such that prototypes that are near one another in the grid are associated with data points that are near one another in the pattern space. Given a training token, the SOM algorithm adapts not only the nearest prototype vector, but also its neighbors in the grid. This can be related to a neural representation of prototypes where lateral control of plasticity is a Gaussian kernel function defining how local activation determines the learning rate in its neighborhood. The LSM is a discriminative approach to classification, in which categories are modeled by a set of basis vectors spanning a subspace of the overall pattern space. Classification is performed by projecting a pattern vector onto each subspace, and choosing the subspace in which the projected vector is nearest to the original pattern vector. The training algorithm rotates the subspaces so as to minimize the number of misclassifications, essentially rotating each category subspace so that it is closer to training vectors for that category, and rotating subspaces for incorrect categories so that they are further away [18]. In sum, both SOM and LSM illustrate the basic aspects of prototype-based learning; in addition, LSM illustrates the principle of discriminative training. By combining these aspects, LVQ provides a good, simple illustration of prototype-based pattern recognition.

Originally, several versions of LVQ were presented [17, 18]; here we focus on the more discriminant of them, LVQ2. In LVQ2, each category to be learned is assigned a number of reference (prototype) vectors, each of which has the same number of dimensions as the input vectors of the categories. In the recognition stage, an unknown input vector is classified as the category of the reference vector that has the smallest Euclidean distance to that input vector. This classification scheme means partitioning the vector space into regions, or cells, defined by individual reference vectors. LVQ training adjusts the reference vectors so that each input vector has a reference vector of the right category as its closest reference vector.

Kohonen, in his formulation of the LVQ2 algorithm, is particularly concerned with the problem of approximating decision lines corresponding to the minimum error rate classification. This becomes particularly relevant when the class joint density functions, $P(C_j)p(\mathbf{x}|C_j)$, overlap. If there is overlap, it is impossible to separate the classes perfectly; the task becomes that of finding the decision lines which minimize the number of misclassifications, i.e., the decision lines that implement the Bayes decision rule. The optimal lines, corresponding to the Bayes decision rule, will be along the hypersurfaces where the class joint density functions cross. Ideally, a classifier should generate decision lines that approximate the optimal lines. This is the motivation for LVQ2.

Figure 5.3 helps to illustrate vector adaptation in LVQ2, for a simple one-dimensional situation. For a given training vector $\mathbf{x}$, three conditions must be met for learning to occur: (1) the nearest class must be incorrect; (2) the next-nearest class (found by searching the reference

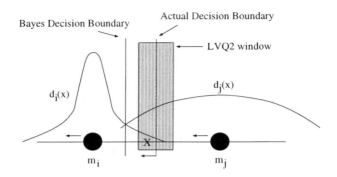

Figure 5.3 LVQ2 adaptation, one-dimensional, two-class problem.

vectors in the remaining classes) must be correct; (3) the training vector must fall inside a small, symmetric window defined around the midpoint of the reference vectors $\mathbf{m}_i$ and $\mathbf{m}_j$ (corresponding to categories C_i and C_j)—this midpoint (in higher dimensions, midplane) being the decision boundary realized by the two vectors. If these conditions are met, the incorrect reference vector is moved farther away from the input, while the correct reference vector is moved closer, according to:

$$\mathbf{m}_i(t+1) = \mathbf{m}_i(t) - \alpha(t)(\mathbf{x}(t) - \mathbf{m}_i(t)), \tag{5.40}$$

$$\mathbf{m}_j(t+1) = \mathbf{m}_j(t) + \alpha(t)(\mathbf{x}(t) - \mathbf{m}_j(t)), \tag{5.41}$$

where $\mathbf{x}$ is a training vector belonging to category C_j, $\mathbf{m}_i$ and $\mathbf{m}_j$ denote the closest reference vectors within categories C_i and C_j, respectively, and $\alpha(t)$ is a monotonically decreasing function of time, e.g., $\alpha(t) = t/T$, where T is a preset total number of token presentations.

These requirements, taken together, are aimed at assuring that the decision line between the two vectors will eventually approximate the Bayes decision boundary that corresponds to the minimum classification error probability situation, at the place where the joint density functions cross. Training samples on the wrong side of the boundary, but close enough to the boundary to be within the window, pull the reference vectors, and hence the boundary, closer to them. This occurs for incorrectly classified tokens on both sides of the boundary, resulting in a kind of tug of war. If there are more such incorrectly classified samples on one side of the boundary than on the other, the net effect will be to displace the boundary toward the larger set of samples. If the number of incorrectly classified samples is the same on either side, there will be no net movement and the system may be said to have converged.

Clearly, the window plays an essential role. If there are enough samples to allow a very small window size, the equilibrium situation of equal pull on either side of the boundary will be attained very near the point where the true joint probability density functions cross, i.e., the Bayes boundary. If there are fewer samples, a larger window must be used, and the resulting decision boundary then applies to a smoothed version of the true joint probability density functions. As with the k-NN and Parzen window methods above, the need for smoothing is related to the number of available training samples; the fewer the samples, the greater the need for smoothing.

The above justification for the LVQ2 training procedure is of course only an intuitive one. Indeed, at the time of its proposal, there was no proof that LVQ2 did in fact converge to the optimal Bayes configuration. We will see later that the MCE framework can provide a theoretical basis for an adaptation procedure very similar to LVQ2.

In a statistical benchmark comparing LVQ, back-propagation, and Boltzmann machines, it was found that the simple, quickly trained LVQ performed extremely well [17]. The ease of implementation of this algorithm, and its high classification power, has spawned a tremendous number of applications, in speech recognition as well as other areas.

5.6.1   Shift-Tolerant LVQ for Speech Recognition

As we have described it, the LVQ algorithm in its basic form is a method for static pattern recognition. In applying it to problems in speech recognition, one must somehow expand its scope of application to handle a stream of dynamically varying patterns. This issue is the same issue facing MLPs; it spawned a variety of solutions in the MLP context [20–22]. Among the extremely large number of applications of LVQ to problems in speech recognition, we here focus on the shift-tolerant LVQ (STLVQ) architecture [23, 24]. This method closely follows the approach in the highly influential time-delay neural network (TDNN) model. Figure 5.4 shows the architecture of the recognition system. Each category is assigned a number of reference vectors. The LVQ training procedure is then applied to speech patterns that are stepped through in time, thus providing the system with a measure of shift-tolerance similar to that of TDNN. To achieve this effect while at the same time capturing a sufficient amount of temporal context for the recognizer, STLVQ uses a 7-frame window that is shifted, one frame at a time, over the 15-frame speech token. Each window position yields an input vector of 112

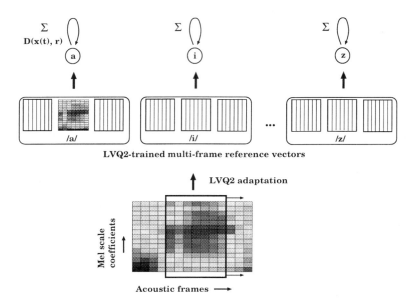

LVQ2-trained multi-frame reference vectors

Figure 5.4 Shift-tolerant LVQ system architecture.

dimensions (7 frames × 16 channels). Given this input vector, LVQ2 is applied as described above.

This moving window scheme requires a different recognition procedure than simply finding the closest vector, as there are now several closest vectors, one for each window position. Departing slightly from the original STLVQ scheme used in [23, 24], the distances to the closest reference vector in each category can simply be summed over all 15 frames of the input, yielding an overall distance for each category that can then be used for categorization. The discriminant function in STLVQ is thus a sum of Euclidean distances:

$$g_k(\mathbf{x}_1^T, \Lambda) = \sum_{t=1}^{T} D(\mathbf{x}_t, \mathbf{r}_{z_{k,t}}^k) \qquad (5.42)$$

where $\mathbf{x}_1^T = (\mathbf{x}_1, ..., \mathbf{x}_T)$ denotes a sequence of T feature vectors.

The STLVQ architecture was successfully trained and tested on the same phoneme recognition tasks used in the evaluation of TDNN. Though much simpler than the TDNN architecture, STLVQ yielded very good results on these tasks, comparable with those obtained for TDNN. These results are one set out of many results showing the power of discriminative methods based on prototype modeling.

5.6.1.1   HMM interpretation of STLVQ

As discussed earlier, Euclidean distance can be seen to derive from a Gaussian pdf. If the feature variances are taken to be uniform across all feature vector dimensions and all categories, taking the logarithm of the Gaussian pdf and dropping the terms that do not affect categorization gives us negative squared Euclidean distance. In a one-state continuous HMM of the kind suggested in Figure 5.4, the overall log likelihood would be found by summing the log likelihoods for each frame. Hence we see a similarity between STLVQ and an HMM using one state per category (and feature vectors taken from large, overlapping windows). The similarity can be deepened by examining how the nearest-neighbor-based recognition procedure in STLVQ can be seen as a particular HMM implementation. The Gaussian mixture typically used in an HMM state s,

$$b_s(\mathbf{x}_t) = \sum_{i=1}^{I_s} c_{s,i} N(\mathbf{x}_t, \mu_{s,i}, \Sigma_{s,i}) \qquad (5.43)$$

(where I_s denotes the number of Gaussians in the mixture) will behave in a nearest-neighbor, winner-takes-all manner if the variances are assumed to be very small. If the mixing weights $c_{s,i}$ are assumed equal, they too can be dropped from the log likelihood. Thus, STLVQ can be seen to correspond to an HMM using one state per category, uniform mixing weights, and small, uniform variances. (The remaining difference between the two architectures is that between summing Euclidean distances and squared Euclidean distances.) Of course, this similarity applies only to the classifier structure, and not to the manner in which the unknown parameters are determined; in STLVQ, via the LVQ algorithm, in an HMM, using the Baum-Welch or Viterbi algorithms [11]. This brings up the interesting topic of how an LVQ-like algorithm for discriminative training could be applied to an HMM without making the above limiting assumptions. It will be seen later that the MCE framework allows just that.

5.6.1.2   Limitations and Strengths of STLVQ Architecture and Training

The primary limitation of the architecture described here is that it assumes an input token that is known to contain a single phoneme. Both training and testing datasets used to evaluate STLVQ were obtained from manually labeled speech databases. Practical speech recognition of course does not allow this luxury. Furthermore, it is not clear how the phoneme recognition achieved with STLVQ can be extended to word or sentence recognition (though the similarities with a simplified HMM are highly suggestive). A limitation of the training method is that LVQ is only applied locally, at the level of the multiframe window. In effect, optimization is directed at the subtask of classifying a multiframe window into one of the phoneme classes; ideally, it would be directed at the overall task of categorizing the entire sequence of feature vectors using the STLVQ discriminant function (5.42). Nonetheless, in spite of these shortcomings, the positive results for STLVQ do show that prototypes spanning multiple frames of speech can successfully be trained to capture the speech dynamics relevant for phoneme classification, and that the classifier can be given a measure of robustness to shifts in the input simply by training the prototypes on shifted versions of the training tokens [24, 25].

5.6.2 Expanding the Scope of LVQ for Speech Recognition: Incorporation into Hidden Markov Modeling

Summarizing the essential aspects of STLVQ, we see that the architecture provides (partial) solutions to the central issues of:

- Representation of local speech dynamics important for discrimination;
- Tolerance to shifts in time.

The issues not addressed satisfactorily in this architecture include:

- Representation of longer speech sequences such as entire utterances;
- Global optimization;
- Application to continuous speech recognition.

These issues are general to prototype-based methods. Prototypes are usually of fixed dimension; it is not obvious how they can be used as the basis for a system that is to classify time-varying inputs of different lengths. The issue of representing longer speech sequences spanning multiple phonemes or words is particularly crucial, and made more difficult by the fact that the timing of speech events varies considerably, even for the same phonetic content. This introduces the need for some kind of time warping or normalization in the model.

Two methods for extending the scope of STLVQ to address these questions have been proposed. Essentially, they attempt to merge the discriminative power of LVQ-designed prototypes with the sequential modeling abilities of HMMs. Though neither of these hybrid approaches, as originally proposed, are in common use today, they have historical significance and illuminate the nature of the challenges facing prototype-based speech recognition.

5.6.2.1 LVQ-HMM

The most straightforward way in which to combine the discriminant power of LVQ with the dynamic modeling capability of an HMM is to use LVQ to design a discriminative codebook for a discrete density HMM. This was the approach taken in a number of studies [26–30]. In particular, in [26], the multiframe LVQ reference vectors designed for STLVQ were grouped together to form a single HMM codebook. The

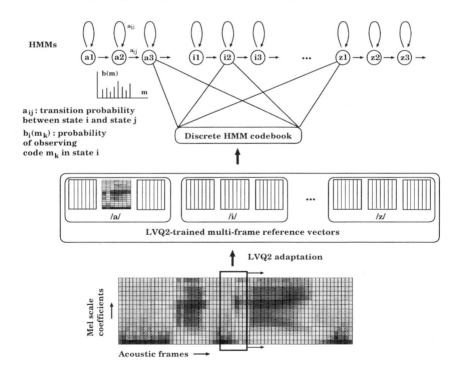

Figure 5.5 LVQ-HMM architecture. LVQ is used to design reference vectors that discriminate well between the categories. The reference vectors are pooled and used as the codebook for a discrete HMM that is capable of modeling variable-length temporal sequences.

resulting hybrid LVQ-HMM system is illustrated in Figure 5.5. It was shown that this approach yielded improvements over the k-means clustering approach to HMM codebook design and could be used for general speech recognition tasks such as word or phrase recognition. This strategy follows many of the ideas described for the integration of MLPs with DP or hidden Markov modeling, in particular the proposals in which the more discriminative MLP is trained locally, and then used in the recognition phase to provide local distances and probabilities to an HMM or DP process that integrates the local scores over time [21, 22]. Note that in this hybrid LVQ-HMM design, the LVQ training phase and the HMM training phase are completely separate and do not interact, leading one to question the optimality of the overall system and to wonder whether some form of global optimization might not be possible.

5.6.2.2   HMM-LVQ

An interesting reversal of the above LVQ-HMM proposal was that suggested in [31], illustrated in Figure 5.6. In this proposal, dubbed HMM-LVQ, a set of n-state HMMs (one HMM per word category) is first used to segment acoustic feature vectors for a speech token. If there are M word categories, M segmentations are generated. For each segmentation, the feature vectors assigned to each HMM state are averaged, and the averages concatenated for each HMM, yielding M sequences, $x_1, ..., x_M$, of n vectors each. Each one of these time-normalized, n-frame sequences is assigned the same label as the training token. LVQ training is performed on reference vectors that attempt to classify the time-normalized sequences. Since M sequences are generated for each training and testing token, a slightly different LVQ discriminant function must be defined. For each category C_k, the reference vector nearest each sequence x_i is found, and the corresponding distance $D(i, k)$ is stored. $D(i, k)$ is either Euclidean distance or a likelihood-based distance that incorporates prototype variances. The discriminant function for each category is then defined to be a weighted sum of the $D(i, k)$:

$$g_k(\mathbf{x}_1^T) = \sum_{i=1}^{M} v_i D(i, k) \qquad (5.44)$$

The category with the smallest discriminant function is chosen as the classification output. During the training phase, reference vectors are adapted according to the LVQ2 training rule, modified here to account for the new definition of discriminant function.

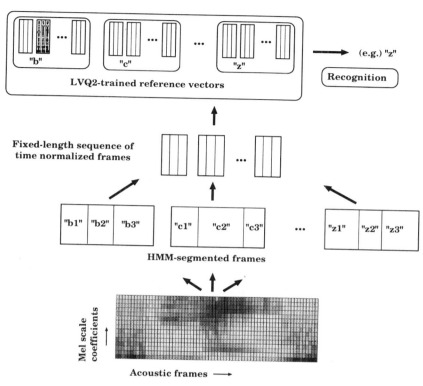

Figure 5.6 HMM-LVQ architecture. In a first stage of processing, HMMs for each category in the "E-set" are used to segment acoustic feature vectors for the speech utterance. The feature vectors assigned to each state are averaged, resulting in a fixed-length sequence of time-normalized frames. In the second stage, LVQ is used to design reference vectors that will discriminate between time-normalized frame sequences from different categories.

One may question the two-stage nature of this proposal and ask about the optimality of the HMM segmentations with respect to the LVQ decisions. There is also, of course, a loss of information incurred in the time-normalization method. Nonetheless, good results for this approach were obtained on the difficult Bell Labs E-set task. More importantly, the method illustrates how prototypes can be used directly at the whole-utterance level via a time-normalization of the acoustic feature frames in a first stage of processing.

Though HMM-LVQ, as described here, addresses only isolated-word recognition, it has been applied to continuous speech recognition is described in [31].

5.6.3 Minimum Classification Error Interpretation of LVQ

The LVQ algorithm can be seen as a prototype-based implementation of the MCE framework. The LVQ classification rule is based on the Euclidean distance between a pattern vector and each category's reference vectors. The category of the nearest reference vector is given as the classification decision. This corresponds to the following definition of discriminant function:

$$g_k(\mathbf{x}, \Lambda) = \left[\sum_{i=1}^{I_k} \|\mathbf{x} - \mathbf{r}_{ik}\|^{-\varsigma} \right]^{-1/\varsigma} \tag{5.45}$$

Here $\mathbf{r}_{ik} \in \Re^N$ is the (adaptable) i-th reference vector of category C_k, I_k is the number of reference vectors of category C_k, $\| \cdot \|$ is the Euclidean norm, ς is a positive scalar, and Λ as above is a general notation for the model parameters, in this case the category reference vectors. If one uses a large ς, (5.45) can be seen to correspond to the distance between observation $\mathbf{x}$ and the closest reference vector of category C_k, which is the distance used in the LVQ classification rule. (Note that we have to handle separately the case where one of the $\|\mathbf{x} - \mathbf{r}_{ik}\|$ equals zero. This is a minor problem that shall not be mentioned when it recurs in slightly different contexts below.)

Given this definition of the LVQ discriminant function, the MCE misclassification measure (5.27) and loss function (5.28) defined previously can be used unchanged. An overall loss function can then be defined as above (5.29) and the GPD method can be used to minimize it. Applying the chain rule to the loss function in (5.28), and chaining through the misclassification measure in (5.27) and the discriminant function in (5.45) (and using large values for ψ and ς), the parameter modification (5.32) becomes

$$\Delta \mathbf{r}_{z_k}^k(\Lambda) = 2\epsilon \ell'(d_k(\mathbf{x}, \Lambda))(\mathbf{x} - \mathbf{r}_{z_k}^k(\Lambda)) \tag{5.46}$$

and

$$\Delta \mathbf{r}_{z_j}^j(\Lambda) = -2\epsilon \ell'(d_k(\mathbf{x}, \Lambda))(\mathbf{x} - \mathbf{r}_{z_j}^j(\Lambda)) \tag{5.47}$$

where C_k is the correct class, C_j is the best matching incorrect class, and z_k and z_j, respectively, denote the closest reference vectors in categories C_k and C_j. The shape of the derivative of $\ell(\ )$ entails that training is only performed for a small number of tokens near the class boundary.

Instead of the sigmoidal loss in (5.28), a piece-wise linear loss can be chosen such that the loss derivative will be zero for correct classifications (implying no learning) and constant for misclassifications near the border. The resulting reference vector adaptation rules in (5.46) and (5.47) correspond very closely to the LVQ2 learning rule expressed in (5.40) and (5.41). The only difference is that LVQ2 imposes an additional condition for adaptation to occur: the next-nearest category must be correct. This is not the case for MCE/GPD learning, where any category could be in second place. The study described in this chapter is based on LVQ2, but it is interesting to note that later versions of LVQ2, for example the version described in [32], were found to be more effective if they dropped the second LVQ2 condition.

To summarize, one can view LVQ as a practical implementation of the GDP method used with an MCE criterion [4, 33].

5.6.4 Smoothness of MCE Loss

In addition to its effect on estimation quality and generalization, it was mentioned earlier that the smoothness of the MCE loss function has an impact on optimization. This can be illustrated by examining the

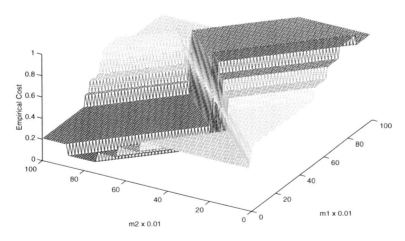

Figure 5.7 Average empirical loss measured over 10 samples from a one-dimensional, two class classification problem. Each class was represented with a single reference vector ($m1$ and $m2$ in the figure). The overall loss was plotted for a range of combinations of $m1$ and $m2$. Here, the ideal zero-one loss was used in calculating the overall loss.

error surfaces that result from different choices for α, the parameter controlling the steepness of the sigmoidal loss function, in the context of a simple, one-dimensional LVQ classifier.

The task we consider is a two class problem, that of classifying one-dimensional data from two overlapping Gaussian distributions (with means at 0.3 and 0.7, and standard deviations of 0.2). The problem is similar to that illustrated in Figure 5.3. One reference vector was assigned to each class, and the overall empirical loss (5.33) for a given sample size was calculated for a grid of positions of the two reference vectors. Figure 5.7 shows the overall loss for a sample size of ten, using the (noncontinuous) *ideal* zero-one loss function. One can see how the resulting error surface is not tractable to gradient search, as it consists of large plateaus of zero gradient, with abrupt, discontinuous transitions between plateaus. However, using a sigmoidal MCE loss function, with $\alpha = 0.1$, to calculate the empirical loss over the same training sample, one can see (Figure 5.8) that the error surface is now smooth and suitable for gradient-based search. With a steeper sigmoidal MCE loss ($\alpha = 1$), one now has an error surface of intermediate smoothness that may be less suited to gradient-based optimization (Figure 5.9). The effect of α on the smoothness of the MCE loss is clearly similar to the effect of h_1 on the smoothness of the Parzen window estimate illustrated in Figure 5.1.

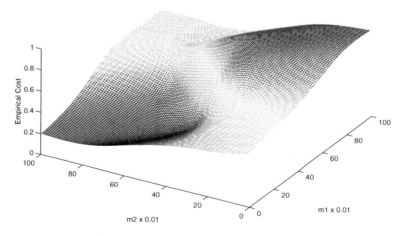

Figure 5.8 Average empirical loss measured over ten samples from a one-dimensional, two-class classification problem. Here, a sigmoidal MCE loss, $\alpha = 0.1$, was used in calculating the overall loss.

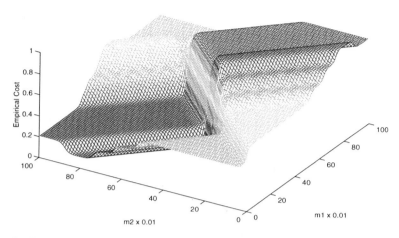

Figure 5.9 Average empirical loss measured over ten samples from a one-dimensional, two-class classification problem. Here, a sigmoidal MCE loss, $\alpha = 1.0$, was used in calculating the overall loss.

5.6.5 LVQ Summary

In the preceding, it was seen that the LVQ algorithm provides both a good example of prototype-based pattern recognition and productive grounds for different applications to speech recognition. The simple vector adaptations aimed at optimal category discrimination illustrate well the concept of discriminative training. The fact that LVQ can be seen as a special, prototype-based instantiation of the MCE framework shows that the intuitively appealing LVQ training procedure can be formally linked to the minimization of a loss function that reflects a classifier's recognition error rate. In the next sections, we will see how MCE can be used to design prototype-based classifiers that are more general than LVQ and better suited to tasks in speech recognition than the architectures considered so far.

5.7 Prototype-Based Methods Using Dynamic Programming

Clearly, the issue of time-normalization is crucial in template- or prototype-based approaches to speech recognition. One of the crucial advances in the field of automatic speech recognition was the advent of dynamic programming (DP) algorithms that allow an efficient nonlinear matching between a sample utterance and a speech prototype [1, 2, 34]. Initially, most template matching methods used one reference per word, and operated in speaker-dependent, isolated word-recognition mode [35]. Later studies showed how DP could be used to perform connected word recognition [2], and furthermore how clustering algorithms (e.g., the modified K-means algorithm) could be used to design several references per word category so as to reflect variants in pronunciation, in particular variants found across different speakers [3].

Given a reference pattern and an input test sample, DP can be used to find the path through a grid of local matches between prototype and test sample frames that has the best overall score. The local matches are typically computed using a distance measure between a feature vector belonging to the reference and a feature vector belonging to the test sample; the local distances corresponding to any given grid path are cumulated, resulting in an overall path distance. The grid path with the best overall match is that which has the smallest cumulated distance. Using DP techniques, it is possible to find the best path efficiently. The power of DP, applied to speech recognition, lies in its ability to nonlinearly compress and stretch prototypes so as to best match an unknown input, yet in a manner that preserves the distinctiveness of the prototype. This is clearly attractive in speech recognition, where speaking rate is rarely uniform, even for the same speaker speaking the same utterance. This aspect of DP applications to speech recognition is the basis of the appellation "dynamic time warping" (DTW). (DTW usually refers to a specific application of DP techniques to speech recognition, one in which whole-utterance templates are used.)

We here focus on the topic of discriminative design of DP prototypes using MCE. The reader is referred to sources such as [3] and [36] for an overview of DP techniques in general.

5.7.1 MCE-Trained Prototypes for DTW-Based Speech Recognition

The approach adopted in [37] is to define the MCE loss in terms of a discriminant function that reflects the structure of a straightforward DTW-based recognizer, yet which can be treated as a continuous function of the input data and the recognizer parameters. Specifically, the discriminant function is defined to be:

$$g_k(\mathbf{x}_1^T, \Lambda) = \left[\sum_{b=1}^{B_k} D(\mathbf{x}_1^T, \mathbf{r}_k^b)^{-\zeta} \right]^{-1/\zeta} \tag{5.48}$$

$D(\mathbf{x}_1^T, \mathbf{r}_k^b)$ is the reference distance between the input utterance

$\mathbf{x}_1^T = (\mathbf{x}_1, ..., \mathbf{x}_T)$ and the reference utterance $\mathbf{r}_k^b = (\mathbf{r}_1, ..., \mathbf{r}_{T_k^b})$. B_k denotes the number of reference assigned to category C_k; T_k^b is the number of frames in the b-th reference of C_k; ξ is a positive number that controls the extent to which the smallest reference distance dominates the discriminant function (a similar type of control was used in the MCE formulation of LVQ described previously). Each reference distance is defined to be:

$$D(\mathbf{x}_1^T, \mathbf{r}_k^b) = \left[\sum_{\theta=1}^{\Theta} D_\theta(\mathbf{x}_1^T, \mathbf{r}_k^b)^{-\xi} \right]^{-1/\xi} \tag{5.49}$$

where $D_\theta(\mathbf{x}_1^T, \mathbf{r}_k^b)$ is a *path distance* accumulated along the θ-th best (i.e., smallest distance) path used in matching the input utterance $\mathbf{x}_1^T$ and the reference $\mathbf{r}_k^b$; Θ is the total number of distinct paths. Each path distance is defined as

$$D_\theta(\mathbf{x}_1^T, \mathbf{r}_k^b) = \sum_{t=1}^{T} w_{k,t}^b \delta_{k,\theta_t}^b \tag{5.50}$$

where $w_{k,t}^b$ is a weighting factor for the t-th frame of $\mathbf{r}_k^b$, and δ_{k,θ_t}^b is a local distance between the t-th frame of $\mathbf{r}_k^b$ and the corresponding frame θ_t of $\mathbf{x}_1^T$ along the θ-th path. The local distance is defined to be:

$$\delta_{k,\theta_t}^b = \|\mathbf{r}_{k,t}^b - \mathbf{x}_{\theta_t}\|^2 \tag{5.51}$$

where $\|\cdot\|$ denotes the Euclidean norm.

Given this definition of $g_k(\mathbf{x}_1^T, \Lambda)$, we can adopt the MCE misclassification measure

$$d_k(\mathbf{x}_1^T, \Lambda) = g_k(\mathbf{x}_1^T, \Lambda) - \left[\frac{1}{M-1} \sum_{j\neq k} g_j(\mathbf{x}_1^T, \Lambda)^{-\psi} \right]^{-1/\psi} \tag{5.52}$$

where the only change compared with the previous definition in (5.27) is the replacement of the single pattern vector $\mathbf{x}$ by a sequence of pattern vectors $\mathbf{x}_1^T$. With a similar change to (5.28), we have an MCE loss function for $\mathbf{x}_1^T$:

$$\ell(d_k(\mathbf{x}_1^T, \Lambda)) = \frac{1}{1 + e^{-\alpha d_k(\mathbf{x}_1^T, \Lambda)}} \tag{5.53}$$

An overall loss function for $\mathbf{x}_1^T$ can also be defined, along the lines of

(5.29). (The reader should refer to [4] for the formal basis of MCE/GPD application to samples that are variable-length vector sequences.)

As in the general description of the MCE loss function, the crucial aspect of this particular definition of loss is that it is a continuous, differentiable function of the system parameters Λ, and reflects classification performance. The particular definition is such that it embodies the structure of a prototype-based recognizer that uses a number of references for each category, and that bases its classifications on the results of matching the category references and the input using DTW. The loss function is a mathematical expression for the performance of this particular classifier, for any given set of parameters values and any given body of training data. The reference parameters $\mathbf{x}_k^b$ can now be adapted, using GPD (or any other gradient-based optimization technique), so as to minimize overall loss. The MCE/GPD parameter updates can be found easily by applying the chain rule to the MCE loss function to find the loss gradient, $\nabla \ell_k(\mathbf{x}_1^T, \Lambda)$, and applying the GPD update procedure [37].

5.7.1.1 Practical Implementation of MCE/GPD

Rather than use every single DP path possible in the calculation of the reference distance (5.49), it is practical to use only the top path or the top few paths. This is possible by simply assuming a very large value of ξ. The resulting reference distance will thus be dominated by the best path(s); using only the top path(s) to calculate (5.49) can then be viewed as an approximation of the original, continuous definition. Further simplifications, which may or may not be as important in practical terms, are to assume a very large value for ζ in the definition of discriminant function (5.48), resulting in the use of only the nearest references, and a very large value of ψ in the misclassification measure (5.52), resulting in the use of only the top competing categories. These assumptions considerably simplify the application of the chain rule to the loss function, and hence the optimization of the reference parameters. One should bear in mind that less extreme choices for ζ, ξ, and ψ result in a smoother loss function, which may result in better generalization to unknown data.

Given these simplifying assumptions, the MCE/GPD parameter updates are found as before by applying the chain rule to the MCE loss function and applying the GPD update procedure. The resulting adaptation procedure is strikingly similar to the MCE-LVQ procedure: the nearest incorrect reference is moved away from the training sample

x_1^T, and the nearest correct reference is moved closer. In the context of this DTW classifier, "move" refers to changing the reference parameters along the best DP path between reference and input so as to reduce or increase the path distance. The reader should refer to [37, 38] for the exact parameter updates and their derivation (they closely follow those for HMMs, given later in this chapter).

5.7.1.2   MCE-DTW Results

This application of MCE-based adaptation to speech recognition using DTW was evaluated on the Bell Labs E-set task, and on phoneme recognition tasks for the ATR 5240 word databases, with good results [37, 38]. A very similar application of MCE to DTW-based speech recognition was described in [39, 40], which yielded similar results for the E-set task.

5.7.2   Prototype-Based Minimum Error Classifier

A different approach to DP-based speech recognition using speech pattern prototypes is the prototype-based minimum error classifier (PBMEC) architecture proposed in [41, 42]. The difference between PBMEC and the MCE-trained DTW examined above is that PBMEC models prototypes at a finer grain. Rather than store templates for entire words, the PBMEC prototypes are modeled within phonetic or subphonetic *states*. Word models are formed by connecting different states together. The phonetic states are usually shared across word models, though this does not have to be so. In the sense that a collection of templates for a given word can be represented by several parallel state-connecting paths, PBMEC can be viewed as a more general structure than the usual DTW template architecture. Indeed, the PBMEC architecture is that of an HMM. However, PBMEC uses the kind of distance measure used in DTW-based procedures in comparing input frames to reference frames, and not the Gaussian pdfs usually assumed in an HMM. Distances such as the Itakura-Saito distance [3], which are not appropriate in the context of HMMs, could be used with PBMEC.

In PBMEC, a category is represented as a particular sequence of subphonemic or subword states. The allowable state sequences thus define the categories of the task. Phonemes, words, and phrases can therefore be represented in a uniform manner, as state sequences. A DP procedure performed over the allowable state sequences is then embedded into the discriminant function for each category.

5.7.2.1 PBMEC State Distance and Discriminant Function

The PBMEC discriminant function is defined for each category C_j in the task using an L_p norm-based state distance e_s, which is a function of the reference distances between a single feature vector $\mathbf{x}_t$ at time t and reference vectors belonging to a state s:

$$e_s(\mathbf{x}_t) = \left[\sum_{i=1}^{I_s} [(\mathbf{x}_t - \mathbf{r}_{s,i})^t (\Sigma_{s,i})^{-1} (\mathbf{x}_t - \mathbf{r}_{s,i})]^{-\zeta} \right]^{-1/\zeta} \tag{5.54}$$

where $\mathbf{r}_{s,i}$ denotes the (adaptable) i-th reference vector of state s, $\Sigma_{s,i}$ is an adaptable positive definite covariance matrix, and I_s is the number of reference vectors assigned to s. The form of the state distance e_s corresponds to an L_p norm of Mahalanobis distances. (Since the Mahalanobis distance derives from a Gaussian pdf under the assumption of a covariance matrix that is shared across all categories, $\Sigma_{s,i}$ should be shared across all states if one wishes to preserve the link to the Gaussian pdf.) As above, a large value for ζ means that the closest reference vector will dominate this expression. Next, an accumulated distance is defined as the sum of the state distances for a given state sequence $\Theta^\rho = (\theta_1^\rho, \theta_2^\rho, ..., \theta_T^\rho)$ permitted by the DP procedure for category C_j, given a training token $\mathbf{x}_1^T = (\mathbf{x}_1, ..., \mathbf{x}_T)$ and the phonemic transcription of category C_j:

$$V(\mathbf{x}_1^T, \Theta^\rho, \Lambda) = \sum_{t=1}^{T} e_{\theta_t^\rho}(\mathbf{x}_t) \tag{5.55}$$

The discriminant function for each target category C_j is then defined to be:

$$g_j(\mathbf{x}_1^T, \Lambda) = \left[\sum_\rho [V(\mathbf{x}_1^T, \Theta^\rho, \Lambda)]^{-\xi} \right]^{1/-\xi} \tag{5.56}$$

Note that a large value for ξ will mean that the state sequence with the smallest accumulated distance will dominate the discriminant function. In this case, the best state sequence $\Theta = (\theta_1, \theta_2, ..., \theta_T)$ is used to define the discriminant function:

$$g_j(\mathbf{x}_1^T, \Lambda) = V(\mathbf{x}_1^T, \Theta, \Lambda) \tag{5.57}$$

Lp norm of state distances propagated through network

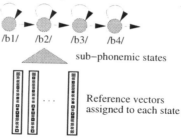

Figure 5.10 Multi-state PBMEC architecture.

(Here and in the following, in order to simplify notation, the depen-dence on category C_j of the state sequences Θ^ρ and the best state sequence Θ is not indicated explicitly, but it should be kept in mind.)

Figure 5.10 illustrates the structure of the model for a single phoneme. Note that training will be performed at the level of the category C_j, not of the subcategory states s.

The full DP score grid for a word utterance and the correct word category, with the best state sequence overlaid, is shown in Figure 5.11.

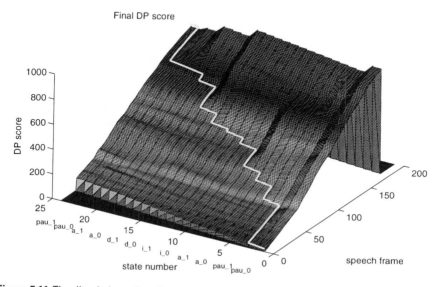

Figure 5.11 The discriminant function for a category is defined as the final accumulated score of the best DP path for that category. The state numbers correspond to PBMEC state models strung together to form a word model.

The final DP score for the entire utterance is used as the value of the discriminant function for that word category. Another representation of the discriminant function is shown in Figure 5.12.

Given this discriminant function, the misclassification measure (5.52) and loss function (5.53) used earlier can be adopted. As before, the gradient of the loss function with respect to all classifier parameters can be calculated using the chain rule. The reader should refer to [15] for the exact parameter updates and their derivation (they closely follow those for HMMs, given later in this chapter). As in the MCE-trained DTW classifier examined earlier, the parameter update becomes significantly simpler by assuming large values for ζ (entailing that only the nearest prototype in a state is used), ξ (entailing that only the top DP state sequence is used), and ψ (entailing that only the top competing category is used). Under these simplifying assumptions, the MCE/GPD update rule for PBMEC is again very similar to the MCE-LVQ update rule, pulling the nearest reference vectors for the correct category closer to the input and pushing the nearest reference vectors for the incorrect category away. As in the MCE-DTW application examined earlier, this is done along the best DP state sequence for correct and incorrect categories.

5.7.2.2   MCE/GPD in the Context of Speech Recognition Using Phoneme Models

The essence of MCE/GPD adaptation is to push the incorrect categories away and pull the correct category closer. When the categories are word models built out of shared phoneme models, the same phoneme model may be part of both correct and incorrect categories. In this case, the amount of adaptation for that model will often cancel out precisely. This is illustrated in Figure 5.13. In this figure, the word "aida" is recognized as "taira." The best DP path for each word segments the utterance into a sequence of start and end times for each phonemic substate in the word. The gradient of the MCE loss function will be zero for phonemic states that are on the shared portion of the segmentations, shown in gray. Only models corresponding to different state-to-frame assignments, shown in black, will have a nonzero gradient. In this manner, discriminative training based on MCE/GPD focuses on the phoneme models and speech frames that caused the misclassification to occur.

5.7.2.3   PBMEC Results

PBMEC has been evaluated on several speech databases, most notably the Bell Labs E-set task and the ATR 5240 word tasks. The results show clear improvements over nondiscriminative methods. Most

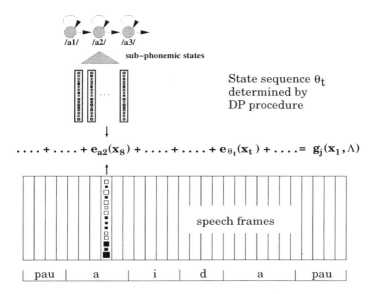

$$\ldots + \ldots + \mathbf{e}_{a2}(\mathbf{x}_8) + \ldots + \ldots + \mathbf{e}_{\theta_t}(\mathbf{x}_t) + \ldots = \mathbf{g}_j(\mathbf{x}_1, \Lambda)$$

Figure 5.12 The discriminative function is a sum of local scores over the best DP path.

striking were the results for the ATR 5240 word tasks, where PBMEC, using 20 times fewer parameters, achieved error rates less than half those of an HMM system with a carefully designed topology [15, 43].

5.7.3 Summary of Prototype-Based Methods Using DP

Using the MCE framework for discriminative training, it is possible to embed the dynamic programming procedure into the definition of loss. Both of the DP-embedding, prototype-based approaches surveyed here can be seen as generalizations of the LVQ algorithm to tasks that involve recognition of dynamic or sequential patterns. The results obtained for both methods show that trained discriminatively with MCE, they can compete favorably with other pattern recognition methods, including HMMs. In their use of MCE as the basis for parameter adaptation, both methods illustrate very clearly that the goal of Bayes decision theory can be pursued without having to explicitly represent a probability density or posterior probability function. Loosely rephrasing the discussion concerning the link between MCE and the Bayes error, if the proto-type-based discriminant functions lead to the same classification decisions as would knowledge of the true posterior probabilities and application of the Bayes decision rule, the error rate for the prototype-based system will be the Bayes error. MCE offers a way of striving for this

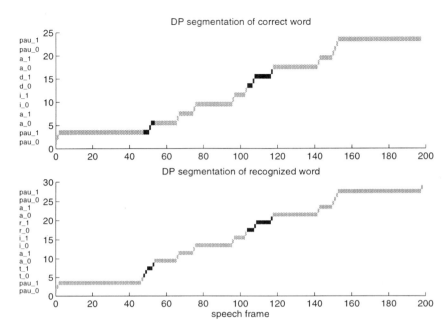

Figure 5.13 DP segmentations for the correct word, "aida," and the incorrectly recognized word, "taira," for a two-state/phoneme PBMEC model. (The states are duplicated to implement a durational constraint, so a phoneme is effectively modeled as the sequence of states phon-0 phon-0 phon-1 phon-1.) Discriminative training based on MCE/GPD will focus on the regions where there is a difference between correct and incorrect segmentations.

goal even for classifiers that do not explicitly model the distributions of the categories, such as prototype-based classifiers. Of course, the goal of minimum-error-rate classification will only be attained when enough system parameters (e.g., references) are used and an infinite amount of training data is available.

5.8   Hidden Markov Model Design Based on MCE

Prototype-based methods, almost by definition, use a distance measure to match a pattern vector to a prototype. Typically, Euclidean distance is used to calculate the distance between prototype and pattern vector. We saw that Euclidean distance, and the related Mahalanobis distance, can be seen to derive from a Gaussian pdf under certain assumptions. The L_p norm of Mahalanobis distances used in the calculation of the local

PBMEC state distance clearly resembles the mixture of Gaussian pdfs used to model HMM state observation probabilities. Furthermore, the use of DP in the prototype-based methods described here is very similar to the use of the Viterbi algorithm to find the best HMM state sequence. Thus there are deep similarities between the prototype methods described here and HMMs.

This can lead us to question any strong division between prototype method and HMMs as being overly artificial. The similarity goes both ways, and it should be clear that HMMs can themselves be viewed as prototype-based methods. The mean vectors that are the centers of an HMM's Gaussian kernels correspond directly to a prototype in a nearest-neighbor classifier based on, for example, Euclidean distance. The difference, compared with a classifier using Euclidean distance, is that the Gaussian mixture used in HMMs offers a more general probabilistic model, particularly in its ability to represent variance in a category-dependent, state-dependent, or kernel-dependent manner. There is clearly a parallel with the contrast (and underlying similarity) between the k-NN method and Parzen windows. While the former uses Euclidean distance to find and count nearest neighbors, the latter defines a kernel function around each example, and attempts to model a density directly by summing the kernels. It should not come as a surprise that a similar choice exists for prototype-based methods. The advent of MCE shows how both choices can be used as the basis for a minimum-error classifier. The greater generality of the Gaussian model used in HMMs does not imply that they will necessarily outperform prototype-based methods using Euclidean distance. The issue depends on many factors, including the number of parameters used, the amount of training data that is available, and the nature of the task. There are also practical issues to consider: classifiers using Euclidean distance are generally easier to implement than their Gaussian counterparts.

The similarity between the prototype-based models we have described so far and HMMs relies on the link between Euclidean or Mahalanobis distance and the Gaussian pdf. It should be kept in mind that there are other distance measures, such as the Itakura-Saito acoustic distance, that could be appropriate in the context of a prototype-based method, but that cannot easily be related to an HMM pdf. Conversely, it should also be pointed out that HMMs do not have to use Gaussian pdfs; other choices are possible.

The MCE framework can be applied to HMMs in a manner very similar to that described for the PBMEC model [15, 44]. In light of the preceding discussion concerning the prototype-like nature of HMMs, a summary of the application of MCE to HMMs completes our survey of discriminatively trained prototype-based models.

5.8.1   HMM State Likelihood and Discriminant Function

The HMM state observation probability density corresponds to the PBMEC state distance. The observation probability density function of observation $\mathbf{x}_t$ at time t, given the mean vectors $\mu_{s,i}$ and covariance matrices $\Sigma_{s,i}$ of an HMM state s, is typically a Gaussian mixture density:

$$b_s(\mathbf{x}_t) = \sum_{i=1}^{I_s} c_{s,i} N(\mathbf{x}_t, \mu_{s,i}, \Sigma_{s,i}) \qquad (5.58)$$

where I_s is the number of mixture components in state s and $c_{s,i}$ are mixing weights satisfying the constraint:

$$\sum_{i=1}^{I_s} c_{s,i} = 1 \qquad (5.59)$$

$N(\mathbf{x}_t, \mu_{s,i}, \Sigma_{s,i})$ is the multivariate Gaussian density:

$$N(\mathbf{x}, \mu, \Sigma) = \frac{1}{(2\pi)^{d/2}|\Sigma|^{1/2}} \exp(-\tfrac{1}{2}(\mathbf{x} - \mu)'\Sigma^{-1}(\mathbf{x} - \mu)) \qquad (5.60)$$

An HMM discriminant function $g_j(\mathbf{x}_1^T, \Lambda)$ for each category C_j can be defined along the same lines as the PBMEC discriminant function. The global probability density of the whole utterance $\mathbf{x}_1^T = (\mathbf{x}_1, ..., \mathbf{x}_T)$ for a given state sequence $\Theta^\rho = (\theta_1^\rho, \theta_2^\rho, ..., \theta_T^\rho)$ allowed for category C_j is defined in terms of the observation probabilities $b_s(\mathbf{x}_t)$ and the state-to-state transition probabilities a_{ij}:

$$f(\mathbf{x}_1^T, \Theta^\rho | \Lambda) = \prod_{t=1}^{T} a_{\theta_{t-1}^\rho \theta_t^\rho} b_{\theta_t^\rho}(\mathbf{x}_t) \qquad (5.61)$$

The HMM discriminant function is defined as the log of an L_p norm of probability densities corresponding to all possible state sequences Θ^ρ:

$$g_j(\mathbf{x}_1^T, \Lambda) = \log\left[\sum_\rho [f(\mathbf{x}_1^T, \Theta^\rho | \Lambda)]^\xi\right]^{1/\xi} \qquad (5.62)$$

A large value for ξ will entail that the path with the highest probability will dominate the discriminant function. For such a choice of ξ, the HMM discriminant function becomes:

$$g_j(\mathbf{x}_1^T, \Lambda) = \log f(\mathbf{x}_1^T, \Theta|\Lambda) = \sum_{t=1}^{T} \log a_{\theta_{t-1}\theta_t} + \sum_{t=1}^{T} \log b_{\theta_t}(\mathbf{x}_t) \quad (5.63)$$

where $\Theta = (\theta_1, \theta_2, ..., \theta_T)$ denotes the best overall state sequence for category C_j.

5.8.2 MCE Misclassification Measure and Loss

The nature of the HMM discriminant function requires a definition of misclassification measure different from (5.52). Following [45], the following definition was used:

$$d_k(\mathbf{x}_1^T, \Lambda) = -g_k(\mathbf{x}_1^T, \Lambda) + \log \left[\frac{1}{M-1} \sum_{j \neq k} e^{g_j(\mathbf{x}_1^T, \Lambda)\psi} \right]^{1/\psi} \quad (5.64)$$

For a large ψ, the bracketed expression is approximately the value of the discriminant function of the best incorrect category.

The same MCE loss function as (5.28) can be applied to this misclassification measure.

5.8.3 Calculation of MCE Gradient for HMMs

The application of MCE to HMMs is very similar to its application to the other prototype-based methods described in this chapter. Given the definitions of the HMM discriminant function, misclassification measure, and loss function, the overall MCE loss function can be minimized using the GPD method, quickprop, or any other gradient-based method. The main task in doing so is to use the chain rule to write out the expression for the loss gradient, $\nabla \ell_k(\mathbf{x}_1^T, \Lambda)$, for HMMs. Let us now derive the expression for this gradient [15].

5.8.3.1 Derivative of Loss with Respect to Misclassification Measure

The derivative of the loss $\ell(d_k(\mathbf{x}_1^T, \Lambda))$ with respect to the misclassification measure $d_k(\mathbf{x}_1^T, \Lambda)$ is

$$\frac{\mathrm{d}\ell(d_k(\mathbf{x}_1^T, \Lambda))}{\mathrm{d}d_k(\mathbf{x}_1^T, \Lambda)} = \ell'(d_k(\mathbf{x}_1^T, \Lambda)) = \alpha \ell(d_k(\mathbf{x}_1^T, \Lambda))(1 - \ell(d_k(\mathbf{x}_1^T, \Lambda)))$$

$$(5.65)$$

5.8.3.2 Derivative of Misclassification Measure with Respect to Discriminant Functions

It can easily be verified that the derivative of the misclassification measure (5.64) with respect to the correct category's discriminant function is simply -1:

$$\frac{\partial d_k(\mathbf{x}_1^T, \Lambda)}{\partial g_k(\mathbf{x}_1^T, \Lambda)} = -1 \tag{5.66}$$

With respect to an incorrect category C_i's discriminant function, the derivative is

$$\frac{\partial d_k(\mathbf{x}_1^T, \Lambda)}{\partial g_i(\mathbf{x}_1^T, \Lambda)} = \frac{e^{g_i(\mathbf{x}_1^T, \Lambda)\psi}}{\displaystyle\sum_{j \neq k} e^{g_j(\mathbf{x}_1^T, \Lambda)\psi}} \tag{5.67}$$

Assuming a very large value of ψ, this expression reduces to unity for the best incorrect category, and zero for all other incorrect categories.

5.8.3.3 Derivative of Discriminant Function with Respect to Observation Probability Density Function

As the best DP path was used to define the discriminant function, the derivative of this function with respect to a given state's pdf will depend on the particular path sequence $\Theta = (\theta_1, \theta_2, ..., \theta_T)$. The partial derivative with respect to each $b_{\theta_t}(\mathbf{x}_t)$ along the best state sequence is

$$\frac{\partial g_k(\mathbf{x}_1^T, \Lambda)}{\partial b_{\theta_t}(\mathbf{x}_t)} = \frac{1}{b_{\theta_t}(\mathbf{x}_t)} \tag{5.68}$$

For states not on the best path at a particular time instant t, the derivative is zero.

As the same state can occur several times on the best DP path, the partial derivative of the discriminant function with respect to the observation probability parameters (the mean vectors and covariances) is summed over the path time instants corresponding to that state. Specifically, using $\mathbf{m}_s$ to denote a parameter vector within a state s, the partial derivative is summed over the time points t where the best path state at t corresponds to s, i.e., over all t such that θ_t is equal to s:

$$\frac{\partial g_k(\mathbf{x}_1^T, \Lambda)}{\partial \mathbf{m}_s} = \sum_{t|\theta_t = s} \frac{\partial g_k(\mathbf{x}_1^T, \Lambda)}{\partial b_{\theta_t}(\mathbf{x}_t)} \frac{\partial b_{\theta_t}(\mathbf{x}_t)}{\partial \mathbf{m}_{\theta_t}} \tag{5.69}$$

5.8.3.4   Derivative of Observation Probability with Respect to Mixing Weights

Recalling that $b_s(\mathbf{x}_t)$ is a Gaussian mixture density:

$$b_s(\mathbf{x}_t) = \sum_{i=1}^{I_s} c_{s,i} N(\mathbf{x}_t, \mu_{s,i}, \Sigma_{s,i}) \tag{5.70}$$

the partial derivative of the observation probability with respect to a mixing weight $c_{s,i}$ is

$$\frac{\partial b_s(\mathbf{x}_t)}{\partial c_{s,i}} = N(\mathbf{x}_t, \mu_{s,i}, \Sigma_{s,i}) \tag{5.71}$$

5.8.3.5   Derivative of Observation Probability with Respect to Mean Vectors

Assuming that $\mu_{s,i}$ is a vector of D components $\mu_{s,i,d}$, and that $\Sigma_{s,i}$ is a diagonal matrix of D components $\sigma_{s,i,d}^2$, the partial derivative of the observation probability with respect to a mean vector component $\mu_{s,i,d}$ is:

$$\frac{\partial b_s(\mathbf{x}_t)}{\partial \mu_{s,i,d}} = c_{s,i} \frac{\partial N(\mathbf{x}_t, \mu_{s,i}, \Sigma_{s,i})}{\partial \mu_{s,i,d}} \tag{5.72}$$

Expanding this expression produces:

$$\frac{\partial b_s(\mathbf{x}_t)}{\partial \mu_{s,i,d}} = c_{s,i} N(\mathbf{x}_t, \mu_{s,i}, \Sigma_{s,i}) \frac{(x_{t,d} - \mu_{s,i,d})}{\sigma_{s,i,d}^2} \tag{5.73}$$

An important issue arises when adapting the mean vectors using the above expression: the derivative can be extremely sensitive to differences in the variance term $\sigma_{s,i,d}^2$. In practice, this term was dropped from the gradient of the mean vectors [45].

5.8.3.6 Derivative of Observation Probability with Respect to Covariances

The partial derivative of the observation probability with respect to a covariance component $\sigma_{s,i,d}$ is:

$$\frac{\partial b_s(\mathbf{x}_t)}{\partial \sigma_{s,i,d}} = c_{s,i} \frac{\partial N(\mathbf{x}_t, \mu_{s,i}, \Sigma_{s,i})}{\partial \sigma_{s,i,d}} \tag{5.74}$$

Expanding this expression produces:

$$\frac{\partial b_s(\mathbf{x}_t)}{\partial \sigma_{s,i,d}} = c_{s,i} N(\mathbf{x}_t, \mu_{s,i}, \Sigma_{s,i}) \frac{1}{\sigma} \left[\frac{(x_{t,d} - \mu_{s,i,d})^2}{\sigma_{s,i,d}^2} - 1 \right] \tag{5.75}$$

Adapting $\sigma_{s,i,d}$ rather than $\sigma_{s,i,d}^2$ ensures that $\sigma_{s,i,d}^2$ will be positive, which preserves the positive definiteness of $\Sigma_{s,i}$.

5.8.3.7 Application of the Chain Rule

Now that the relevant partial derivatives have been found, they can be chained together to yield the partial derivative of the loss with respect to each individual component of the parameter vector Λ_n. This process is performed in the top-down manner outlined above. For example, the derivative of the loss $\ell(d_k(\mathbf{x}_1^T, \Lambda))$ w.r.t. a mean vector component $\mu_{s,i,d}$ belonging to the correct category C_k is:

$$\frac{\partial \ell(d_k(\mathbf{x}_1^T, \Lambda))}{\partial \mu_{s,i,d}} = \frac{\partial \ell(d_k(\mathbf{x}_1^T, \Lambda))}{\partial d_k(\mathbf{x}_1^T, \Lambda)} \frac{\partial d_k(\mathbf{x}_1^T, \Lambda)}{\partial g_k(\mathbf{x}_1^T, \Lambda)} \sum_{t|\theta_t=s} \frac{\partial g_k(\mathbf{x}_1^T, \Lambda)}{\partial b_{\theta_t}(\mathbf{x}_t)} \frac{\partial b_{\theta_t}(\mathbf{x}_t)}{\partial \mu_{\theta_t,i,d}} \tag{5.76}$$

Expanding the derivatives produces:

$$\frac{\partial \ell(d_k(\mathbf{x}_1^T, \Lambda))}{\partial \mu_{s,i,d}} = -\ell'(d_k(\mathbf{x}_1^T, \Lambda))$$

$$\times \sum_{t|\theta_t=s} \frac{1}{b_{\theta_t}(\mathbf{x}_t)} c_{\theta_t,i} N(\mathbf{x}_t, \mu_{\theta_t,i}, \Sigma_{\theta_t,i}) \frac{(x_{t,d} - \mu_{\theta_t,i,d})}{\sigma_{\theta_t,i,d}^2} \tag{5.77}$$

If a large ψ is used in the misclassification measure (5.64), the derivative of the loss $\ell(d_k(\mathbf{x}_1^T, \Lambda))$ w.r.t. a mean vector component $\mu_{s,i,d}$ belonging to the best incorrect category with best DP path sequence Θ

is:

$$\frac{\partial \ell(d_k(\mathbf{x}_1^T, \Lambda))}{\partial \mu_{s,i,d}} = \ell'(d_k(\mathbf{x}_1^T, \Lambda))$$

$$\times \sum_{t|\theta_t=s} \frac{1}{b_{\theta_t}(\mathbf{x}_t)} c_{\theta_t,i} N(\mathbf{x}_t, \mu_{\theta_t,i}, \Sigma_{\theta_t,i}) \frac{(x_{t,d} - \mu_{\theta_t,i,d})}{\sigma_{\theta_t,i,d}^2}. \quad (5.78)$$

Thus the gradient term has the same form as that for the correct category, with the sign reversed (from the derivative of the mis-classification measure). Use of these derivatives in the GPD adaptation procedure means that mean vectors belonging to the correct category will be pulled closer to the input, and mean vectors belonging to incorrect categories will be pushed away, in a manner strongly reminiscent of the LVQ adaptation rule.

If phoneme models are used to represent the categories, it is natural that many categories will share model parameters. In this case, the above derivative terms are summed over all shared models.

In MCE-based optimization, the above derivative calculations are used to produce a total gradient vector, $\nabla \ell_k(X_n, \Lambda_n)$, containing in each position the partial derivative of the final loss $\ell_k(X_n, \Lambda_n)$ with respect to an individual component of the parameter vector Λ_n. Given this total gradient, GPD, Quickprop, or any other gradient-based method can be used to minimize the overall MCE loss function.

5.8.4 MCE-HMM Results

Since the initial proposal of MCE [4, 44, 45], a very large number of applications of MCE-trained HMMs have been reported. To cite but two results, using MCE-trained HMMs, some of the best context-independent HMM results have been reported for the Texas Instruments-Massachusetts Institute of Technology TIMIT database [15, 43], and recognition improvements were reported for the resource management database [46]. In general, MCE seems to nearly always outperform the nondiscriminative MLE when used for the HMM training procedure. (The typical practice is to first use MLE to design the HMM, and to the refine the HMM using MCE.) However, the extent of the improvement afforded by MCE depends greatly on the quality of the model. A carefully designed HMM topology, with context-dependent models and an ample number of Gaussians and mixtures, trained with MLE, usually cannot be improved upon very much by MCE. However, when only a crude model topology is used and only a small number of parameters are available, the difference between MCE and MLE can be striking. There is clearly a strong practical implementational advantage to using MCE, as one can afford to use a simpler model with relatively light computational needs and still attain good performance. This is offset by the long training time that a discriminative training method like MCE requires compared with the nondiscriminative MLE. Whereas the latter designs the parameters for a given category using only data from that category, MCE (and other discriminative training methods such as MMI [6]) use data from all categories, not just the correct category, in designing each category's model parameters.

5.9 Conclusion

The central theme in this chapter was the use of pattern prototypes as the basis for a pattern classification system. The nature of prototype-based methods was highlighted by contrasting them with classic example-based methods, k-NN and Parzen windows. Both types of approach can be related to Bayes decision theory and to the goal of minimum-error-rate classification. In the case of the prototype-based methods, the MCE framework is instrumental in establishing the link to the Bayes error target. This is particularly so since prototype-based methods typically do not use an explicitly probabilistic model, making their evaluation in terms of quality of density estimation difficult if not impossible. MCE directly uses classification performance as the criterion for evaluation; it directly evaluates the decisions made by the classifier. The decision-oriented MCE loss function is particularly appropriate for designing prototype-based methods.

The prototype-based methods described here cover a substantial range. Starting with the static pattern recognition case in which patterns are represented as single vectors of fixed dimensionality, the simple but powerful LVQ method was introduced. In its straightforward use of reference vectors and nearest-neighbor classification based on Euclidean distance, LVQ exemplifies prototype-based pattern recognition. It was seen that LVQ can be viewed as an application of the MCE framework for discriminative training. Moving to the recognition of dynamic or sequential patterns, such as speech patterns, several LVQ-based architectures, including two LVQ/HMM hybrids, were described. These architectures illustrate some of the important issues in prototype-based speech recognition, primarily, the need to handle the nonlinear compressions and expansions characteristic of speech patterns, and the desirability of global classifier optimization. (None of the LVQ-based architectures allowed for global optimization.) These issues were addressed much more satisfactorily in the MCE-trained DP-embedding prototype-based methods considered next. In those methods, the dynamic programming procedure is embedded into the definition of discriminant function. An MCE loss function can then be defined in terms of those discriminant functions, and GPD optimization (or any other gradient-based optimization methods, such as the quick-prop algorithm) can be applied to learn the classifier parameters, i.e., the prototypes or templates. These methods are very similar in architecture to a hidden Markov model, leading us to view HMMs themselves as prototype-based methods, and to present the fundamentals of the application of MCE to HMM design.

References

[1] Sakoe, H., and S. Chiba, "A Dynamic Programming Algorithm Optimization for Spoken Word Recognition," *IEEE Trans. on Acoustics, Speech, and Signal Processing*, Vol. ASSP-26, no. 1, February 1978, pp. 43–49.

[2] Ney, H., "The Use of a One-Stage Dynamic Programming Algorithm for Connected Word Recognition," *IEEE Trans. on Acoustics, Speech, and Signal Processing*, Vol. ASSP-32(2), 1984, pp. 263–271.

[3] Rabiner, L. R., and B.-H. Juang, *Fundamentals of Speech Recognition*, Englewood Cliffs, N.J.: Prentice Hall, 1993.

[4] Katagiri, S., C.-H. Lee, and B.-H. Juang, "A Generalized Probabilistic Descent Method," *Proc. Acoustical Society of Japan*, fall meeting, September, 1990, pp. 141–142.

[5] Juang, B.-H., and S. Katagiri, "Discriminative Learning for Minimum Error Classification," *IEEE Trans. on Acoustics, Speech, and Signal Processing*, Vol. 40, no. 12, December 1992, pp. 3043–3054.

[6] Normandin, Y., "Hidden Markov Models, Maximum Mutual Information Estimation, and the Speech Recognition Problem," Ph.D. thesis, McGill University, Montreal, Department of Electrical Engineering, 1991.

[7] Duda, R. O. and P. E. Hart, *Pattern Classification and Scene Analysis*, New York: Wiley, 1973.

[8] Fukunaga, K. K., and P. M. Narendra, "A Branch and Bound Algorithm for Computing *k*-nearest Neighbors," *IEEE Trans. on Computers*, Vol. 24, 1975, pp. 750–753.

[9] Bishop, C. M., *Neural Networks for Pattern Recognition*, Oxford and New York: Oxford University Press, 1995.

[10] Niles, L. T., H. F. Silverman, G. Tajchman, and M. Bush, "How Limited Training Data Can Allow a Neural Network to Outperform an 'Optimal' Statistical Classifier," in *Proc. ICASSP '89*, Glasgow, Scotland, May 1989, pp. 17–20.

[11] Huang, X. D., Y. Ariki, and M. A. Jack, *Hidden Markov Models For Speech Recognition*, Edinburgh University Press, 1990.

[12] Amari, S., "A Theory of Adaptive Pattern Classifiers," *IEEE Trans. on Electronic Computers*, Vol. EC-16, no. 3, 1967, pp. 299–307.

[13] Battiti, R., "First- and Second-Order Methods for Learning: Between Steepest Descent and Newton's Method," *Neural Computation*, Vol. 4, 1992, pp. 141–166.

[14] Fahlman, S. E., "An Empirical Study of Learning Speech in Back-Propagation Networks," Technical report, Carnegie Mellon University, 1988.

[15] McDermott, E., "Discriminative Training for Speech Recognition," Ph.D. thesis, Waseda University, Tokyo, Japan, March 1997.

[16] Bottou, L. "Une Approche théorique de l'apprentissage connectionniste: Application `a la Reconnaissane de la Parole," Ph.D. thesis, Université de Paris Sud, 1991.

[17] Kohonen, T., G. Barna, and R. Chrisley, "Statistical Pattern Recognition with Neural Networks: Benchmarking Studies," *Proc. IEEE Int. Conf. Neural Networks*, Vol. 1, 1988, pp. 61–68.

[18] Kohonen, T., *Self-Organizing Maps*, 2d ed., Berlin and New York: Springer, 1997.

[19] Oja, E., *Subspace Method of Pattern Recognition*, New York: Wiley, 1983.

[20] Waibel, A., et al., "Phoneme Recognition Using Time-Delay Neural Networks," *IEEE Trans. on Acoustics, Speech, and Signal Processing*, Vol. ASSP-37, no. 3, 1989, pp. 328–339.

[21] Morgan, N., and H. Bourlard, "Continuous Speech Recognition Using Multilayer Perceptrons with Hidden Markov Models," *Proc. IEEE ICASSP*, 1990, pp. 413–416.

[22] Robinson, A.J., "Application of Recurrent Nets to Phone Probability Estimation," *IEEE Transactions on Neural Networks*, Vol. 5, no. 2, March 1994, pp. 298–305.

[23] McDermott, E., and S. Katagiri, "Shift-Invariant, Multi-Category Phoneme Recognition using Kohonen's LVQ2," *Proc. IEEE ICASSP*, Vol. 1, 1989, pp. 81–84.

[24] McDermott, E., and S. Katagiri, "LVQ-Based Shift-Tolerant Phoneme Recognition," *IEEE Trans. on Acoustics, Speech, and Signal Processing*, Vol. 39, no. 6, 1991, pp. 1398–1411.

[25] Minami, Y., et al., "On the Robustness of HMM and ANN Speech Recognition Algorithms, *International Conference on Spoken Language Processing*, Vol. 2, 1990, pp. 1345–1348.

[26] Iwamida, H., S. Katagiri, E. McDermott, and Y. Tohkura, "A Hybrid Speech Recognition System Using HMMs with an LVQ-Based Codebook," *Proc. ICASSP '90*, Alburquerque, N.M., April 1990, pp. 489–492.

[27] Iwamida, H., S. Katagiri, and E. McDermott, "Speaker-Independent Large Vocabulary Word Recognition Using an LVQ/HMM Hybrid Algorithm," *Proc. ICASSP '91*, Toronto, Canada, May 1991, pp. 553–556.

[28] Kimber, D. G., M. A. Bush, and G. N. Tajchman, "Speaker Independent Vowel

Classification Using Hidden Markov Models and LVQ2," *Proc. ICASSP '90*, Alburquerque, N.M., April 1990, pp. 497–500.

[29] Yu, G., W. Russel, R. Schwartz, and J. Makhoul, "Discriminant Analysis and Supervised Vector Quantization for Continuous Speech Recognition," *Proc. ICASSP '90*, Alburquerque, N.M., April 1990, pp. 685–688.

[30] Kurimo, M., *Using Self-Organizing Maps and Learning Vector Quantization for Mixture Density Hidden Markov Models*," Ph.D. thesis, Helsinki University of Technology, 1997.

[31] Katagiri, S., and C.-H. Lee, "A New Hybrid Algorithm for Speech Recognition Based on HMM Segmentation and Learning Vector Quantization," *IEEE Transactions on Speech and Audio Processing*, Vol. 1, no. 4, 1993, pp. 421–430.

[32] McDermott, E., and S. Katagiri, "LVQ3 for Phoneme Recognition," *Proc. Acoustical Society of Japan*, spring meeting, 1990, pp. 151–152.

[33] Katagiri, S., C-H. Lee, and B.-H. Juang, "Discriminative Multilayer Feed-Forward Networks," *Proc. IEEE Workshop on Neural Networks for Signal Processing*, 1991, pp. 309–318.

[34] Vintsyuk, T.K., "Element-Wise Recognition of Continuous Speech Composed of Words from a Specified Dictionary," *Kibernetika*, Vol. 7, March–April 1971, pp. 133–143.

[35] Itakura, F., "Minimum Prediction Residual Principle Applied to Speech Recognition," *IEEE Trans. on Acoustics, Speech, and Signal Processing*, Vol. 26, no. 1, 1975, pp. 52–59.

[36] Waibel, A., and K.-F. Lee, eds., *Readings in Speech Recognition*, San Mateo, Calif.: Morgan Kaufmann, 1990.

[37] Komori, T. and S. Katagiri, "Application of a Generalized Probabilistic Descent Method to Dynamic Time Warping-based Speech Recognition," *Proc. IEEE ICASSP*, Vol. 1, March 1992, pp. 497–500.

[38] Komori, T., and S. Katagiri, "GPD Training of Dynamic Programming-based Speech Recognizers," *jasj*, Vol. 13, no. 6, 1992, pp. 341–349.

[39] Chang, P.-C., and B.-H. Juang, "Discriminative Template Training for Dynamic Programming Speech Recognition," *Proc. IEEE ICASSP*, Vol. 1, 1992, pp. 493–496.

[40] Chang, P.-C., and B.-H. Juang, "Discriminative Training of Dynamic Programming Based Speech Recognizers," *IEEE Transactions on Speech and Audio Processing*, Vol. 1, no. 2, 1993, pp. 135–143.

[41] McDermott, E., and S. Katagiri, "Prototype-Based Discriminative Training for Various Speech Units," *Proc. IEEE ICASSP*, Vol. 1, March 1992, pp. 417–420.

[42] McDermott, E., and S. Katagiri, "Prototype Based Discriminative Training for Various Speech Units," *Computer Speech and Language*, Vol. 8, 1994, pp. 351–368.

[43] McDermott, E., and S. Katagiri, "String-Level MCE for Continuous Phoneme Recognition," *Proc. Eurospeech '97*, Rhodes, Greece, September 1997, pp. 123–126.

[44] Katagiri, S., C-H. Lee, and B.-H. Juang, "New Discriminative Training Algorithms Based on the Generalized Descent Method," *Proc. IEEE Workshop on Neural Networks for Signal Processing*, 1991, pp. 299–308.

[45] Chou, W., B.-H. Juang, and C.-H. Lee, "Segmental GPD Training of HMM Based Speech Recognizer," *Proc. IEEE ICASSP*, Vol. 1, March 1992, pp. 473–476.

[46] Chou, W., C.-H. Lee, and B.-H. Juang, "Minimum Error Rate Training Based on N-best String Models," *Proc. IEEE ICASSP*, Vol. 2, 1993, pp. 652–655.

6

Recurrent Neural Networks for Speech Recognition
Mike Schuster

6. 1 Introduction

6. 1. 1 Background and Motivation

Speech recognition by machines, commonly defined as the automatic transcription of spoken words or sentences (utterances), is a challenging engineering problem. Not only all of today's state-of-the-art speech recognition systems, but a whole category of applications, which include speaker and language recognition as well as problems occurring in speech synthesis, image recognition, time series prediction, and character recognition, are based on the theory of pattern recognition. This category of problems is often solved by statistical approaches using the simple but powerful principle of supervised learning from examples, which has been used successfully for speech recognition since about 1975, and has been applied to many other problems of engineering interest as well.

A finite amount of examples, or training data, for speech recognition, a number of recorded waveforms reduced to an observation feature vector sequence $\mathbf{x}_1^T$ of length T frames plus their known and correct transcriptions, are used to train a model $\mathbf{M}$, that can later be used to transcribe new, previously unseen waveforms (test data). The model corresponds to a given structure and a number of W parameters to maximize some predefined optimality criterion, which in the case of speech recognition is ideally the percentage of correct words.

In almost all of today's systems, the model **M** is a hidden Markov model (HMM) modeling the observation vector sequence. Often conventional Gaussian mixture distributions are used to model a feature vector given a certain phone (or a part of it called a "state"), because this greatly simplifies training and use of the model and makes it possible to use hundreds of hours of speech data for training, which is necessary to build state-of-the-art models given the current state of research. The disadvantage of using conventional Gaussian mixture distributions is that

- Their use requires a number of assumptions about the speech data;
- Their evaluation for a given feature vector to calculate its likelihood $P_k(\mathbf{x})$ for all possible states K can be time-consuming, if many mixture components and states are used;
- They are in general not trained to maximize the ability to discriminate between different phones or states.

This last point will be discussed in more detail in a later section.

Since around 1988, neural networks (NNs) have been used as general tools or black boxes to solve statistical pattern recognition problems by learning from examples, either solely or for a part of the problem. Speech recognition for an arbitrary task is a very complex process and currently (1999) cannot be expected to be solved solely by neural networks. The term *neural networks* includes historically a large number of different architectures, models, and algorithms, many of them without an explicit objective function to maximize. The NNs to be discussed here are based on a sound theory that requires to maximize the likelihood measured on the training data. They are applied to solve one specific subproblem, that is, the estimation of a local score to substitute the observation likelihood calculation replacing the conventional Gaussian mixture distributions. Based on these local phone (or state) score estimates, further steps can be taken to search for more useful outputs like words or sentences. These systems are often referred to as hybrid systems, because NNs are used in combination with other techniques like HMMs [1, 2].

Different from that of many other problems, the task in speech recognition is to map sequences of observation vectors to sequences of class labels (phonemes or words), and not single vectors to a number of different classes. For this, sequential data, especially one type of NN model, the recurrent neural network (RNN), which is only a structural variation of nonrecurrent NNs, has been shown to be useful and has been used successfully in speech recognition engines [3, 4].

The use of NNs, compared with that of Gaussian mixture distribution strained to maximize the likelihood, has a number of advantages.

- They can be used to generate likelihood-like scores that are discriminative on the state level.
- They can be used with an exceptionally simple pruning strategy called *posterior pruning*, to speed up the search for the best word sequence.
- They in general need less free parameters to achieve similar word recognition performance.
- They can automatically share parameters, which has to be done explicitly for conventional Gaussian mixture distributions.
- They can be organized in a tree structure to speed up the likelihood calculation.
- They can, using recurrent neural networks, make use of an arbitrary amount of input information not bounded by ad hoc sized input windows.

These advantages will be discussed in more detail in the sections below. The main disadvantages of neural networks is that their training is time-consuming and often, depending on the implementation, difficult to converge, which limits in many cases the amount of data that can be used for training, and therefore also the number of potential applications and users.

6.1.2 Chapter Overview

The first part of the introduction has loosely defined the speech recognition problem and has listed advantages and disadvantages of using neural networks and competing approaches to solve the speech recognition problem. In Section 6.2 the basic theory of the speech recognition problem for a simple case using context-independent acoustic models is reviewed. Section 6.3 explains the necessary (for an understanding of this chapter) basics of neural networks, covering the types of problems to be solved, NN elements, architectures, and training algorithms. One architecture especially useful for speech recognition, the RNN, its specific problems and merits as well as a useful extension of the basic unidirectional RNN to a bidirectional structure, is discussed in Section 6.4. Section 6.5 explains issues that arise when NNs are used to build context-dependent acoustic models, which are necessary to achieve state-of-the-art recognition results. The more practical aspects of training and complete systems using neural networks are addressed in Section 6.6. Finally, a discussion and some potential improvements for the field are given in Section 6.7.

6.2 Speech Recognition Theory

Speech recognition using standard statistical methods (e.g., HMMs) is well documented in several books [5, 6]. To introduce some necessary notation within the context of using neural networks, the speech recognition problem can be written as

$$\mathbf{Y}^* = \arg\max_{\mathcal{Y}} P(\mathbf{Y}|\mathbf{X}) \qquad (6.1)$$

with

$$\mathbf{X} = \mathbf{x}_1^T = \{\mathbf{x}_1, \mathbf{x}_2, \ldots, \mathbf{x}_T\} \qquad (6.2)$$

being the input observation feature vector sequence (frames) of dimensionality D calculated from the observed waveform, in practical systems around 100 40-dimensional vectors/sec input speech (for a description of common feature extraction methods see [7]), and

$$\mathbf{Y} = \mathbf{y}_1^T = \{\mathbf{y}_1, \mathbf{y}_2, \ldots, \mathbf{y}_T\} \qquad (6.3)$$

being any associated valid symbol sequence coded in K-dimensional vectors with the component corresponding to the symbol index being one and all others being zero. $\mathbf{Y}^*$ is the recognized symbol sequence with the highest probability[1] among all possible sequences from the space $\mathcal{Y}$. Valid symbol classes y_t at time t are in the case of word

---

1. Throughout this chapter there is no distinction made between probability mass and density, usually denoted as P and p, respectively. If the variable to model is categorical, a probability mass is assumed; if it is continuous, a probability density is assumed.

recognition any words that are listed in a pronunciation dictionary which contains all words to be recognized as phoneme sequences, and in the case of phoneme recognition all possible phonemes, which usually number, depending on the language to recognize, around 50. Since the principle usage of neural networks is the same for word and phoneme recognition, discussion is here limited to the latter in order to simplify notation.

Training a speech recognition system in principle corresponds to estimating the probability distribution[2] $P(\mathbf{Y}|\mathbf{X})$, which includes (a) defining an appropriate model $\mathbf{M}$ and (b) estimating its parameters $\mathbf{w}$ maximizing some predefined optimality criterion. In practice the model $\mathbf{M}$ consists of several modules with each one being responsible for a different part of $P(\mathbf{Y}|\mathbf{X})$. Usage of the trained system or recognition for a given observation sequence $\mathbf{X}$ corresponds principally to the evaluation of $P(\mathbf{Y}|\mathbf{X})$ for all possible symbol sequences to find the best one $\mathbf{Y}^*$. This procedure is called the search, for which efficient algorithms are known [8–12].

It is important to notice that although $\mathbf{Y}^*$ is the symbol sequence with the highest probability, it is not guaranteed that it corresponds to the symbol sequence with the highest word or phoneme recognition rate, which is defined by $r = (N - I - D - S)/N$ (N: number of symbols in correct sequence; I: number of insertions in recognized sequence; D: number of deletions; S: number of substitutions) and has to be found by a dynamic programming procedure [6]. This mismatch is a well-accepted fact and doesn't seem to be a problem in practical systems.

Using the Bayes rule $P(B|A) = P(A|B)P(B)/P(A)$ and the product rule of probability $P(A, B) = P(A)P(B|A)$, the conditional sequence probability $P(\mathbf{Y}|\mathbf{X})$ can for a simple example be broken down to three terms as:

$$\mathbf{Y}^* = \arg\max_{\mathcal{Y}} P(\mathbf{Y}|\mathbf{X})$$

$$= \arg\max_{\mathcal{Y}} P(\mathbf{X}|\mathbf{Y})P(\mathbf{Y})$$

$$= \arg\max_{\mathcal{Y}} P(\mathbf{x}_1^T|\mathbf{y}_1^T)P(\mathbf{y}_1^T)$$

---

2. To simplify notation, throughout this chapter, random variables and their values are often *not* denoted as different symbols when their identity is obvious from the context. This means, $P(\mathbf{x}) = P(X = \mathbf{x})$.

$$= \arg\max_{\mathscr{Y}} \left[\prod_{t=1}^{T} P(\mathbf{x}_t | \mathbf{x}_1, \mathbf{x}_2, \ldots, \mathbf{x}_{t-1}, \mathbf{y}_1^T) \right]$$

$$\times \left[\prod_{t=1}^{T} P(\mathbf{y}_t | \mathbf{y}_1, \mathbf{y}_2, \ldots, \mathbf{y}_{t-1}) \right] \tag{6.4}$$

$$\approx \arg\max_{\mathscr{Y}} \left[\prod_{t=1}^{T} P(\mathbf{x}_t | \mathbf{y}_t) \right] \left[\prod_{t=1}^{T} P(\mathbf{y}_t | \mathbf{y}_{t-1}) \right] \tag{6.5}$$

$$= \arg\max_{\mathscr{Y}} \left[\prod_{t=1}^{T} \frac{P(\mathbf{y}_t | \mathbf{x}_t)}{P(\mathbf{y}_t)} P(\mathbf{y}_t | \mathbf{y}_{t-1}) \right] \tag{6.6}$$

making some simplifying approximations. These were, for this example:

1. Every output state class $\mathbf{y}_t$ depends only on the previous state $\mathbf{y}_{t-1}$ and not on all previous state classes, making it a first order Markov model:

$$P(\mathbf{y}_t | \mathbf{y}_1, \mathbf{y}_2, \ldots, \mathbf{y}_{t-1}) \Rightarrow P(\mathbf{y}_t | \mathbf{y}_{t-1}) \tag{6.7}$$

2a. The feature vectors are assumed to be statistically independent in time:

$$P(\mathbf{x}_t | \mathbf{x}_1, \mathbf{x}_2, \ldots, \mathbf{x}_{t-1}, \mathbf{y}_1^T) \Rightarrow P(\mathbf{x}_t | \mathbf{y}_1^T) \tag{6.8}$$

This assumption is called the *independence assumption*, and it is considered to be a strong assumption, because feature vectors close in the observation sequence are often in similar ranges and therefore not independent.

2b. The likelihood of feature vector $\mathbf{x}_t$ given the complete symbol sequence $\mathbf{y}_1^T$ is assumed to depend only on the symbol found at t and not on any other ones, making it a context-independent model or monophone, here with only one state per HMM:

$$P(\mathbf{x}_t | \mathbf{y}_1^T) \Rightarrow P(\mathbf{x}_t | \mathbf{y}_t) \tag{6.9}$$

The equations (6.5) and (6.6) or very similar expressions are the basis

for many frequently used approaches to speech recognition systems—
in general (6.5), if a Gaussian mixture system is used, and (6.6), if a
system based on neural networks is used. Note that they are both only
approximations to expression (6.4), which is optimal given the original
problem (6.1). The four probability expressions in (6.5) and (6.6) to be
estimated are:

1. $P(\mathbf{x}_t|\mathbf{y}_t)$—the likelihood of input observation vector $\mathbf{x}_t$ given
   phone $\mathbf{y}_t$, in general modeled by conventional Gaussian
   mixture distributions, but can be modeled by a neural network
   also as done in [12];
2. $P(\mathbf{y}_t|\mathbf{x}_t)$—the posterior probability of phoneme c_t given input
   vector $\mathbf{x}_t$, to be estimated by a neural network;
3. $P(\mathbf{y}_t)$—the prior probability of phoneme frame c_t, to be
   approximated by the relative frequency of observing phoneme
   frame c_t in the training data;
4. $P(\mathbf{y}_t|\mathbf{y}_{t-1})$—the transition probability for the HMMs, also to be
   estimated by counting from the aligned training data.

Often the expression

$$\tilde{P}(\mathbf{y}_t|\mathbf{x}_t) = \frac{P(\mathbf{y}_t|\mathbf{x}_t)}{P(\mathbf{y}_t)} \tag{6.10}$$

is referred to as the scaled likelihood because it is proportional to the
real likelihood $P(\mathbf{x}_t|\mathbf{y}_t)$.

6.3 Basics of Neural Networks

Artificial neural networks (see [13] for an excellent introduction) can be
used for many supervised learning tasks. Given as training data N input/
target data vector pairs $\mathbf{D} = \{\mathbf{x}_n, \mathbf{t}_n\}$ (a mapping from input to a target
data), with dimensions D and K, respectively, the aim of a supervised
learning process is to learn how to predict output data given new input
data, which is written as a K-dimensional function $y^k(\mathbf{x}_n; \mathbf{w})$ depending
on the current input vector $\mathbf{x}$ and the NN parameter value vector $\mathbf{w}$ with
W weights. The weights are combined in structures, whose different
types are discussed next.

6.3.1 Parameter Estimation by Maximum Likelihood

The estimation of the parameters of the model is often guided by the *maximum likelihood* principle, which can be stated as follows:

There is given training data $\mathscr{D}$ and a model structure **M**, which is characterized by the parameter vector **w** out of all possible parameter vectors in the space $\mathscr{W}$, and all parameter value sets are assumed to be equally probable

$$P(\mathbf{w}) = const. \tag{6.11}$$

Then the goal of the parameter estimation process is to find a single set of parameter values $\mathbf{w}^*$ that maximize the probability of the model parameters given the data, which is called MAP (maximum a posteriori) estimation. Given assumption (6.11) and the fact that the unconditional probability of the data $\mathscr{D}$ is independent of the parameters **w** this can be simplified to

$$\mathbf{w}^* = \arg\max_{\mathscr{W}}\{P(\mathbf{w}|\mathscr{D})\} \tag{6.12}$$

$$= \arg\max_{\mathscr{W}}\left\{\frac{P(\mathscr{D}|\mathbf{w})P(\mathbf{w})}{P(\mathscr{D})}\right\} \tag{6.13}$$

$$= \arg\max_{\mathscr{W}}\{P(\mathscr{D}|\mathbf{w})P(\mathbf{w})\} \tag{6.14}$$

$$= \arg\max_{\mathscr{W}}\{P(\mathscr{D}|\mathbf{w})\} \tag{6.15}$$

such that the problem now becomes to maximize the *likelihood* of the data given the model with its parameters **w** without having a preference for certain parameter sets that seem to be more plausible than others ($\Rightarrow p(\mathbf{w}) = const.$).

In the case of supervised learning with training data $\mathbf{D} = \{\mathbf{x}_n, \mathbf{t}_n\}$, whose data pairs are assumed to be conditionally independent given **w**, the likelihood $\mathscr{L}$ becomes

$$\mathscr{L} = P(\mathscr{D}|\mathbf{w}) \tag{6.16}$$

$$= \prod_N P(\mathbf{t}_n, \mathbf{x}_n|\mathbf{w}) \tag{6.17}$$

$$= \prod_N P(\mathbf{t}_n|\mathbf{x}_n, \mathbf{w})P(\mathbf{x}_n|\mathbf{w}) \tag{6.18}$$

$$\propto N_N P(\mathbf{t}_n|\mathbf{x}_n, \mathbf{w}), \tag{6.19}$$

because for neural networks as considered in this chapter the inputs are not modeled and do not depend on $\mathbf{w}$, the term $\prod_N P(\mathbf{x}_n | \mathbf{w})$ is constant for all $\mathbf{w}$ and therefore vanishes during maximization of $\mathcal{L}$. With the negative logarithm taken, the function becomes

$$E(\mathcal{D}, \mathbf{w}) = -\ln \prod_N P(\mathbf{t}_n | \mathbf{x}_n, \mathbf{w}) \qquad (6.20)$$

$$= -\sum_N \ln P(\mathbf{t}_n | \mathbf{x}_n, \mathbf{w}) \qquad (6.21)$$

which is called an *objective function* or *error function*, which has to be minimized during training. Making appropriate assumptions for the conditional output distribution $P(\mathbf{t}_n | \mathbf{x}_n, \mathbf{w})$ makes it possible to classify the types of problems to be solved by neural networks, which is discussed in the next section.

6.3.2 Problem Classification

Problems suitable to be solved with neural networks can be divided into three groups depending on the type of input and target variables. Inputs and targets can, in general, be continuous and/or categorical variables, which defines the two categories of supervised learning problems. When targets are continuous, the problem is known as a regression problem; when they are categorical (class labels), the problem is known as a classification problem. Within this chapter, the term *prediction* is used as a general term that includes classification and both types of regression.

6.3.2.1 Regression

For unimodal regression or *function approximation*, the components of the output vectors are continuous variables. The NN parameters are estimated to minimize some predefined error criterion, e.g., maximize the likelihood of the target data given the model and the input data as discussed above. When it is assumed that (a) the distribution of the errors between the desired target and the estimated output vectors is a single Gaussian with zero mean and a global data-dependent variance, and (b) all K outputs are conditionally independent, then the likelihood criterion reduces to the convenient Euclidean distance measure between the desired and the estimated output vectors or the *squared-error function,*

$$E = \sum_N \sum_K (y^k(\mathbf{x}_n; \mathbf{w}) - t_n^k)^2 \qquad (6.22)$$

which has to be minimized during training. It has been shown several times [13] that neural networks can estimate the conditional average of the desired target vectors at their network outputs; i.e., $y^k(\mathbf{x}; \mathbf{w}^*) = \langle t^k | \mathbf{x} \rangle$, where $\langle \cdot \rangle$ is an expectation operator and $\mathbf{w}^*$ is the parameter (weight) vector at the minimum of the error function. For unimodal regression the network outputs can be interpreted as the mean of a Gaussian distribution, which varies depending on the current input.

6.3.2.2 Classification

In the case of a classification problem, one seeks the most probable class out of a given pool of K classes for each input vector $\mathbf{x}_n$. To make this kind of problem suitable to be solved by an NN, the categorical target variables are usually coded as vectors as follows. Consider that k is the desired class label for an input vector $\mathbf{x}$. Then construct a K-dimensional target vector $\mathbf{t}$ such that its kth component is one and all other components are zero. The kth component can be interpreted as the probability of $\mathbf{x}$ belonging to class k. The target vectors $\mathbf{t}_n$ constructed in this manner along with the input vectors $\mathbf{x}_n$ can be used to train the NN under some optimality criterion, usually the cross-entropy function,

$$E = -\sum_N \sum_K t_n^k \, ln(y^k(\mathbf{x}_n; \mathbf{w})) \qquad (6.23)$$

which results from a maximum likelihood estimation assuming a multi-nomial target distribution [13]. It has been shown [11] that the kth network output can be interpreted as an estimate of the conditional posterior probability of class membership $(y^k(\mathbf{x}; \mathbf{w}^*) = P(C = k | \mathbf{x}))$, with the quality of the estimate depending on the amount of training data and the complexity of the network.

6.3.3 Neural Network Training

Training of neural networks as discussed here is equivalent to adjusting the weights $\mathbf{w}$ iteratively such that an error function is minimized, which is, depending on the type of problem, either the squared error or the cross-entropy measured on the training data. Function minimization is a problem occurring in many disciplines of science, and standard procedures are well documented [13, 15–16]. Usual approaches for neural networks are

1. First-order methods, which use the first derivative of the error function $(\partial/\partial\mathbf{w})E$ to be minimized, for example, gradient descent, gradient descent with momentum, resilient propagation (RPROP) algorithm;

2. Second-order methods [17], which also use the second derivative (Hessian) or approximations to it, for example quasi-Newton, conjugate gradient, Levenberg-Marquardt, Back-Propagation based on partial Quasi-Newton (BPQ) [18], or quickprop [19].

The first (and also second) derivative of the error function in feedforward neural networks can be calculated efficiently with a procedure called back-propagation [13, 20], which requires a forward pass (calculate $y^k(\mathbf{x}_n; \mathbf{w})\ \forall\ k$) and a backward pass (calculate $(\partial/\partial\mathbf{w})E$ vector) through the network for each of the N training vector pairs. All training procedures can be (1) offline or batch methods, for which the weights $\mathbf{w}$ are updated after all N training samples have been used to calculate the first derivative, or (2) online methods, for which only a part of the training samples is used to get an estimate of the first derivative which is then used to update the weights.

The use of neural networks for large-scale real-world problems such as, for example, speech recognition adds two practical problems to training: (1) the number of parameters W (weights) is considerably high, often in the range of 10,000–2,000,000, and (2) the number of training data vectors N is often in the range of 1 million to 100 million, being also much higher than for the average NN application (medical applications, etc.). These two problems rule out many of the theoretically superior and more sophisticated second-order training algorithms because of insufficient memory resources and/or a too complicated implementation. Algorithms used in practice for large-scale problems are currently mostly first-order methods, i.e., online gradient descent and online RPROP procedures.

6.3.3.1 Gradient Descent Training

Gradient descent training refers to adjusting the weight vector $\mathbf{w}$ after each iteration i by a small vector $\Delta\mathbf{w}$ proportional to the negative gradient $-(\partial/\partial \mathbf{w})E^{(i)} = -(\partial/\partial \mathbf{w})E(\mathbf{D}, \mathbf{w}^{(i)})$:

$$\Delta \mathbf{w}^{(i)} = -\eta \frac{\partial}{\partial \mathbf{w}} E^{(i)} \tag{6.24}$$

$$\mathbf{w}^{(i+1)} = \mathbf{w}^{(i)} + \Delta \mathbf{w}^{(i)} \tag{6.25}$$

This procedure can be refined by making the weight change $\Delta \mathbf{w}$ linearly dependent on the previous change

$$\Delta \mathbf{w}^{(i)} = -\eta \frac{\partial}{\partial \mathbf{w}} E^{(i)} + \rho \cdot \Delta \mathbf{w}^{(i-1)}, \tag{6.26}$$

which often leads to a considerable speedup. Good values for η and ρ heavily depend on the used NN structure, the training data, and the initialization of $\mathbf{w}$ (which is often random using small values) and have to be found by experiments.

   If an online procedure is used, then the estimated gradient to be used for one update depends only on a small part of the available training data, which might lead to a large fluctuation of the gradient and therefore to slower training. In this case the local gradient estimate may be smoothed and improved by

$$\frac{\partial}{\partial \mathbf{w}} E^{(i)} := (1 - \alpha) \cdot \frac{\partial}{\partial \mathbf{w}} E^{(i-1)} + \alpha \cdot \frac{\partial}{\partial \mathbf{w}} E^{(i)} \tag{6.27}$$

with $0 \leq \alpha \leq 1$ controlling the amount of smoothing. For an additional improvement, α can be a variable slowly increasing toward one during training.

6.3.3.2 RPROP Training

RPROP in [21], is a simple, heuristic first-order procedure that has been proposed in many variations by various researchers (see [13]), and works reasonably well also for large-scale problems. The idea is to keep a step-size δ_w for each weight individually and make the update dependent only on the sign of the wth component of the gradient $(\partial/\partial w)E^{(i)}$ as:

$$\text{IF} \quad \frac{\partial}{\partial \mathbf{w}} w E^{(i)} > 0 \quad \text{THEN} \quad w_w^{(i+1)} := w_w^{(i)} - \delta_w^{(i)}$$

$$\text{ELSE IF} \quad \frac{\partial}{\partial \mathbf{w}} w E^{(i)} < 0 \quad \text{THEN} \quad w_w^{(i+1)} := w_w^{(i)} + \delta_w^{(i)}$$

The step-size itself is updated depending on the gradient component change as

$$\text{IF} \quad \frac{\partial}{\partial \mathbf{w}} wE^{(i)} \cdot \frac{\partial}{\partial \mathbf{w}} wE^{(i-1)} > 0 \quad \text{THEN} \quad \delta_w^{(i+1)} = \delta_w^{(i)} \cdot \tau_+$$
$$\text{ELSE} \qquad\qquad\qquad\qquad\qquad\qquad\qquad \delta_w^{(i+1)} = \delta_w^{(i)} \cdot \tau_-$$

with good values being $\tau_+ = 1.2$ and $\tau_- = 0.5$ for many problems. It is useful to limit δ_w to not exceed a certain range, which is not very critical and often set to $0.000001 < \delta_w < 50$. A good initial start value for δ_w is often $\delta_w = J/10$, with J being the number of input weights to a certain neuron.

6.3.3.3 ARPROP Training

A simple adaptive refinement of the RPROP procedure exists, which was found to be very robust against variations of the data block sizes used for an update of the weights and therefore well-suited for online training, is explained in [12]. The problem of RPROP for online training, which is necessary to train large networks with lots of data provided in small chunks, is that in general the step-sizes approach zero too quickly. This can be avoided by heuristically increasing the step-sizes after a certain number of iterations, or by an automatic procedure, here called Automatic RPROP or ARPROP. An efficient method is described below.

An average step-size $\delta_{average}$ is calculated after each update of the weights as

$$\delta_{average} = \frac{1}{W} \sum_{w=1}^{W} \delta_w.$$

If the average step-size of the current update $\delta_{average}^{(i)}$ is below the average step-size from the previous update $\delta_{average}^{(i-1)}$, the parameter τ_+ controlling the step-size increase is decreased by a small value C, otherwise it is increased; that is,

$$\text{IF} \quad \delta_{average}^{(i)} > \delta_{average}^{(i-1)} \quad \text{AND} \quad \tau_+^{(i)} > 1.0 \quad \text{THEN} \quad \tau_+^{(i+1)} = \tau_+^{(i)} - C$$
$$\text{ELSE} \qquad\qquad\qquad\qquad\qquad\qquad\qquad \tau_+^{(i+1)} = \tau_+^{(i)} + C,$$

with $C = 0.01$, $\tau_+^{(0)} = [1.1; 1.2]$ and $\tau_- = [0.5; 0.9]$ being useful values to achieve fast convergence. In this way the average step-size decreases very slowly and a complete training until convergence can be run without any human-induced restarts. ARPROP was found to be an efficient training procedure for large recurrent and nonrecurrent networks (50,000–300,000 weights), as necessary to achieve good speech recognition results, using a large amount of data (10^6–10^8 observation vectors).

6.3.4 Neural Network Architectures

For supervised learning (from sequences), several neural network architectures are in use. The most common are multilayer perceptrons (MLPs), time-delay neural networks (TDNNs), and recurrent neural networks (RNNs); these are briefly discussed below. Other common architectures such as radial basis functions (RBFs) and hierarchical mixtures of experts (HMEs) [22] have interesting properties but are commonly not used for speech recognition applications and their explanations are therefore omitted here.

The type of neural networks discussed here have as elements neurons connected by directed connection weights representing scalar parameters w, which are combined in a structure to provide a D- (input) to K-dimensional (output) mapping. Each neuron has one output o and many (for example J) inputs connected to outputs of other neurons or the input vector itself (Figure 6.1). The output o of each neuron is a

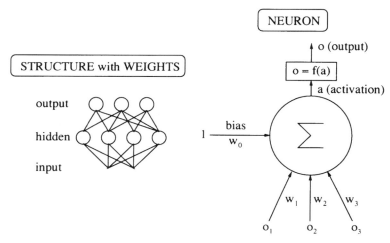

Figure 6.1 Elements of feedforward neural networks as used in this chapter are *neurons* and *connection weights* in a *structure*.

function of its activation a, so $o = f_{act}(a)$, with the activation calculated as a sum of all inputs to the neuron multiplied by its corresponding weight, $a = \sum_J o_j w_j$. Usually there is also a bias weight, which acts as an additional input constantly set to one and in general treated like one of the J inputs.

The neurons are often organized in layers as groups of neurons, with consecutive layers usually being fully connected, meaning that each neuron of a layer is connected to all neurons of the next layer. When neurons' outputs are used as one of the K neural network outputs, they belong to the output layer; otherwise they belong to one of the hidden layers. As activation functions for hidden-layer neurons, commonly the *sigmoid* function $f_{act}(a) = 1/(1 + e^{-a})$ or its equivalent by a linear transformation, the *tanh*-function $f_{act}(a) = (e^a - e^{-a})/(e^a + e^{-a})$ is used (note that $2f_{act}^{sigmoid}(2a) - 1 = f_{act}^{tanh}(a)$), with the latter often leading to slightly faster convergence using commonly used training procedures. The choice of the sigmoid activation function is motivated by its distinct property of being the discriminant function for a two-class classification problem that makes the output the posterior probability of class membership, if the input distributions are Gaussian with equal covariance matrices [13]. The choice of activation functions for the output layer depends on the problem to be solved. If it is a unimodal regression problem, the output of the network represents the mean of a Gaussian distribution, which shouldn't be bounded. Therefore, usually the linear activation function $f_{act}(a) = a$ is used. If it is a classification problem, the softmax [23] function $f_{act}(a) = e^a / \sum_K e^{a_k}$ is used, which can be interpreted as the generalized sigmoid for the K-class classification problem [13]. The combinations of objective function and output activation function depending on the problem to be solved are summarized in Table 6.1.

Table 6.1
Common setups for neural networks

Problem	Objective Function	Output Activation Function
unimodal regression	squared error	linear
classification	cross-entropy	softmax

6.3.4.1 Multilayer Perceptrons

Multilayer perceptrons (MLPs) are the most common type of architecture, in many practical applications only with two layers of weights, one hidden and the output layer as shown in the left part of Figure 6.1. More layers are possible but not necessary, since there are proofs that any mapping can be approximated with arbitrary accuracy with only two layers ([13] and references therein), although using more layers can be a more efficient realization of a certain mapping. In practice, however, more than two layers are rarely used because of little expected performance gain and practical problems during training.

For sequence processing with neighboring vector pairs being correlated, it is common to use, besides the current input vector $\mathbf{x}_t$ information from its $2L$ neighboring vectors $\mathbf{x}_{t-L}, \mathbf{x}_{t-L+1}, \ldots, \mathbf{x}_{t-1}$ and $\mathbf{x}_{t+1}, \mathbf{x}_{t+2}, \ldots, \mathbf{x}_{t+L}$ from a window as input to the MLP to improve prediction. The optimal width of the window in order to reach a good performance on unseen test data drawn from the same distribution obviously depends on the type of data and the predefined structure. If the window is too small, not enough information will be present to provide an optimal prediction. If the window is very large, the performance on the training data will be almost perfect, but the performance on test data will be poor. Hence, an optimal window size must be found by trial and error on development data.

6.3.4.2 Time-Delay Neural Networks

Time-delay neural networks (TDNNs) [24] have the same structure as a regular MLP but a reduced number of total weight parameters; they have proven to be a useful improvement over regular MLPs in many applications. This is achieved by a user-defined hard-tying of parameters, meaning forcing certain parameters to have the same values. Which parameters are useful to tie depends heavily on the used data and can only be found by experiments.

6.4 Recurrent Neural Networks

For many applications the data $\mathscr{D}$ is not a collection of vector pairs in arbitrary order, but rather the data comes in sequences of vector pairs, with the order being not arbitrary. Speech recognition is a typical example for this case—every preprocessed waveform is an array of vectors $\mathbf{x}_1^T$, that is to be mapped to an array of target classes c_1^T in the form of K-dimensional vectors $\mathbf{t}_1^T$.

6.4.1 Unidirectional Recurrent Neural Network

One type of RNN, the common unidirectional RNN, provides an elegant way of dealing with (time) sequential data that embodies correlations between data points that are close in the sequence.

6.4.1.1 RNN Architecture

Figure 6.2 shows a basic RNN architecture with a delay line and unfolded in time for two time steps. In this structure, the input vectors x_t are fed one at a time into the RNN. Instead of using a fixed number of input vectors as done in the MLP and TDNN structures, this architecture can make use of all the available input information up to the *current* time frame t_c (i.e., $\{x_t, t = 1, 2, \ldots, t_c\}$) to predict y_{t_c}. How much of this information is captured by a particular RNN depends on its structure and the training algorithm. An illustration of the amount of input information used for prediction with different kinds of NNs is given in Figure 6.3.

Future input information coming up later than t_c is usually also useful for prediction. With an RNN, this can be partially achieved by delaying the output by a certain number of S time frames to include future information up to x_{t_c+S} to predict y_{t_c} (Figure 6.3). Theoretically S could be made very large to capture all the available future information, but in practice it is found that prediction results drop if S is too large. A rough explanation for this could be that with rising S the modeling power of the RNN is more and more concentrated on remembering the input information up to x_{t_c+S} for the prediction of y_{t_c}, leaving less modeling power for combining the prediction knowledge from different input vectors.

While delaying the output by some frames (between 30 and 60 ms)

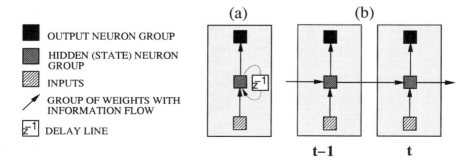

Figure 6.2 General structure of a regular unidirectional RNN shown (a) with a delay line, and (b) unfolded in time for two time steps.

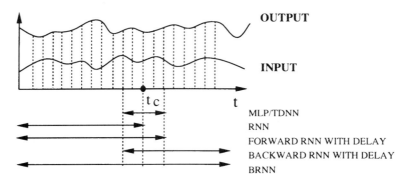

Figure 6.3 Visualization of the amount of input information used for prediction by different network structures.

has been used successfully to improve results in a practical speech recognition system [25] the optimal delay is task dependent and has to be found by the trial-and-error method on a validation test set. One possibility to get around this user-defined delay is to use bidirectional recurrent neural networks (BRNNs), which are discussed in the next section.

To use all available input information, it is possible to use two separate networks—one for each time direction—and then somehow merge the results. Both networks can then be called experts for the specific problem the networks are trained on. One way of merging the opinions of different experts is to assume that the opinions are statistically independent and hyperdistributed. Making certain simplifying assumptions about these distributions leads to the often used arithmetic averaging for unimodal regression and to geometric averaging (what corresponds to an arithmetic averaging in the log-domain) for classification. These merging procedures are referred to as linear opinion pooling and logarithmic opinion pooling, respectively [26–27]. Although simple merging of network outputs has been applied successfully in practice [3], it is generally not clear how to merge network outputs in an optimal way, since different networks trained on the same data can no longer be regarded as independent.

At this point it should be noted that the inclusion of more input information to enhance performance given the original problem (6.1) is a questionable procedure, although it has led to very good results in phoneme recognition experiments [12, 25] and continuous speech recognition [3, 10, 28, 29]. Although more input information will in general improve the estimation of the frame class posterior probabilities, it alters implicitly one of the expressions in the approximation to the original problem (6.6),

$$P(\mathbf{y}_t|\mathbf{x}_t) \Rightarrow P(\mathbf{y}_t|\mathbf{x}_1, \mathbf{x}_2, \ldots, \mathbf{x}_{t-1}, \mathbf{x}_t) \qquad (6.28)$$

which cannot be justified by the theory and needs further investigation.

6.4.1.2 RNN Training

Because of their recurrent connections, training of RNNs is slightly more complicated than for feedforward neural networks like MLPs. An often used training procedure is back-propagation through time (BPTT). For BPTT, first the RNN structure is unfolded up to the length of the training sequence as shown for two time-steps in Figure 6.2, which transforms the RNN in a large feedforward neural network. Now regular back-propagation can be applied, only at the beginning and the end of the training data sequence some special treatment is necessary. The state inputs at $t = 1$ are not known and can be set to an arbitrary, but fixed, value in practice. Also, the local state derivatives at $t = T$ are not known and can be set to zero, assuming that input information beyond that point is not important for the current update, which is for the boundaries certainly the case.

It should be noted that training of recurrent neural networks is not more time-consuming than nonrecurrent NNs like MLPs for a similar number of total free parameters. In fact, in many cases RNNs need less parameters than MLPs to achieve the same performance, and are therefore faster to train.

6.4.2   Bidirectional Recurrent Neural Network

To overcome the limitations of a regular RNN outlined in the previous section, here a bidirectional recurrent neural network (BRNN) which can be trained using all available input information in the past and future of a specific time frame, is explained [30].

6.4.2.1   BRNN Architecture

The idea is to split the state neurons of a regular RNN in one part, which is responsible for the positive time direction (forward states), and a second part for the negative time direction (backward states). Outputs from forward states are not connected to inputs of backward states and vice versa. This leads to the general structure that can be seen in

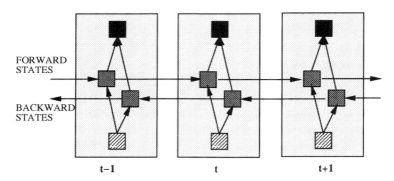

Figure 6.4 General structure of the bidirectional recurrent neural network (BRNN) with hidden states in forward and backward time direction, shown unfolded in time for three time steps.

Figure 6.4, where it is unfolded over three time-steps. It is not possible to display the BRNN structure in a figure similar to Figure 6.3 with the delay line, since the delay would have to be positive and negative in time. Note that without the backward states this structures implifies to a regular unidirectional forward RNN, as shown in Figure 6.3. If the forward states are taken out, it results in a regular RNN with a reversed time axis. With both time directions taken care of in the same network, input information in the past and the future of the currently evaluated time frame can directly be used to minimize the objective function, without the need of delays to include future information as for the regular unidirectional RNN discussed above.

6.4.2.2 BRNN Training

The BRNN can in principle be trained with the same algorithms as a regular unidirectional RNN because there are no interactions between the two types of state neurons, and therefore can be unfolded into a general feedforward network. However, if, for example, any form of back-propagation through time (BPTT) is used, the forward and backward pass procedure is slightly more complicated, because the update of state and output neurons can no longer be done one at a time. If BPTT is used, the forward and backward pass over the unfolded BRNN over time is done almost in the same way as for a regular MLP—only at the beginning and the end of the training data some special treatment is necessary. The forward state inputs at $t = 1$ and the backward state inputs at $t = T$ are not known. Setting these could be made part of the learning process, but are in practice often set to an arbitrary but fixed value. Also, the local state derivatives at $t = T$ for the forward states and at $t = 1$ for the backward states are not known, and are set to zero. The training procedure for the unfolded bidirectional network over time can be summarized as follows:

1. *FORWARD PASS*
   Feed all input data for one time slice $1 \leqslant t \leqslant T$ into the BRNN and determine all predicted outputs.
   (a) Do forward pass just for forward states (from $t = 1$ to $t = T$) and backward states (from $t = T$ to $t = 1$).
   (b) Do forward pass for output neurons.

2. *BACKWARD PASS*
   Calculate the part of the objective function derivative for the time slice $1 \leqslant t \leqslant T$ used in the forward pass.
   (a) Do backward pass for output neurons.
   (b) Do backward pass just for forward states (from $t = T$ to $t = 1$) and backward states (from $t = 1$ to $t = T$).

3. *UPDATE WEIGHTS*

One obvious disadvantage of the bidirectional structure is the fact that, theoretically, for its usage, the complete input sequence must be known, which prohibits any online processing for time-sequential data. This disadvantage can be partially removed by cutting the sequences in shorter chunks, which are used for training and testing. If these chunks are long enough to include all context effects that can be used by the BRNN, then this representation with many short sequences is equivalent to the original one with one long sequence. In this case, only a time-lag equal to the length of the chunks would occur, if an online procedure is used.

6.5 Modeling Phonetic Context

Using assumption (6.9) made the models context-independent one-state models, which are valid models for simple tasks and to introduce basic concepts. State-of-the-art speech recognition systems usually make less severe assumptions by introducing context-dependent models (depending on a context class ϕ) and more than one HMM state per model denoted by s. How to determine the optimal set of context classes and number of states per model for a given task is a current research issue and beyond the scope of this chapter. Detailed procedures can be found for example in [31] and in references in there. The scaled likelihood with time t dropped in notation becomes then $\tilde{P}(c, \phi, s|\mathbf{x}) = P(c, \phi, s|\mathbf{x})/P(c, \phi, s)$ instead of $\tilde{P}(c|\mathbf{x}) = P(c|\mathbf{x})/P(c)$. This representation is not useful for use in a neural-network-based system, since the number of different output classes, for all combinations of phonemes, context classes and states is in general large ($5{,}000$ to $30{,}000$) and would lead to an NN with a huge output layer that couldn't be trained in practice. It is possible to decompose the scaled likelihood, for example, as

$$\tilde{P}(c, \phi, s|\mathbf{x}) = \frac{P(c, \phi, s|\mathbf{x})}{P(c, \phi, s)} \tag{6.29}$$

$$= \frac{P(\phi, s|c, \mathbf{x})}{P(\phi, s|c)} \cdot \frac{P(c|\mathbf{x})}{P(c)} \tag{6.30}$$

$$= \frac{P(s|\phi, c, \mathbf{x})}{P(s|\phi, c)} \cdot \frac{P(\phi|c, \mathbf{x})}{P(\phi|c)} \cdot \frac{P(c|\mathbf{x})}{P(c)}, \tag{6.31}$$

which results in several terms that can be estimated independently. The last term $P(c|\mathbf{x})/P(c)$ is the regular monophone-scaled likelihood. The denominator of the middle term $P(\phi|c)$ and the first term $P(s|\phi, c)$ can be estimated by the relative frequencies of the events in the training data. The numerators $P(s|\phi, c, \mathbf{x})$ and $P(\phi|c, \mathbf{x})$ represent like $P(c|\mathbf{x})$ classification problems conditioned on a continuous input $\mathbf{x}$, but depend also on the discrete inputs c and ϕ, which could be treated as additional input vector components that could, for example, be set to one and zero depending on their discrete input state. For estimation of each of these terms there are two possibilities: (1) with one NN that takes also discrete inputs as part of an enlarged input vector allowing parameter sharing between different context-dependent models, or (2) many smaller NNs for each discrete possibility occurring on the right-hand side of the terms (for example, K networks for the estimation of $P(\phi|c, \mathbf{x})$ if there are K monophone classes c), which allows greater control over the encapsulated context-dependent models and faster execution. Currently common is the latter approach and is, for example, discussed in [32, 33]. A very appealing method to discriminate thousands of classes by automatically growing a tree structure, where decomposition at each node in the tree is based on the basic idea of decomposing posterior probabilities, has also been applied successfully to a speech recognition problem and is described in [34, 35].

6.6 System Training and Usage

6.6.1 Training

In the discussion up to now, it has been assumed that frame-labeled training data is available, meaning that for each input vector **x** there is a known target class, which is usually not the case. Instead, there is often only a transcription of the utterance, which might include word boundary or phoneme boundary information but not complete state alignments, which have to be built in incremental steps. Training all acoustic parameters of a complete system (NN weights, transition probabilities, and prior weights) involves a number of iterative steps, summarized below.

1. Assign a target class c (phone or state) to each frame of the training data; this is done by aligning the known word transcriptions to the waveforms using the acoustic models from the previous iteration. In the beginning there are no acoustic models available, and the initial state alignment has to be done by hand (or by using another existing speech recognizer) for at least a few sentences to bootstrap the system.

2. Calculate the state priors $P(c^{(i)}) = N(i)/N_{all}$ and the transition probabilities $P(c^{(i)}|c^{(j)}) = N(j,i)/N(j)$, with $N(\cdot)$ denoting the occurrences of the corresponding frames (or frame pairs) in the training data.

3. Train the NN minimizing the cross-entropy objective function (6.23), set up for a classification problem with softmax output activation function, using the assigned target classes.

4. Go to 1, until there is no significant change in the alignments anymore. In [4] it was found that around four iterations are sufficient.

This procedure is called Viterbi training, because a distinct target class is assigned to each frame. It is also possible to perform a more general but also more memory-consuming forward-backward training [36], where each frame gets assigned to all target classes with a certain probability.

6.6.2 Usage

Usage of the trained system involves, according to (6.6) in principle, the following steps:

1. Estimate the posterior frame probability vector $P(\mathbf{y}_t|\mathbf{x}_t)$ (for all K phones or states) and all times t using the trained NN.

2. Apply the heuristic posterior pruning [29] to the estimated probabilities as

$$\text{IF} \quad P(y_t^k|\mathbf{x}) < threshold \quad \text{THEN} \quad P(y_t^k|\mathbf{x}) := 0 \qquad (6.32)$$

   to limit the number of states that have to be considered for the search. The threshold depends heavily on the number of output classes and is often in the range of around 10^{-5}.

3. Calculate the scaled likelihoods according to (6.10) by dividing the posterior probability estimates by the associated frame class prior probability. This likelihood is used by the search engine to find the sequence with the highest probability $\mathbf{Y}^*$.

In a practical online system these two steps will be applied incrementally according to the current time index. Also, in general it is only required to generate the scaled likelihoods that are requested by the search engine. This can help to save computation time if the NN is organized in a tree structure as mentioned in Section 6.5.

6.7 Conclusions

6.7.1 Discussion

There are a number of issues raised in using (6.6) with neural networks versus (6.5) for recognition (training criterion, discriminative training, distribution of model complexity), which are summarized here:

6.7.1.1 Training Criterion

The training criterion for (6.5) is to maximize the likelihood $P(\mathbf{x}_t|\mathbf{y}_t)$ for all samples, for (6.6) the maximization of the posterior probabilities $P(\mathbf{y}_t|\mathbf{x}_t)$, which leads to a maximum frame classification rate. These two training criteria are for the purpose of recognition using (6.5) or (6.6) equivalent, if an infinite amount of data and arbitrarily complex models are available, because (6.5) is equal to (6.6) using the Bayes rule.

However, for the practical case of not having an infinite amount of data available, using (6.5) or (6.6) will make a considerable difference, as discussed below.

6.7.1.2 Discriminative Training

The use of (6.5) requires the estimation of the distributions $P(\mathbf{x}_t|\mathbf{y}_t)$, which are often chosen as mixtures of Gaussians. It might be the case that the true distributions are quite different from those and can be approximated only poorly by mixtures of Gaussians. Also, because parameter estimation is performed independently for each class, the models' complexity might be used to model areas of the distribution which are useless for the discrimination between them.

If one neural network instead of K mixtures of Gaussians is used, discrimination ability between frame classes is maximized because only the differences between classes are modeled.

6.7.1.3 Distribution of Model Complexity

If Gaussian mixtures are used to model the observation distributions, the model complexity spent on each class is easily controlled by adjusting the number of mixtures for that class.

If one neural network is used to model all observation distributions, the distribution of complexity spent on each class is not as easily controlled. The maximization of the posterior probabilities guarantees a maximum frame classification rate incorporating the class prior probabilities during training. This has the effect of reserving more available model complexity to classes with high prior probabilities, which in turn leads to poor estimation of the posterior probabilities of classes with a low prior probability. This behavior can be avoided partially by using the tree-based NN setup briefly described in Section 6.5, with a clustering procedure for the classes that lead to a similar prior probability for all output classes of each network.

6.7.2 Summary

This chapter showed how to use recurrent neural networks for speech recognition. Advantages and disadvantages of using (recurrent) neural networks instead of conventional techniques for modeling the observation likelihood distributions in HMM-based speech recognition systems were discussed. The chapter outlined the underlying theory, basic architectures, their problems and merits, as well as procedures to train large networks with large amounts of data. Finally, modeling of phonetic context in the neural network framework was discussed and a practical recipe of how to use NNs for speech recognition was given.

References

[1] Bourlard, H., and C. Wellekens, "Links between Markov Models and Multilayer Perceptrons," *IEEE Transactions on Pattern Analysis and Machine Intelligence*, Vol. 12, no. 12, pp. 1167–1178.

[2] Bourlard, H., and M. Morgan, *Connectionist Speech Recognition: A Hybrid Approach*. Boston; Kluwer Academic Press, 1994.

[3] Robinson, T., M. Hochberg, and S. Renals. "Improved Phone Modeling with Recurrent Neural Networks," *Proc. IEEE Int. Conf. Acoustics, Speech, and Signal Processing*, Adelaine, Australia, 1994, Vol. 1, pp. 37–40.

[4] Robinson, T., M. Hochberg, and S. Renals, "The Use of Recurrent Neural Networks in Continuous Speech Recognition," in Chin-Hui Lee, Frank K. Soong, and Kuldip K. Paliwal, eds., *Automatic Speech and Speaker Recognition, Advanced Topics*, Boston; Kluwer Academic Publishers, 1996, pp. 233–258.

[5] Huang, X. D., Y. Ariki, and M. A. Jack, *Hidden Markov Models for Speech Recognition*, Edinburgh: Edinburgh University Press, 1990.

[6] Rabiner, L., and B. H. Juang, *Fundamentals of Speech Recognition*, Englewood Cliffs, N.J.: Prentice Hall, 1993.

[7] Young, S., "A Review of large vocabulary speech recognition," *IEEE Signal Processing Magazine*, Vol. 15, no. 5, pp. 45–57.

[8] Ney, H., and X. Aubert, "Dynamic Programming Search: From Digit Strings to Large Vocabulary Speech Recognition," in Chin-Hui Lee, Frank K. Soong, and Kuldip K. Paliwal, eds., *Automatic Speech and Speaker Recognition: Advanced Topics*, Boston: Kluwer Academic Publishers, 1996, pp. 385–412.

[9] Schwartz, R., L. Nguyen, and J. Makhoul, "Multiple-Pass Search Strategies," in Chin-Hui Lee, Frank K. Soong, and Kuldip K. Paliwal, eds., *Automatic Speech and Speaker Recognition: Advanced Topics*, Boston: Kluwer Academic Publishers, 1996, pp. 429–456.

[10] Renals, S., and M. Hochberg, "Decoder Technology for Connectionist Large Vocabulary Speech Recognition," Technical Report CUED/F-INGENG/TR186, Cambridge University Engineering Department, Cambridge, England, 1995.

[11] Ortmanns, S., H. Ney, and X. Aubert, "A Word Graph Algorithm for Large Vocabulary Continuous Speech Recognition," *Computer, Speech and Language*, Vol. 11, 1999, pp. 43–72.

[12] Schuster, M., "On Supervised Learning from Sequential Data with Applications for Speech Recognition," Ph.D. thesis, Nara Institute of Science and Technology, Nara, Japan, 1999.

[13] Bishop, C. M., *Neural Networks for Pattern Recognition*, Oxford, England: Clarendon Press, 1995.

[14] Richard, M. D., and R. P. Lippman, "Neural Network Classifiers Estimate Bayesian A Posteriori Probabilities," *Neural Computation*, Vol. 3, no. 4, 1991, pp. 461–483.

[15] Press, W. H., S. A. Teukolsky, W. T. Vetterling, and B. P. Flannery, *Numerical Recipes in C*, 2d ed., Cambridge: Cambridge University Press, 1992.

[16] Battiti, R. "First- and Second-Order Methods for Learning: Between Steepest Descent and Newton's Method," *Neural Computation*, Vol. 4, 1992, pp. 141–166.

[17] Shepherd, Adrain, *Second-Order Methods for Neural Networks*, London: Springer, 1997.

[18] Saito, K., and R. Nakano, "Partial BFGS Update and Efficient Step-Length Calculation for Three-Layer Neural Networks," *Neural Computation*, Vol. 9, no. 1, 1997, pp. 123–141.

[19] Fahlmann, E. F., "An Empirical Study of Learning Speed in Back-Propagation Networks," Technical Report CMU-CS-88-162, Carnegie Mellon University, Pittsburgh, Penn., 1988.

[20] Rumelhart, D. E., G. E. Hinton, and R. J. Williams, "Learning Internal Representations by Error Backpropagation," in D. E. Rumelhart and J. L. McClelland, eds., *Parallel Distributed Processing*, Vol. 1, Cambridge, Mass.: MIT Press, 1986, pp. 318–362.

[21] Riedmiller, M., and H. Braun, "A Direct Adaptive Method for Faster Back-Propagation Learning: The RPROP Algorithm," *Proc. IEEE Int. Conf. Neural Networks*, 1993, pp. 586–591.

[22] Jordan, M. I., and R. A. Jacobs, "Hierarchical Mixtures of Experts and the EM algorithm," *Neural Computation*, Vol. 6, no. 2, 1994, pp. 181–214.

[23] Bridle, J. S., "Probabilistic Interpretation of Feed-Forward Classification Network Outputs, with Relationships to Statistical Pattern Recognition," in F. Fougelman-Soulie and J. Herault, eds., *Neurocomputing: Algorithms, Architectures and Applications*, Vol. F68, NATO ASI Series, Berlin: Springer-Verlag, 1989, pp. 227–236.

[24] Waibel, A., "Phoneme Recognition Using Time-Delay Neural Networks," *IEEE Trans. on Acoustics, Speech, and Signal Processing*, Vol. 37, no. 3, 1989, pp. 328–339.

[25] Robinson, A. J., "An Application of Recurrent Neural Nets to Phone Probability Estimation," *IEEE Trans. on Neural Networks*, Vol. 5, no. 2, 1994, pp. 298–305.

[26] Berger, J. O., *Statistical Decision Theory and Bayesian Analysis*, Berlin: Springer-Verlag, 1985.

[27] Jacobs, R. A., "Methods for Combining Experts' Probability Assessments," *Neural Computation*, Vol. 7, no. 5, 1995, pp. 867–888.

[28] Renals, S., and M. Hochberg, "Efficient Search Using Posterior Phone Probability Estimates," *Proc. IEEE Int. Conf. Acoustics, Speech and Signal Processing*, Vol. 1, Detroit, Mich., 1995, pp. 596–599.

[29] Renals, S., and M. Hochberg, "Efficient Evaluation of the Search Space Using the NOWAY Decoder," *Proc. IEEE Int. Conf. Acoustics, Speech, and Signal Processing*, Vol. 1, Atlanta, Ga., 1996, pp. 149–153.

[30] Schuster, M., and K. K. Paliwal, "Bidirectional Recurrent Neural Networks," *IEEE Trans. on Neural Networks*, Vol. 45, no. 11, 1997, pp. 2673–2681.

[31] Odell, J. J., "The Use of Context in Large Vocabulary Speech Recognition," Ph.D. thesis, Cambridge University, Cambridge, England, 1995.

[32] Franco, H., et al., "Context-Dependent Connectionist Probability Estimation in a Hybrid Hidden Markov Model Speech Recognition," *Computer, Speech and Language*, Vol. 8, 1994, pp. 211–222.

[33] Kershaw, D. J., M. M. Hochberg, and A. J. Robinson, "Context-Dependent Classes in a Hybrid Recurrent Network-HMM Speech Recognition System," Technical Report CUED/F-INGENG/TR217, Cambridge University Engineering Department, Cambridge, England, 1995.

[34] Fritsch, J., "ACID/HNN: Clustering Hierarchies of Neural Networks for Context-Dependent Connectionist Acoustic Modeling," *Proc. IEEE Int. Conf. Acoustics, Speech, and Signal Processing*, Vol. 1, Seattle, Wash., 1998, pp. 505–508.

[35] Fritsch, J., "Applying Divide and Conquer to Large Scale Pattern Recognition Tasks," in Genevieve B. Orr and Klaus-Robert Müller, eds., *Neural Networks: Tricks of the Trade*, Berlin: Springer-Verlag, 1998, pp. 315–342.

[36] Senior, A., and A. J. Robinson, "Forward-Backward Retraining of Recurrent Neural Networks," in D. S. Touretzky, M. C. Mozer, and M. E. Hasselmo, eds., *Advances in Neural Information Processing Systems*, Cambridge, Mass.: MIT Press, 1996, pp. 743–749.

7

Time-Delay Neural Networks and NN/HMM Hybrids: A Family of Connectionist Continuous-Speech Recognition Systems

Jürgen Fritsch, Hermann Hild, Uwe Meier, and Alex Waibel

7.1 Introduction

A decade has passed since early connectionist models were first applied to the speech recognition problem. Early work on neural modeling for speech perception led to a great number of very insightful and innovative ideas for modeling speech and for developing speech recognition systems. While the excitement led to a flurry of activity and to many creative approaches that differed from traditional speech recognition systems, it was soon also complemented by theoretical and experimental rigor aiming to understand connectionist models' theoretical underpinnings as well as possible links to classical recognition theory and practice and their relative contribution to modern speech recognition systems.

The efforts have been rewarded with a thorough understanding of connectionist speech technology performance and theory. Early purist attempts at all-connectionist speech systems and at achieving biological plausibility have mostly given way to various hybrid strategies that have blended the best of connectionist and nonconnectionist techniques into full system solutions with best overall performance in mind. Connectionist models have since become a powerful tool in the toolbox of the practitioner, ready to be applied wherever appropriate. They have continued in their role as superior classifiers and predictors, easy to implement and to train, parsimonious in their use of parameters (and hence storage and run-time requirements) and easy to fuse with other signals or features of varying meaning and origin. Together with classical techniques and processing strategies, they have led to most successful state-of-the-art systems.

Our own early explorations, the time delay neural network (TDNN), modular hierarchical neural networks and gating networks, neural network hidden Markov Model (NN/HMM) hybrid systems, and connectionist signal fusion have all been extended further. The extensions represent practical high-performance speech recognition solutions for a number of current challenging recognition tasks. Below we review these models and two of their extensions: the MS-TDNN (a multistate extension of the TDNN), and ACID/HNN, a hierarchical modular network architecture for hybrid NN/HMM-based systems. Both have achieved excellent recognition performance and high run-time efficiency. We will explore their operation and performance on three challenging recognition problems:

1. Continuous spelling recognition (high confusability);
2. Lipreading (fusing speech and visual signals);
3. Large-vocabulary continuous conversational speech recognition over telephone lines (large vocabularies, poor signal quality, and poor speaking style).

7.2 MS-TDNNs and NN/HMM Hybrid Approaches

With the revival of the connectionist paradigm in the 1980s, neural nets also became interesting for speech recognition. First approaches used static net architecture for small tasks such as phoneme or digit recognition. Later, dynamic network structures were developed to cope with speech input of variable length and positioning. Today, basically all connectionist speech recognition systems that perform isolated or continuous word recognition take advantage of some form of conventional nonlinear time alignment procedures, such as hybrid connectionist NN/HMM systems, linear predictive neural networks (LPNN), or MS-TDNNs.

The reader is referred to a detailed summary of early applications of neural nets to speech recognition [1]. In this chapter, we will discuss two of the approaches that continue to be used successfully today: the TDNN/MS-TDNN and hybrid NN/HMM architectures.

The TDNN was one of the first dynamic networks at its time, featuring a shift-invariant architecture with time delays that can be trained for phoneme recognition. The multistate TDNN (MS-TDNN) extends the TDNN philosophy by integration of a nonlinear time alignment in its architecture, thus allowing discriminative back-propagation training not only on the phoneme, but also on the word and sentence, level.

While the MS-TDNN solves the word recognition problem from a connectionist point of view, hybrid NN/HMM systems approach the problem from the HMM perspective: instead of the Gaussian mixture densities of the HMMs, the emission (class-conditional) probabilities $p(\mathbf{x}|s)$ to observe speech vector $\mathbf{x}$, given the system is in state s, are estimated using neural networks. As neural networks will learn a posteriori probabilities $p(s|\mathbf{x})$, these probabilities must first be converted to class-conditional probabilities before they can be fed into an HMM. This is achieved using the Bayes rule:

$$p(\mathbf{x}|s) = \frac{P(s|\mathbf{x})p(\mathbf{x})}{P(s)} \qquad (7.1)$$

These two approaches are now described in more detail.

7.2.1   The Time-Delay Neural Network (TDNN)

The time-delay neural network (TDNN) was developed for phoneme classification in 1987 by Waibel and Lang [2–5]. The TDNN computes phoneme scores for a short speech segment, representing one of the phonemes /b/, /d/, or /g/. These scores are then integrated (accumulated) over time into one score for each phoneme, as shown in Figure 7.1.

The time delays allow the TDNN to discover temporal relationship both directly in the speech input and in the more abstract representation of the hidden layer. By using the same set of weights for each time-step, the TDNN is shift-invariant: features in the speech signal can be detected independently of their position. Also, the shared weights allow for a much smaller set of parameters as compared with a fully connected net.

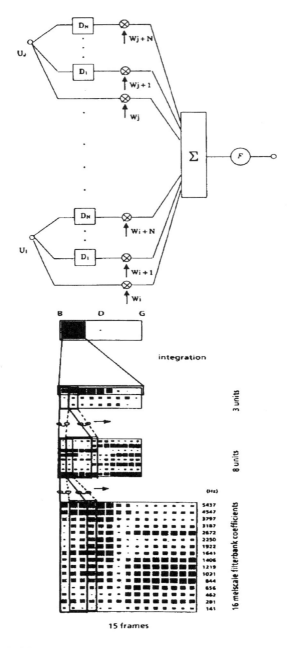

Figure 7.1 The TDNN architecture: A single neuron with time-delayed connections $(D_1, \ldots, D_N)$ and the complete system (from Waibel et al., "Phoneme Recognition Using Time-Delay Neural Networks.")

7.2.2 Multistate TDNN

The multistate TDNN extends the TDNN philosophy. Instead of single phonemes, phoneme sequences, i.e., words, are integrated over time. For each word to be recognized, a nonlinear time-alignment between the phoneme scores of the speech input and the corresponding phoneme sequence of the word model is computed. Like in HMM algorithms, an optimal path can be efficiently computed with the dynamic time-warping algorithm.

7.2.3 MS-TDNN Variants

The MS-TDNN as an extension of the TDNN was first described by Haffner et al. in [6]. Since then, many architectures and applications have emerged from this approach, some of which are summarized below. Two applications, alphabet recognition and multimodal lipreading, are described in more detail in the next sections.

Tebelskis experiments with connectionist speech recognition for large vocabularies, using linear predictive neural networks (LPNNs) and MS-TDNNs [7–9]. His works show that the performance of MS-TDNNs is clearly superior to that of LPNNs. Bregler [10, 11], Duchnowski [12], and Meier [13] extended the MS-TDNN approach to a lipreading system, using visual input as a second modality to support more robust alphabet recognition. Zeppenfeld [14] uses the MS-TDNN for keyword spotting. The task is to detect a small group of keywords in an arbitrary text, for example to determine the topic of a spoken dialogue.

Besides speech recognition, the MS-TDNN is being applied to online cursive handwriting recognition [15]. In contrast to optical character recognition (OCR), online data is not recorded as bitmaps but as writing trajectories. Thus a written word is a sequence of coordinates (and other derived features such as writing speed, etc.), similar to the sequence of frequency coefficients in speech.

Bodenhausen [16–18] derives algorithms for automatic structuring of the network architecture on the example of the MS-TDNN architecture. His algorithm automatically finds a good parameterization of the number of hidden neurons, the width of the time delays, and the number-of-states-per-word model.

7.2.4 Hybrid NN/HMM Variants

Hybrid NN/HMM architectures are often termed connectionist acoustic models since they differ from standard HMM-based statistical speech recognizers in the way that acoustic observation likelihoods are estimated. While traditional HMM recognizers mostly rely on parametric densities such as mixtures of Gaussians to model state emission probabilities, hybrid NN/HMM systems employ connectionist models to estimate state posterior probabilities. Using Bayes rule and estimates of state prior probabilities, state posteriors can be converted to (scaled) likelihoods required by the HMM framework. The advantage of taking such a detour in computing state emission probabilities lies in the discriminative power of connectionist models. Since mainstream HMM speech recognizers are mostly trained in a maximum likelihood framework using the expectation-maximization (EM) algorithm, incorporation of discriminatively trained neural networks that focus on modeling of class boundaries instead of class distributions is often observed to be beneficial. Also, compared with mixtures of Gaussian-based acoustic models, connectionist acoustic models are often reported to achieve the same accuracy with far less parameters.

A variety of connectionist estimators of posterior probabilities have been applied to the task of connectionist acoustic modeling. The most popular connectionist architecture for hybrid speech recognition seems to be the multilayer perceptron (MLP) and variants of it [19]. The MLP was shown to be a reasonably good estimator of posterior probabilities that, in addition, can be trained quite efficiently using the back-propagation algorithm. In principle, any feedforward network structure can be applied to the estimation of posterior probabilities. Mostly, networks with a single hidden layer of rather large size (up to 4,000 units) have been used for context-independent single-state phone modeling.

Attempts at applying a more dynamic connectionist model led to a hybrid system based on a recurrent neural network (RNN) [20]. Such an architecture consists of a set of state units that have recurrent connections from their outputs back to their inputs (these units also have connections to the input nodes). State units and input nodes are connected to the output layer. It is argued that the recurrent structure allows for the training of acoustic models that are sensitive to contextual phenomena present in the speech signal. Also, an RNN somewhat alleviates the independence assumption of HMMs, which surely does not hold for speech signals. However, RNN modeling suffers from an extremely expensive training algorithm (back-propagation through time) that seems to require parallel implementation on special purpose hardware.

Another connectionist model for hybrid NN/HMM systems that has only recently attracted interest in the speech community is the hierarchical-mixtures-of-experts model [21], which we will discuss in more detail later.

7.3 Alphabet Recognition with the MS-TDNN

One of the major advantages of the MS-TDNN is that its architecture allows for discriminative training not only on the phoneme, but also on the word and sentence, level. Thus, the MS-TDNN is especially well suited for tasks with an easily confused vocabulary, such as spoken letter recognition.

In the following sections, we will describe the training procedures of the MS-TDNN, several techniques for searching in large name lists, as well as the many speech recognition experiments that have accompanied the development of the MS-TDNN.

7.3.1 Training Procedures

The training of the MS-TDNN starts on the phoneme level and then moves on to the more complex task of word and finally sentence (letter sequence) recognition.

On the phoneme level, the MS-TDNN is trained for frame-by-frame phoneme classification, using only its first three layers (input, hidden, and phoneme; see Figure 7.2). Each frame is assigned a phoneme target, and a standard error back-propagation algorithm adapted to account for the shared weights and time delays is used to iteratively update the network's parameters. For each time frame t (one every 10 milliseconds), a vector of n phoneme scores $\mathbf{y}_t = (y_{1,t}, \ldots, y_{n,t})$ is computed and compared with the corresponding target vector $\mathbf{d}_t$.

For a high-dimensional "1-out-of-n" coding ($n > 50$ phonemes in our case), the classical mean square error function ($E_{MSE} = \sum_{i=1}^{n} (y_{i,t} - d_{i,t})^2$) turned out to be a suboptimal choice. To illustrate this, consider for example that for a target $(1, 0, \ldots, 0)$, the "zero" output $(0, \ldots, 0)$ has only half the error of the more desirable output $(1, 0.2, \ldots, 0.2)$.

Both the cross entropy and the McClelland error function

$$E_{McClelland}(T, Y) = \sum_{i=1}^{n} \log(1 - (y_i - d_i)^2) \qquad (7.)$$

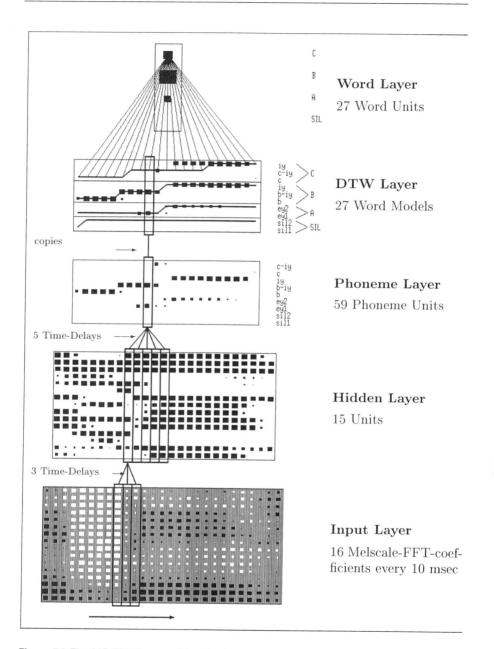

C

B

A

SIL

Word Layer
27 Word Units

iy
c-iy
c   > C
iy
b-iy   > B
b
ey2
ey1   > A
sil2
sil1   > SIL

DTW Layer
27 Word Models

copies

c-iy
c
iy
b-iy
b
ey2
ey1
sil2
sil1

Phoneme Layer
59 Phoneme Units

5 Time-Delays

Hidden Layer
15 Units

3 Time-Delays

Input Layer
16 Melscale-FFT-coef-
ficients every 10 msec

Figure 7.2 The MS-TDNN recognizing the letter B.

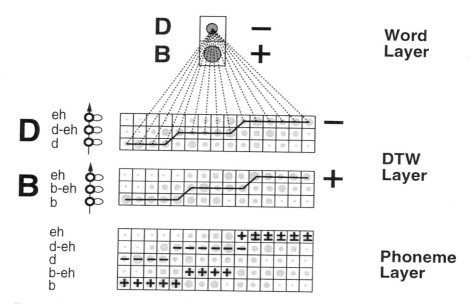

Figure 7.3 Error back-propagation through the alignment paths of the correct and highest incorrect word model.

avoid this problem by punishing "outliers" with an error approaching infinity for $|d_i - y_i| \to 1.0$.

Once the network has learned to classify phonemes, the word level training is entered. Like in an HMM, a word is modeled as a sequence of phonemes. Note that, for consistency with other speech recognition tasks, we call an element of the recognition vocabulary a word, and a sequence of words a sentence. Thus, in the case of alphabet recognition, a word represents a spoken letter, and a sentence is a string of spelled letters.

During word-level training, one score is computed for each letter to be recognized by accumulating the scores of all phonemes of the corresponding word model. In contrast to phoneme-level training, phoneme boundaries are no longer fixed. Instead, an alignment is searched which maximizes the sum of the phoneme scores on the corresponding alignment path, resulting in the optimal score for each word in the output layer.

The training error is now a function of the word output, and the error derivatives are back-propagated from the word units through the alignment paths and the front-end TDNN, as illustrated in an example Figure 7.3: *D* is the correct output, and the phonemes along the alignment path of the *B*-model receive positive training, while the phonemes along the path through *D* receive negative training. The error signals are neutralized in the common phone 'eh,' hence training focuses on the critical differences between the two models.

For the word-level training, we have achieved best results with an objective function similar to the classification figure of merit (CFM) [22], which is closely related to the later published, more general formulation of generalized probabilistic descent (GPD) [23]. CFM tries to maximize the distance *d* between the correct score y_c and the incorrect scores y_i instead of using absolute targets 1.0 and 0.0 for correct and incorrect word units. We have achieved best results by focusing only on the highest incorrect output y_{bi}. In this case, the error is a function of the distance or misclassification measure $d = y_c - y_{bi}$:

$$E_{CFM} = f(y_c - y_{bi}) = f(d) \tag{7.3}$$

We have been successfully using a quadratic error function

$$f(d) = (1 - d)^2 \tag{7.4}$$

which results in an error of 0 for the ideal case $d = y_c - y_{bi} = 1$, an error of 1 if y_c and y_{bi} are indistinguishable ($d = 0$), and an increasing error for misclassifications, i.e., $y_c < y_{bi}$. The classical sigmoid function

$$f(d) = \frac{1}{1 + e^{-\alpha * - d + \beta}} \tag{7.5}$$

will work as well, if appropriate values for the slope α and shift β are used.

For sentence level training, a similar scheme is applied. If a sentence, i.e., a sequence of letters, is incorrectly recognized, a correct alignment path is found using a forced alignment on the known transcription. As on the word level, positive and negative training can then be applied along the correct and incorrect alignment path.

7.3.2 Duration Modeling

In continuous letter recognition, the short and easily confused letters are especially prone to insertion and deletion errors. For example, the letter *E* is often erroneously inserted, resulting in the recognition of, e.g., *TE* instead of *T.*

In continuous speech recognition, the commonly practiced technique to cope with this problem is a global word entrance penalty, which is adjusted to balance insertion and deletion errors.

The small vocabulary of the alphabet allowed us to use more powerful techniques for duration control both on the phoneme and word levels. On the phoneme level, a phoneme-specific minimum duration constraint is introduced. By duplicating a phoneme state n times, a minimum duration of n time frames is enforced. For each phoneme p_i, the corresponding minimum duration n_i is derived from statistics from the training data. On the word level, instead of a global word entrance penalty, a word-specific penalty is used. The individual penalties are determined by a gradient descent method which increases or decreases a word-specific entrance penalty if the corresponding letter is inserted or deleted too often, respectively.

7.3.3   Experiments

Over the course of our research, we have been working with various speech databases. Our first database, CMU-Alph, was clean-speech, speaker-dependent spelling data. We then moved to speaker-independent tasks (RM-Spell, KA-Alph), and finally to telephone speech quality (OGI telephone data, discussed below).

Unless otherwise mentioned, all of the experiments described below are performed on continuous spelling data (no pauses between letters), and no language models are used. Depending on the task, between 15 and 100 units were used for a frame in the hidden layer. With the typical number of 50 hidden units, this results in about 20,000 trainable parameters.

7.3.3.1   Speaker-Dependent Data

Our early research on speaker-dependent letter recognition was performed on the CMU-Alph database, which contains about 1,000 continuously spelled names and random letter sequences from about ten speakers. Two speakers, mjmt and mdbs, have been extensively used by several researches for speaker-dependent letter recognition experiments. The results of these experiments are summarized in Table 7.1.

7.3.3.2   Speaker-Independent Data

For speaker-independent spelling recognition, we experimented with an English and a German speech database. For English, we used the

Table 7.1
Speaker-independent letter recognition on the
CMU-Alph data, in percent letter accuracy

CMU Alph	mjmt	mdbs
SPHINX (HMM) [6]	96.0	83.9
LPNN (Iso [24])	91.2	–
TDNN-HMM (Windheuser [25])	97.9	–
MS-TDNN [6]	97.5	89.7
MS-TDNN (Hild)	**99.3**	**95.3**

Resource-Management spell data. In this database, 120 speakers are spelling about 15 sequences, mostly proper names. In Table 7.2, our results are compared with those achieved by an HMM speech recognizer as reported in [26].

The German database comprises a total of about 10,000/80,000 spelled sequences/letters from 100 speakers. In addition to the special characters in the German alphabet (umlaut on ä, ö, ü; letter ß with three pronunciation variants), words for "double" and pronunciation variants for hyphen ("–") were introduced, resulting in a total vocabulary size of 35 words. On a test set of 20 of the 100 speakers, a letter accuracy of 90.4% was achieved.

7.3.3.3 Telephone Data

The "Oregon Graduate Institute (OGI) Spelled and Spoken Word Telephone Corpus" provides recordings from about 4,000 calls over the public telephone line. Among other prompts, callers were asked to spell their last names with and without short pauses between letters

Table 7.2
Speaker-independent letter recognition
on the Resource-Management Spell
data, in percent letter accuracy

RM-SPELL	%
SPHINX (HMM) [26]	88.7
+ senones	90.4
MS-TDNN (Hild)	90.8
+ gender specific	92.0

(SLP/SLN), their first names with pauses (SFP), and the letters of the alphabet (ALP). About 8% of the spellings contain out-of-alphabet words, e.g., "c h e [sorry] c h a v [as in victor] e z," which were excluded (together with about 1% of cutoff data) from the experiments. Using the partition provided with the database, this resulted in a training and development test set of 4,132 and 2,064 strings (from the SLN, SLP, SFP, and ALP sets) and two test sets, 685 last names from the SLN set and 305 last names with pauses from the SLP set.

Table 7.3 reports results on the SLN and SLP test set achieved by several researches, demonstrating the viability of the MS-TDNN approach.

7.3.4 Searching in Large Name Lists

Without language modeling, the letter accuracy for speaker-independent, continuous letter recognition is in the order of 90%. Naturally, the probability to correctly recognize every letter of a sequence is much lower. Thus, only with strong language models can acceptable recognition rates be achieved for spelled names.

In a first set of experiments [27, 31], we compared several techniques for using language models to improve the recognition of spelled names. Our test set consisted of 1,316 names spelled by 23 German speakers. A name directory with over 110,000 last names or 43,181 unique names was used as a list of legal names.

As seen in Table 7.4, without any language models (a), a correct-name rate of less than 60% was achieved. Conventional language model techniques such as bigrams or trigrams do not provide strong constraints and therefore resulted in only small performance improvements (b).

Table 7.3
Percentage of letter accuracy on the speaker-independent OGI telephone data

Data base	SLN		SLP (with pauses)	
	none	bigrams	none	bigrams
MS-TDNN [27]	88.2	91.0	90.6	92.3
OGI [28]	–	–	87.0	–
Junqua [29]	–	88.0	–	–
Spanias [30]	–	–	–	90.6

Table 7.4

Percentage of names correct for different language models and search techniques using a list of 40,000 unique names. The letter accuracy without language modeling is 90.1%

Language Model	% names correct
(a) No language model	56.4
(b) Bigrams	62.1
(c) Closest match	85.0
(d) Closest match	85.0
(e) N-best	88.1
(f) Fully constrained	92.7

Therefore, we examined several techniques to apply the full constraints of a given list of legal names at different times during the recognition process.

The given list of names can be used to correct a hypothesis h after the recognition process by finding the closest match between h and the list of names (c). Results are improved if the distance calculation for the match considers the confusability of different letter pairs (d). An alternative approach is to compute not just one but a large list of hypotheses and search for the highest ranking hypothesis h^* which at the same time is a legal name (e). However, the best results are achieved if the constraints are applied directly in the search process. This is implemented by constructing a large graph, where each node represents one letter and the entire graph represents all names in the list (f). A minimal search graph can be obtained by exploiting common prefixes and suffixes of all names. However, we found that a tree-structured graph can be searched more efficiently: since each final node uniquely represents one name, no backpointers are needed to recover the search results. An example of such a tree structure is depicted in Figure 7.4.

The tree-structure graph turned out to be the most promising search technique. To explore this approach in more detail, we performed a second set of experiments [27] with larger lists and telephone quality speech. As the list of last names, we extracted a total of 14 million residential telephone listings from the northeast coast of the United States, representing about 807,000 unique last names. Sublists of various sizes were randomly drawn (without replacing) from the 14 million

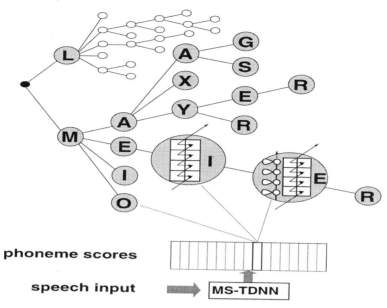

phoneme scores

speech input ➡ | MS-TDNN |

Figure 7.4 Tree search. The entire name dictionary is represented in a big search tree. Each node corresponds to a search hypothesis. With each new frame, new accumulated scores are computed for each node in the tree.

entries. The SLN and SLP test set from the OGI telephone data was used as acoustic test data.

The largest tree represented the 807,000 unique names with about 2 million nodes. Clearly, pruning strategies such as beam search are needed to handle such large search structures. In addition to the full syntactic constraints provided by the list of legal names, there is one more source of knowledge to be exploited: common names like "Smith" are very frequent, while others occur very infrequently. Thus it makes sense to use the a priori probability of the names.

It turned out that the best method to take the a priori probabilities into account is not to assign them to the final nodes of the tree, but to consider them already during the search process in the tree. That is, whenever an internal tree node branches into its subtrees, the relative probability mass of all the remaining final nodes of each subtree is used to guide the search.

Figure 7.5 shows name recognition results as a function of the list sizes (total and unique number of names) on the two test sets of the OGI telephone data, the 685 spelled last names without (SLN) and 305 names with pauses (SLP).

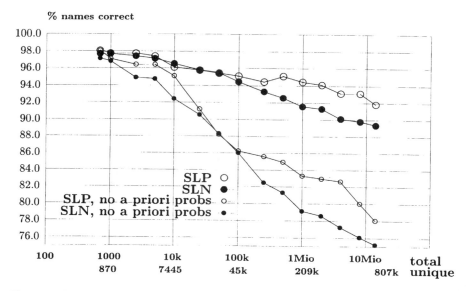

Figure 7.5 Percentage of names correct for trees with and without a priori probabilities on the OGI test sets of spelled last names without (SLP) and with (SLN) pauses. For each name list, the total number of entries and the number of unique names are shown.

7.4 Multimodal Input: Lipreading

7.4.1 Motivation

Most approaches to automated speech recognition that consider solely acoustic information are very sensitive to background noise or fail totally when two or more voices are presented simultaneously ("cocktail party effect"). Humans deal with these distortions by considering additional sources such as directional, contextual, and visual information, primarily lip movements. We are interested in emulating some of these capabilities by combining speech recognition with lipreading to improve robustness and flexibility by offering complementary information [32].

7.4.2 The Recognizer

A modular MS-TDNN is used to perform the recognition. We extended the recognizer described in the previous chapter to perform the audiovisual recognition [33, 34]. Combining visual and acoustic data is done on the phonetic layer (Figure 7.6) or on a lower level (Figures 7.7 and 7.8) [13, 35]. The combination of acoustic and visual

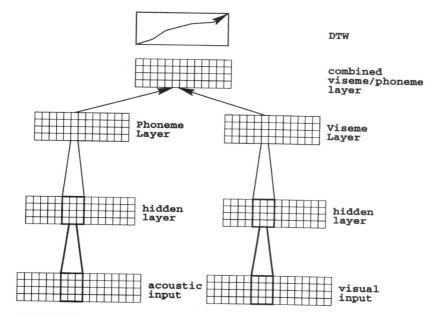

Figure 7.6 MS-TDNN architecture for combining audiovisual data on phonetic layer.

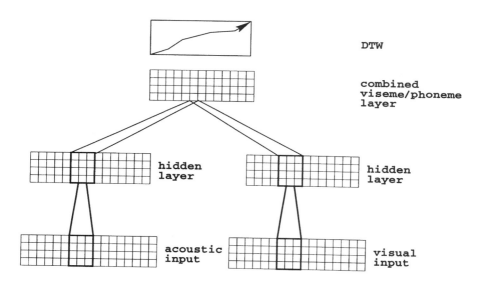

Figure 7.7 MS-TDNN architectures for combining audiovisual data on hidden layer.

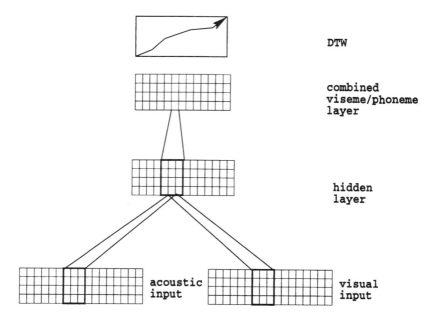

Figure 7.8 MS-TDNN architectures for combining audiovisual data on input layer.

information on the phoneme layer offers several advantages. There is independent control of the two-modality network, allowing for separate training rates and number of training epochs. On the other hand, this method forces us to develop a viseme alphabet for the visual net, as well as a one-to-many correspondence between the visemes and phonemes. A viseme (visual phoneme) is the smallest part of lip movement that can be distinguished. Several phonemes are usually mapped to each viseme, e.g., the phones p, b, m are visually not distinguishable and are mapped to one viseme. Combination on the input or hidden level has the advantage that lower-level correlations between acoustic and visual events can be used in the net.

For the combination on the phonetic layer, in the basic system an acoustic and a visual TDNN are trained separately. The acoustic net is trained on 63 phonemes, the visual on 42 visemes. The audiovisual activation b_{AV} for a phoneme is

$$b_{AV} = \lambda_A b_A + \lambda_V b_V, \qquad \text{where } \lambda_A + \lambda_V = 1 \qquad (7.6)$$

where b_A is the acoustic activation of a phoneme and b_V is the activation of the corresponding viseme. Under nonnoise conditions the acoustic net works much better than a visual-only net, but the additional visual information still can improve the acoustic results. The weight λ_A depends on the quality of the acoustic data. If the quality is high, i.e., in the absence of noise, the weight λ_A should be high, otherwise it should be low. The methods to determine λ_A are described below:

- Entropy Weight
  The entropy is a good measure on the reliability of the net output. The entropy weights λ_A for the acoustic and λ_V for the visual side are given by:

$$\lambda_A = b + \frac{S_V - S_A}{\Delta S_{max-over-data}}, \quad \text{and} \quad \lambda_V = 1 - \lambda_A \quad (7.7)$$

  The entropy quantities S_A and S_V are computed for the acoustic and visual activations by normalizing these to sum to 1.0 (over all phonemes or visemes, respectively) and treating them as probability mass functions. High entropy is found when activations are evenly spread over the units, which indicates high ambiguity of the decision from that particular modality. The bias b preskews the weights to favor one of the modalities. This bias can be determined automatically using the SNR of the acoustic data.

- SNR Weight
  Another measure for the quality of the acoustic data is the signal-to-noise ratio (SNR). High SNR means high quality of the acoustic data. We used a piecewise-linear mapping to adjust λ_A and λ_V as a function of the SNR.

- Neural Net
  In this case a simple back-propagation network without hidden layer is used to combine the visemes and phonemes. Input to this net are the scores of the separately trained acoustic and visual TDNN. These units are fully connected to the output layer, which implements the combined viseme/phoneme layer.

There are two basic MS-TDNN architectures for the lower-level combination. Audio and visual data is either fed directly into the input layer, or the acoustic and visual streams are combined in the hidden layer. The motivation for the latter is that the net can learn a better internal representation for each input modality before the streams are combined. The SNR can be used as additional input for each frame to enable a combination depending on the quality of the acoustic signal.

7.4.3 Results

We have trained the recognizer on 170 sequences of acoustic/visual data from one speaker and tested on 30 sequences of the same person. For each combination method we have trained the nets on clean acoustic data. We separately trained an acoustic TDNN on the same sequences of clean and corrupted data with white noise at 16 dB SNR. For testing we also added different types of artificial noise to the test set of clean data.

Figure 7.9 shows the results for the different combination methods in comparison with the acoustic-only recognition rate in different noise environments. All the nets were trained on clean acoustic data. The recognition rate on the visual data (without acoustic information) was 55%.

With all combination methods we get an improvement compared with the acoustic-only recognition, especially in the case of high background noise. We obtain the best results using the combination on the phonetic layer.

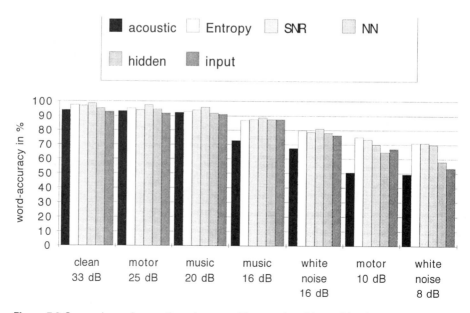

Figure 7.9 Comparison of acoustic-only recognition results with combination on the phonetic layer (entropy, SNR, NN) and combination on lower levels (input, hidden).

Entropy and SNR combination methods have the disadvantage of not taking the inherent confusability of some phonemes and visemes into consideration, but use a single weight in each acoustic/visual time frame depending only on the quality of the acoustic data. The approach that uses a neural network for combination relies on the fact that some phonemes are easier to recognize acoustically while some can be more reliably distinguished from the visual input, by using different weights for each phoneme/viseme pair. As expected, this method delivers the best results, except in the case of high background noise (i.e., motor 10 dB and white noise 8 dB).

Similarly, the hidden- and input-combination recognition performance suffers more in the case of high background noise. However, when evaluating the different approaches one has to remember that the neural net combination, just as the hidden- and input-combination, has no explicit information about the quality of the acoustic input data, as opposed to the combination at the phonetic level with the entropy- and the SNR-weights.

Both under clean and noisy conditions, we get the best performance with the combination on the phonetic level. The advantage of the high-level combination is that the acoustic and visual net can be trained and optimized separately.

7.5 Modular Neural Networks

In 1989, Hampshire and Waibel [36] proposed a modular time delay network architecture consisting of a set of task-specific expert networks that are linearly combined by an expert weighting network—the meta-pi network. Assuming pretrained expert networks, the meta-pi architecture can be trained by back-propagating errors through the linear weighting device into the meta-pi network. Independently and about the same time, Jordan and Jacobs [21, 37] introduced a similar architecture for supervised learning based on the divide-and-conquer strategy. In their architecture, the learning task is divided in sets of overlapping regions by means of a tree-organized hierarchy of gating networks. Learning algorithms capable of jointly estimating gating and expert network parameters were proposed.

Expert networks at the leaves of the tree perform the learning task in their specific region of the input space. Expert outputs are blended by the gating networks and proceed up the tree to yield the final output. Expert and gating network parameters are jointly estimated in order to

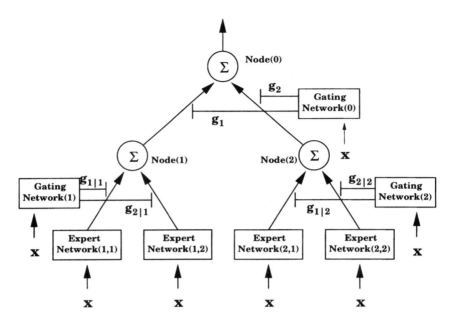

Figure 7.10 Hierarchical mixtures of experts architecture.

maximize the likelihood of a generative model, that is, the construction of overlapping regions in which experts act requires no supervision and is part of the learning algorithm. It was shown that such hierarchical mixtures of expert nets can model discontinuities in the input-output mapping better than monolithic neural networks alone.

7.5.1 Architecture

We restrict the presentation of modular neural networks to a hierarchical-mixtures-of-experts (HME) architecture and its application to classification. Some modifications in the underlying probabilistic model are required to apply the architecture to classification tasks. The architecture is a tree with gating networks at nonterminal nodes and expert networks at its leaves. The gating networks receive the input vectors and divide the input space into a nested set of regions that corresponds to the leaves of the tree. The expert networks also receive the input vectors and produce estimates of the a posteriori class probabilities, which are then blended by the gating network outputs. All networks in the tree are linear, with a softmax nonlinearity as their activation function.

Such networks are also known in statistics as multinomial logit models, a special case of generalized linear models (GLIM) in which the probabilistic component is the multinomial density. This allows for a probabilistic interpretation of the hierarchy in terms of a generative likelihood-based model. For each input vector $\mathbf{x}$, the outputs of the gating networks are interpreted as the input-dependent multinomial probabilities for the decisions about which child nodes are responsible for the generation of the actual target vector $\mathbf{y}$. After a sequence of these decisions, a particular expert network is chosen as the current classifier and computes multinomial probabilities for the output classes. The overall output of the hierarchy is

$$\mu(\mathbf{x}, \Theta) = \sum_{i=1}^{N} g_i(\mathbf{x}, \mathbf{v}_i) \sum_{j=1}^{N} g_{j|i}(\mathbf{x}, \mathbf{v}_{ij}) \mu_{ij}(\mathbf{x}, \theta_{ij}) \tag{7.8}$$

where the g_i and $g_{j|i}$ are outputs of the gating networks and the $\mu_{ij}(\ )$ represent expert networks. The parameters of the first- and second-level gating and the expert networks are denoted by v_i, v_{ij} and θ_{ij}, respectively.

7.5.2 Application to NN/HMM Models

In our case, HMEs are being used in a hybrid NN/HMM speech recognition framework as classifiers, estimating posterior class probabilities. For classification, expert and gating networks in an HME compute multinomial probability models and are therefore parameterized using the softmax nonlinearity ("canonical link" in GLIM theory):

$$z_i(\mathbf{x}) = \frac{\exp y_i(\mathbf{x})}{\sum_j \exp y_j(\mathbf{x})} \tag{7.9}$$

In [21, 37] the $y_i(\mathbf{x})$ are parameterized as linear models, leading to an efficient EM training algorithm (iteratively reweighted weighted least squares) for the hierarchy. However, we discovered [38] that it is sometimes advantageous to use more complex parameterizations for gates and experts, e.g., multilayer feedforward architectures. Such architectures can still be trained efficiently using generalized EM algorithms with online updates [38, 39].

7.5.3 Experiments with a Hybrid HME/HMM System

In [38] we reported on experiments and results with HME-based hybrid NN/HMM systems for connectionist speech recognition. By applying the HME architecture to acoustic modeling for speech recognition, we were aiming at dividing the task of estimating phone posterior probabilities into subtasks of smaller size that can be handled more effectively by standard architectures such as MLPs. Also, in contrast to a single MLP, a modular architecture yields a set of specialized MLPs which are active only in certain regions of the input space. Therefore, evaluation of posterior phone probabilities can be accelerated by pruning the HME tree based on gating network activations.

We trained an HME-based hybrid NN/HMM recognizer on the English Spontaneous Scheduling Task (ESST) in the domain of meeting negotiations [38]. The ESST training database consists of approximately 26 hours of training data, recorded at a sampling rate of 16 kHz. The HME consists of 16 experts and five gates, organized into a two-level tree of branching factor 4. After training the HME, we examined the ability of the architecture to distribute the learning task among several expert networks. Figure 7.11 shows a plot that was generated by computing HME activations along a forced alignment of a recognized hypothesis. It also contains vertical lines indicating word boundaries.

The plot reveals some interesting aspects of our hybrid HME/HMM system. The beginning and ending part of the above utterance contains long noise parts, which coincide with strong activations of just two experts (numbers 10 and 11 from top to bottom). Expert numbers 2, 13,

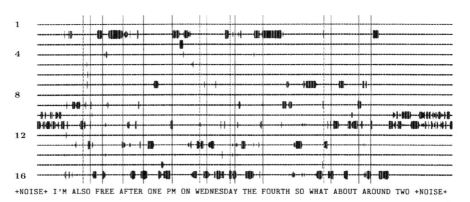

+NOISE+ I'M ALSO FREE AFTER ONE PM ON WEDNESDAY THE FOURTH SO WHAT ABOUT AROUND TWO +NOISE+

Figure 7.11 Gating diagram showing expert activations over time for an HME of depth 2, branching factor 4, trained on the English Spontaneous Scheduling Task (ESST).

and 16 are contributing most during speech segments. There are also some experts that are hardly ever active at all (1, 6, and 8, for instance). However, we found that in other utterances, spoken by different speakers, some of these experts show different behavior and are contributing to the HME's decision. Nevertheless, some experts are subject to pruning, because their contribution, accumulated over a set of test utterances, is too low to be of any significance.

7.6   Context Modeling

Refinement of traditional mixtures of Gaussians-based acoustic modeling [40] using phonetic decision trees for polyphonic context modeling recently led to systems consisting of thousands of HMM states [41]. Significant gains in recognition accuracy have been observed in such systems. Nevertheless, research in context-dependent NN/HMM acoustic models has long concentrated on comparably small systems since it was not clear how to reliably estimate posterior probabilities for thousands of states. Application of a single artificial neural network as in context-independent modeling leads to an unfeasibly large number of output nodes. Similar to the task decomposition scheme applied in modular neural networks (such as meta-pi and HME), the computation of context-dependent posterior state probabilities can be decomposed by a technique called conditional factoring. This technique allows the application of a hierarchically organized set of relatively small neural networks to the (modular) computation of context-dependent posteriors. Below, we discuss the clustering of context classes based on phonetic decision trees, then we present two specific context-dependent hybrid NN/HMM architectures based on hierarchical factoring.

7.6.1   Clustering Context Classes

Typically, a context-dependent system is built by independently clustering the polyphonic contexts of each context-independent state (monophone states). A context-dependent phone consists of a monophone indexed by an identifier for its neighboring context. Since the number of distinct contexts depends exponentially on the number of monophones, the theoretical number of context-dependent phones is extremely large, even when only modeling the dependency on one or two phones to the left and to the right. Furthermore, most of these context-dependent phones will occur rarely or never at all in any training database. Therefore, there is a need for a procedure to reduce the number of context-dependent phone models to a reasonably small number of generalized context-dependent phones.

Currently, the most popular approach is based on binary decision trees. In our system, the splitting criterion for growing the decision trees is based on weighted gain in entropy between the discrete probability distributions (the mixture coefficients in the Gaussian mixtures) before and after a potential split.

$$D(\mathbf{p}, \mathbf{p}_l, \mathbf{p}_r) = n_l H_l(\mathbf{p}_l) + n_r H_r(\mathbf{p}_r) - n H(\mathbf{p}) \qquad (7.10)$$

$$\text{with} \qquad H_l(\mathbf{p}_l) = -\sum_i p_{li} \log p_{li}$$

$$H_r(\mathbf{p}_r) = -\sum_i p_{ri} \log p_{ri}$$

$$H(\mathbf{p}) = -\sum_i p_i \log p_i$$

where $\mathbf{p}$ is the vector of mixture coefficients before and $\mathbf{p}_l$, $\mathbf{p}_r$ are the vectors of mixture coefficients resulting from the separate modeling in the two children nodes after a split. Potential splits are generated by asking phonetic questions in polyphonic contexts, with the restriction that only one phone is considered across word boundaries.

Figure 7.12 shows an example of such a cluster tree for a single monophone. Internal nodes contain phonetic questions (numbers in questions are positions relative to the current monophone); leaves contain model names. The decision tree for the complete system is built by merging the decision trees of all monophone states using questions on the identity of the center monophone state.

7.6.2 Factoring Context-Dependent Posteriors

The set of all leaf nodes in the (final) complete decision tree constitutes the number of distinct states in the context-dependent system. Depending on the size of the task, today's state-of-the-art context-dependent systems use phonetic decision trees with up to 24,000 such leaf nodes. In order to be able to apply connectionist estimators to the estimation of posterior probabilities for such a large amount of

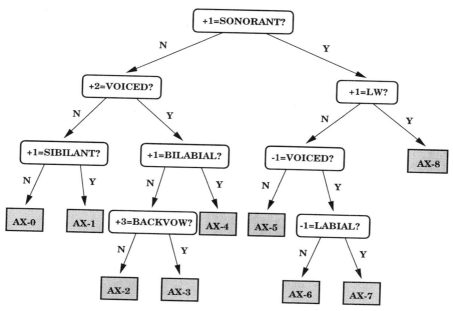

Figure 7.12 Phonetic decision tree clustering context classes of AX-m.

classes, decomposition based on factoring the context-dependent posteriors becomes necessary.

Let S denote the set of all (decision-tree-clustered) HMM states s_k. Consider the partitioning S into M disjoint and nonempty subsets S_i. A particular state s_k will now be a member of S and exactly one of the subsets S_i. Therefore, we can rewrite the posterior probability of state s_k as a joint probability of state and appropriate subset S_i and factor it according to

$$p(s_k|\mathbf{x}) = p(s_k, S_i|\mathbf{x}) \qquad \text{with} \qquad s_k \in S_i \qquad (7.11)$$

$$= p(S_i|\mathbf{x})\, p(s_k|S_i, \mathbf{x})$$

Thus, the global task of discriminating between all the states in S has been converted into (1) discriminating between subsets S_i and (2) independently discriminating between the states s_k contained within each of the subsets S_i. Recursively repeating this process yields a hierarchical tree-organized structure (Figure 7.13).

One of the critical aspects of such a hierarchical decomposition of posteriors is the strategy for partitioning state sets. Typically, posterior

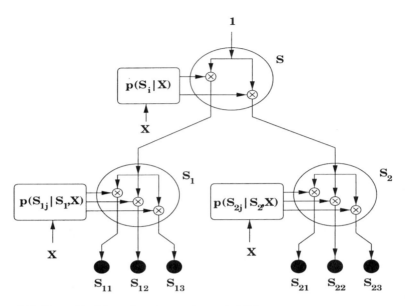

Figure 7.13 Hierarchical factoring of posterior probabilities.

state probabilities are factored according to monophone and context identity. In addition to this manual structuring technique, we present an automatic and more principled approach where factoring is guided by an agglomerative clustering process.

7.6.3   Hierarchies of Neural Networks

Recursively repeating the process of factoring conditional posteriors results in a tree-structured decomposition of the overall posterior probability. Each conditional posterior probability in such a tree structure can be estimated using a feedforward neural network and restricting the training set to match the corresponding dependent variable. Therefore, the hierarchical decomposition of the context-dependent posterior state probabilities maps directly to a hierarchy of neural networks (HNN). In such an architecture, a specific context-dependent posterior probability can be computed by evaluating and multiplying all the conditional posteriors from the root node down to the corresponding leaf node in the factoring tree. The size and shape of the HNN depends on the type of conditioning variables (binary, tertiary, ...) in the tree.

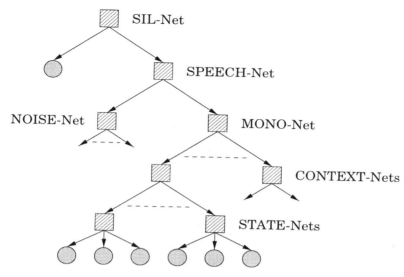

Figure 7.14 Manually constructed hierarchy of neural networks.

7.6.3.1   Manually Structured Hierarchies

The straightforward approach to establishing a hierarchical decomposition of the posterior state probabilities is to adopt the structure of the underlying phonetic decision tree and factor according to the monophone state identity. Additionally, this decomposition scheme can be refined by conditioning on nonspeech sound classes such as silence or noise(s). Figure 7.14 shows the structure of such an HNN [39, 42].

At the top of this hierarchy, we discriminate silence, noise, and speech sounds by means of two networks (SIL-Net and SPEECH-Net). The motivation for this specific partitioning comes from the observation that these three classes are easy to distinguish acoustically. The remainder of the tree structure decomposes the posterior of speech, conditioning on monophone, context, and state identity as these are convenient sound classes modeled by any phone-based HMM speech recognizer. The hierarchy of Figure 7.14 can be decomposed even further, for instance by factoring conditional monophone posteriors (estimated by the MONO-Net) based on linguistic features (e.g., voiced/ unvoiced, vowel/consonantal, fricative, etc.). The motivation behind such a decomposition is twofold. First, it reduces the number of local classes in each node, improving approximation accuracy, and second, it yields a decoupled and specialized set of expert networks having to handle a smaller amount of phonetic variation.

7.6.3.2 Clustering Hierarchies of Neural Networks

The use of prior knowledge for the design of a hierarchy of neural networks as presented in the previous section does not take into account dissimilarity of the observed classes in feature space. We therefore developed an agglomerative clustering algorithm to automatically design a hierarchy of neural networks for the estimation of posteriors for any number of classes. We termed this framework ACID/HNN [43], which stands for agglomerative clustering based on information divergence applied to hierarchies of neural networks.

The agglomerative (bottom-up) clustering scheme in the ACID/HNN framework is based on the symmetric information divergence

$$d(s_i, s_j) = \int_{\mathbf{x}} (p(\mathbf{x}|s_i) - p(\mathbf{x}|s_j)) \log \frac{p(\mathbf{x}|s_i)}{p(\mathbf{x}|s_j)}\, d\mathbf{x} \qquad (7.12)$$

as a measure of acoustic dissimilarity of subphonetic units. Based on this rather inexpensive distance measure, even large amounts of subphonetic units can be clustered efficiently. We typically model the class-conditional likelihoods using single diagonal covariance multivariate Gaussians with mean vectors μ_i and variance vectors σ_i^2. In this case, the symmetric information divergence between two states s_i and s_j amounts to

$$d(s_i, s_j) = \frac{1}{2} \sum_{k=1}^{n} \frac{(\sigma_{jk}^2 - \sigma_{ik}^2)^2 + (\sigma_{ik}^2 + \sigma_{jk}^2)(\mu_{ik} - \mu_{jk})^2}{\sigma_{ik}^2 \sigma_{jk}^2} \qquad (7.13)$$

Making the simplifying assumption of linearity of information divergence, we can define the following distance measure between clusters of states S_k and S_l

$$D(S_k, S_l) = \sum_{s_i \in S_k} p(s_i|S_k) \sum_{s_j \in S_l} p(s_j|S_l) d(s_i, s_j) \qquad (7.14)$$

The ACID algorithm uses the above distance measure in a standard bottom-up agglomerative clustering method. Note that this algorithm clusters HMM states without knowledge of their phonetic identity solely based on acoustic dissimilarity. Figure 7.15 illustrates ACID clustering on a very small subset of initial clusters. The ordinate of the dendrogram plot shows the information divergence at which the merger occurred. Names encode monophone, state (begin, middle, end) and context ID (numeric).

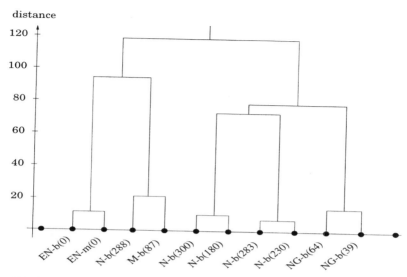

Figure 7.15 Partial dendrogram of ACID clustering.

Each node in an ACID-clustered tree structure represents conditional posteriors when interpreted as a hierarchical decomposition. We are currently experimenting with two-layer MLPs, trained in the framework of a generalized EM algorithm using error back-propagation to estimate these conditional posteriors. Figure 7.16 gives an overview of the resulting HNN.

Challenging aspects of such an architecture are model complexity and adaptation of learning rates during training. While the network in the root node is trained on all of the training data, networks deeper down the tree receive less training data than their predecessors. We found that it is advantageous to reduce the number of networks in an HNN by applying a greedy bottom-up node merging algorithm as a second step of ACID clustering. Using this strategy, we typically increase the average arity of the HNN tree from two to about eight.

7.6.4 Experiments and Results

Experiments with the ACID/HNN approach were carried out on the Switchboard LVCSR corpus. We chose Switchboard because it consists of very noisy spontaneous speech in telephone quality requiring excessive modeling of coarticulation to achieve state-of-the-art performance. Switchboard also is a comparatively hard speech recognition task. Current best systems based on traditional HMM approaches achieve

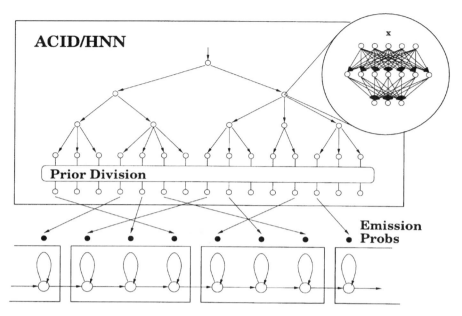

Figure 7.16 ACID clustered hierarchy of neural networks for hybrid NN/HMMs.

word error rates in the vicinity of 30%–40% while running in 150–300 times real time.

Table 7.5 summarizes results for various hybrid NN/HMM models focusing on the ACID/HNN framework. The models were trained on 170 hours of Switchboard training data corresponding to roughly 60 million

Table 7.5
Results for hierarchical and conventional connectionist acoustic models

Acoustic model	# states	# NNs	# params	xRT	word error
CI HME/HMM	166	59	220k	80	58.6%
CD HME/HMM	10,000	224	1.2M	130	37.3%
CD ACID/HNN	**6,000**	**962**	**1.6M**	**120**	**35.7**%
CD ACID/HNN	**24,000**	**4046**	**2.8M**	**145**	**33.3**%
adapted ACID/HNN	**24,000**	**4046**	**2.8M**	**130**	**31.8**%
adapted ACID/HNN	**24,000**	**≈ 500**	**2.8M**	**26**	**33.6**%
CD mixture Gaussians	24,000	–	6.5M	330	31.5%

patterns. Recognition experiments were performed with the Janus-RTk [41] Switchboard recognizer on the 1997 development test set, consisting of 40 unseen speakers. The first two rows give results that we obtained with manually structured HNN models [42]. CI denotes context-independent, CD, context-dependent modeling. Apart from word error rates, the table gives the number of HMM states, effective number of evaluated networks per frame, number of parameters, and real-time factors for each system. The last row gives comparative results obtained with a state-of-the-art mixture Gaussians-based acoustic model. Note that such modeling leads to a significantly larger amount of parameters and a much higher computational burden in terms of decoding speed.

Obviously, context-dependent modeling improves performance vastly. We trained two ACID/HNN acoustic models with 6k and 24k tied states, respectively, to demonstrate the scalability of the proposed approach. Furthermore, our results indicate that going from 6k to 24k HMM states improves performance significantly. Unsupervised speaker adaptation can be applied very easily to our model by simply retraining those networks in the HNN that receive more than a certain amount of adaptation data (the ones at the top of the tree). An additional gain of 1.5% in accuracy was achieved using this simple algorithm. Finally, ACID/HNN models trade off accuracy against decoding speed by simply pruning the evaluation of the HNN tree in each frame based on partial posteriors. This way, a significant amount of network evaluations can be omitted with almost no loss in accuracy. In contrast, traditional acoustic models usually require much more effort to achieve the same goal.

References

[1] Lippmann, R., "Review of Neural Networks for Speech Recognition," *Neural Computation*, Vol. 1, no. 1, March 1989. (Also in [44], pp. 374–392.)

[2] Waibel, A., et al., "Phoneme Recognition Using Time-Delay Neural Networks," Technical Report TR-1–0006, ATR Interpreting Telephony Research Laboratories, October 1987.

[3] Waibel, A., et al., "Phoneme Recognition Using Time-Delay Neural Networks," *IEEE Trans. on Acoustics, Speech, and Signal Processing*, Vol. ASSP-37, no. 3, March 1989, pp. 328–339. (Also in [44], pp. 393–405.)

[4] Lang, K., "A Time-Delay Neural Network Architecture for Speech Recognition," PhD thesis, CMU-CS-89-185, Carnegie Mellon University, Pittsburgh, Penn., July 1989.

[5] Lang, K., A. Waibel, and G. Hinton, "A Time-Delay Neural Network Architecture for Isolated Word Recognition." *Neural Networks*, Vol. 3, no. 1, 1990, pp. 23–43.

[6] Haffner, P., M. Franzini, and A. Waibel, "Integrating Time Alignment and Neural Networks for High Performance Continuous Speech Recognition," *Proc. IEEE Int. Conf. Acoustics, Speech, and Signal Processing*, Vol. 1, Toronto, Canada, May 1991, pp. 105–108.

[7] Tebelskis, J., and A. Waibel, "Performance Through Consistency: MS-TDNN's for Large Vocabulary Continuous Speech Recognition," in *Advances in Neural Information Processing Systems 4 (NIPS*92)*, San Mateo, Calif.: Morgan Kaufmann, 1993, pp. 696–703.

[8] Tebelskis, J., "Performance Through Consistency: Connectionist Large Vocabulary Continuous Speech Recognition," *Proc. IEEE Int. Conf. Acoustics, Speech, and Signal Processing*, Vol. 2, Minneapolis, Minn., April 1993, pp. 259–262.

[9] Tebelskis, J., "Speech Recognition Using Neural Networks," Ph.D. thesis, CMU-CS-95-142, Carnegie Mellon University, Pittsburgh, Penn., May 1995.

[10] Bregler, C., S. Manke, H. Hild, and A. Waibel, "Improving Connected Letter Recognition by Lipreading," *Proc. IEEE Int. Conf. Acoustics, Speech, and Signal Processing*, Vol. 1, Minneapolis, Minn., April 1993, pp. 557–561.

[11] Bregler, C., S. Manke, H. Hild, and A. Waibel, "Bimodal Sensor Integration on the Example of 'Speech-Reading,'" *Proc. IEEE Int. Conf. Neural Networks*, Vol. 2, San Francisco, March 1993, pp. 667–670.

[12] Duchnowski, P., U. Meier, and A. Waibel, "See Me, Hear Me: Integrating Automatic Speech Recognition and Lip-Reading," *Proc. Int. Conf. Speech and Language Processing*, Vol. 2, Yokohama, Japan, September 1994, pp. 547–550.

[13] Meier, U., W. Hürst, and P. Duchnowski, "Adaptive Bimodal Sensor Fusion for Automatic Speechreading," *Proc. IEEE Int. Conf. Acoustics, Speech, and Signal Processing*, Atlanta, Ga., May 1996, pp. 833–836.

[14] Zeppenfeld, T., R. Houghton, and A. Waibel, "Improving the MS-TDNN for Word Spotting," *Proc. IEEE Int. Conf. Acoustics, Speech, and Signal Processing*, Vol. 2, Minneapolis, Minn., April 1993, pp. 475–478.

[15] Manke, S., M. Finke, and A. Waibel, "NPen++: A Writer Independent, Large Vocabulary On-Line Cursive Handwriting Recognition System," *Int. Conf. Document Analysis and Recognition*, Montreal, August 1995, IEEE.

[16] Bodenhausen, U., and A. Waibel, "Learning the Architecture of Neural Networks for Speech Recognition," *Proc. IEEE Int. Conf. Acoustics, Speech, and Signal Processing*, May 1991, pp. 117–120.

[17] Bodenhausen, U., "Automatic Structuring of Neural Networks for Spatio-Temporal Real-World Applications," Ph.D. thesis, Universität Karlsruhe, June 1994.

[18] Bodenhausen, U., and H. Hild, "Automatic Construction of Neural Networks for Special Purpose Speech Recognition Systems," *Proc. IEEE Int. Conf. Acoustics, Speech, and Signal Processing*, Vol. 5, Detroit, Mich., May 1995, pp. 3327–3330.

[19] Bourlard, H., and N. Morgan, *Connectionist Speech Recognition: A Hybrid Approach*. Kluwer Academic Press, 1994.

[20] Hochberg, M. M., et al., "The 1994 ABBOT Hybrid Connectionist-HMM Large-

Vocabulary Recognition System," *Spoken Language Systems Technology Workshop*, 1995.

[21] Jordan, M., and R. Jacobs, "Hierarchical Mixtures of Experts and the EM Algorithm," *Neural Computation*, Vol. 6, no. 2, December 1994, pp. 181–214.

[22] Hampshire II, J. B., and A. Waibel, "A Novel Objective Function for Improved Phoneme Recognition Using Time-Delay Neural Networks," *IEEE Trans. on Neural Networks*, Vol. 1, no. 2, June 1990, pp. 216–228.

[23] Juang, B.-H., and S. Katagiri, "Discriminative Learning for Minimum Error Classification," *IEEE Trans. on Signal Processing*, Vol. 40, no. 12, December 1992, pp. 3043–3054.

[24] Iso, K.-I., "Speech Recognition Using Dynamical Model of Speech Production," Technical Report CMU-CS-92–187, Carnegie Mellon University, Pittsburgh, Penn., July 1992.

[25] Puzrla, P., F. Bimbot, and C. Windheuser. "Distributed Binary Representations for Word Recognition by TDNN-DTW Hybrid Systems," *EUROSPEECH'95 (4th European Conference on Speech Communication and Technology)*, Vol. 3, Madrid, Spain, September 1995, pp. 2175–2178.

[26] Hwang, M.-Y., and X. Huang, "Subphonetic Modeling with Markov States— Senone," *Proc. IEEE Int. Conf. Acoustics, Speech, and Signal Processing*, Vol. 1, San Francisco, March 1992, pp. 33–37.

[27] Hild, H., and A. Waibel, "Recognition of Spelled Names over the Telephone," *Proc. 4th Int. Conf. Speech and Language Processing*, Vol. 1, Philadephia, October 1996, pp. 346–349.

[28] Fanty, M., R. Cole, and K. Roginski, "English Alphabet Recognition with Telephone Speech," J. E. Moody, S. J. Hanson, and R. P. Lippmann, eds., *Advances in Neural Information Processing Systems 5 (NIPS*91)*, Morgan Kaufmann, 1992, pp. 199–206.

[29] Junqua, J.-C., "SmarTspelL: A Multipass Recognition System for Name Retrieval over the Telephone," *IEEE Trans. on Speech and Audio Processing*, Vol. 5, no. 2, March 1997, pp. 173–182.

[30] Loizou, P., and A. Spanias, "High-Performance Alphabet Recognition," *IEEE Trans. on Speech and Audio Processing*, Vol. 4, no. 6, November 1996, pp. 430–445.

[31] Betz, M., and H. Hild, "Language Models for a Spelled Letter Recognizer," *Proc. IEEE Int. Conf. Acoustics, Speech, and Signal Processing*, Vol. 1, Detroit, Mich., May 1995, pp. 856–859.

[32] Stiefelhagen, R., J. Yang, and U. Meier, "Real Time Lip Tracking for Lipreading," *Eurospeech 97*, 1997

[33] Duchnowski, P., U. Meier, and A. Waibel, "See Me, Hear Me: Integrating Automatic Speech Recognition and Lip-Reading," *Int. Conf. on Spoken Language Processing (ICSLP)*, 1994, pp. 547–550.

[34] Duchnowski, P., et al., "Toward Movement-Invariant Automatic Lip-Reading and Speech Recognition," *Proc. ICASSP*, 1995.

[35] Meier, U., R. Stiefelhagen, and J. Yang, "Preprocessing of Visual Speech under Real World Conditions," *Proc. European Tutorial and Research Workshop on Audio-Visual Speech Processing (AVSP 97)*, 1997.

[36] Hampshire, J. B., and A. H. Waibel, "The meta-pi Network: Building Distributed Knowledge Representations for Robust Pattern Recognition," Technical Report CMU-CS-89–166, Carnegie Mellon University, Pittsburgh, Penn., August 1989.

[37] Jordan, M. I., and R. A. Jacobs, "Hierarchies of Adaptive Experts," *Advances in Neural Information Processing*, Vol. 4, 1992, pp. 985–993.

[38] Fritsch, J., "Modular Neural Networks for Speech Recognition," Technical Report CMU-CS-96–203, Carnegie Mellon University, Pittsburgh, Penn., August 1996.

[39] Fritsch, J., M. Finke, and A. Waibel, "Context-Dependent Hybrid HME/HMM Speech Recognition Using Polyphone Clustering Decision Trees," *Proc. ICASSP 97*, Munich, April 1997.

[40] Young, S., "Large Vocabulary Continuous Speech Recognition: A Review," Technical Report, Cambridge University Engineering Department, 1996.

[41] Finke, M., et al., "The Janus-RTk Switchboard/Callhome 1997 Evaluation System," *Proc. LVCSR Hub5-e Workshop*, Baltimore, May 1997.

[42] Fritsch, J., and M. Finke, "Improving Performance on Switchboard by Combining Hybrid HME/HMM and Mixture of Gaussians Acoustic Models," *Proc. Eurospeech 97*, Rhodes, September 1997.

[43] Fritsch, J., "ACID/HNN: A Framework for Hierarchical Connectionist Acoustic Modeling," *Proc. IEEE Workshop on Speech Recognition and Understanding (ASRU97)*, December 1997.

8

Probability-Oriented Neural Networks and Hybrid Connectionist/Stochastic Networks

Jean-Paul Haton

8.1 Introduction

Artificial neural networks (ANNs) have been widely used since the early 1990s in various areas of automatic speech processing. Promising results have been reported in the literature, especially for speech recognition [1, 2], even though neural networks have not yet outperformed classical models like HMM and other stochastic models. A large variety of models have been proposed so far: multilayer perceptrons, self-organizing maps, recurrent networks, predictive networks, etc. These models are described in this book. In this chapter, we are interested in the fundamental and applied aspects of two particular subfields of neural networks, which have received particular attention during the past few years, i.e., probabilistic neural networks for speech pattern classification, and hybrid models that aim at combining neural networks and stochastic models efficiently. The chapter is thus divided into two parts. The first part is devoted to probabilistic neural networks. In Section 8.2, the fundamentals of these models are renamed: principles, different types of models, and their relationships. Section 8.3 addresses the central problem of learning in probabilistic neural networks. Applications of these models to speech and speaker recognition are presented and discussed in Section 8.4. The second part of the chapter concerns hybrid models combining connectionist models (including probabilistic neural networks) and stochastic models. Section 8.5 presents the rationale of such models, together with the different types of solutions that have been envisaged so far.

8.2 Fundamentals of Probability-Oriented Neural Networks

8.2.1 The Bayes Decision Framework

Probability-oriented neural networks (PONNs) are networks that are based on the estimation of statistical distributions of the pattern classes to be classified. In order to formalize the statistical classification problem, let us first briefly recall the optimal (in the sense of minimum error rate) Bayes decision theory. We consider the general case of Bayesian classifiers with Gaussian kernel and mixtures of Gaussian functions for approximating the probability densities.

Let's suppose that we have I mutually exclusive classe of patterns. The realization of a class is an observation vector $\mathbf{x}$ of dimension D which is made up of a set of measurement features. Strictly speaking, this is not totally correct in speech recognition, since the observations are sequences of acoustic vectors whose structure is varying in time. This aspect will be considered in the section on hybrid models. The class-conditional probability for $\mathbf{x}$ to belong to class Ω_i is

$$p(\mathbf{x}/W_i) = \sum_{j=1}^{J_i} f(\mathbf{x}/w_{ij})P(w_{ij}) \tag{8.1}$$

where $f(\mathbf{x}/w_{ij})$ is the component's density and $P(w_{ij})$ the mixture coefficient. Class i is described by J_i mixture components. All component densities are supposed to be Gaussian, i.e.,

$$f(\mathbf{x}/w_{ij}) = N(\mathbf{x}, m_{ij}, \sigma_{ij}) \tag{8.2}$$

Besides, for a particular class

$$\sum_{j=1}^{J_i} P(w_{ij}) = 1 \tag{8.3}$$

The Bayes decision rule can be applied to solve the classification problem of an unknown pattern **x**. This rule of probability gives the class Ω_i with the maximum a posteriori probability by

$$p(\Omega_i/\mathbf{x}) = \frac{p(\mathbf{x}/\Omega_i)P(\Omega_i)}{\sum_k p(\mathbf{x}/\Omega_k)P(\Omega_k)} \qquad (8.4)$$

where $P(\Omega_i)$ represents the a priori probability of class i.

Thus, according to the Bayes decision rule

$$\mathbf{x} \in \Omega_i \text{ iff } i = \text{Arg} \max_k p(\Omega_k/\mathbf{x}) \qquad (8.5)$$

8.2.2 Training Procedures

A major problem associated with the use of a Bayesian classifier is the estimation of probability density functions. Estimation techniques can be classified into two categories: (1) *parametric techniques*: in which case the probability law is known (e.g., a Gaussian function) and the estimation consists in determining the parameters of this law, and (2) *nonparametric techniques*, in which case the probability function is unknown, and the problem is to estimate a probability density function in each point. This is usually the case in most pattern recognition problems. In nonparametric estimation techniques, several approaches can be used. The simplest approach consists in making a multi-dimensional histogram of the learning data by partitioning the sample space into a number of discrete regions. In most practical applications, the data training set cannot be set large enough to give reliable estimates of the probabilities. A more practical approach, proposed by Parzen [3], is to define an interpolation function (or window) centered on an unknown sample **x**. An estimate of $p(\mathbf{x})$ is computed by summing up the value of the window function for each sample in the data set. The assumption underlying this method is that, within a small region r of the feature space, the probability density function does not vary significantly, so that the probability that an input pattern **x** belonging to class Ω falls into region r can be approximated by the number of pattern vectors of class Ω, n_Ω, in r divided by the total number of vectors in the feature space, N. The pdf can thus be approximated as:

$$p(\mathbf{x}/\Omega) = \frac{n_\Omega}{N} * \frac{1}{V} \qquad (8.6)$$

where V is the volume of region r. The Parzen window introduces a window function ϕ for the computation of n_Ω:

$$n_\Omega = \sum_{n=1}^{N_c} \phi(\mathbf{x} - \mathbf{x}(n)) \tag{8.7}$$

where N_c is the total number of patterns $\mathbf{x}(n)$ of class C and

$$\phi(\mathbf{x} - \mathbf{x}(n)) = 1 \text{ iff } |\mathbf{x} - \mathbf{x}(n)| < \theta$$
$$= 0 \text{ elsewhere} \tag{8.8}$$

θ is a threshold that defines the estimation region.

Another approach is based on the decomposition of the probability density function on a basis of orthogonal functions. The main problem raised by this approach is related to the choice of the optimal set of orthogonal functions [4]. Other nonparametric estimation techniques also have to be mentioned, especially the k-nearest-neighbor method and the classification and regression trees [5].

8.2.3 Types of PONNs

8.2.3.1 Radial Basis Function Networks

Radial basis function (RBF) networks are a special case of multilayer perceptrons (MLPs), with cells that compute their activation not through a sigmoid function but according to a Euclidean similarity measure between patterns and to an activation, or basis, function [6, 7]. Suitable basis functions can be Gaussian, Laplacian, or Cauchy-like density functions. Experiments have shown that the form of the basis function is not determinative regarding the overall performance of the network [8, 9]. Gaussian functions are the most commonly used, since they provide a particularly smooth interpolation [10]. RBF networks are related to the potential function method introduced in the 1960s [11]. It has been shown that RBF cells are better descriptors of pattern distributions than inner product cells of classical perceptrons when the pattern classes can be described by mixtures of Gaussian variables [12].

More precisely, an RBF neural network has a two-layer architecture. The first layer maps an input feature vector to a set of hidden nodes (or centers), that constitute a basis for the input pattern space.

The basis functions of this hidden layer produce a localized response to an input pattern, i.e., their response takes a nonzero value only when the input pattern falls within a small region around the center represented by a basis function. It can be noticed that this model is motivated by the locally tuned response characteristics that exist in many parts of biological nervous systems as, for instance, in cochlear cells of the auditory systems. Such nervous cells have response characteristics that are selective for a finite range of the input stimuli.

The second layer implements a linear mapping from the activation of the centers to the output nodes corresponding to the different pattern classes. Thus, the network carries out a nonlinear transformation by forming a linear combination of the basic functions. For each input pattern, the RBF network computes nodal outputs that are estimates of Bayesian probabilities [13]. Figure 8.1 shows the organization of an RBF with Gaussian kernels.

The training of an RBF network is a hybrid one. It is made up of an unsupervised phase followed by a supervised phase:

- The parameters of the Gaussian functions representing the centers are first computed by carrying out a clustering on the training pattern set (for instance, with a K-means algorithm). This phase uses classical training methods of pattern recognition that differ from the methods encountered in the training of neural nets.

- The weights of the connections between the hidden layer and the output layer are then trained by minimizing the mean-square error between actual and desired output values.

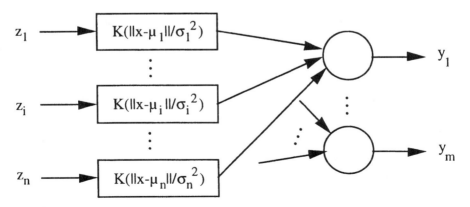

Figure 8.1 An RBF network with Gaussian kernels.

Several variants of RBF learning algorithm have also been proposed [14], as well as various extensions of the initial RBF model [15].

As MLPs, RBF networks can be used for classification and function approximation tasks. Their main advantage is that they keep the pattern classification performances of other ANNs with a much lower training computation cost (the convergence to a solution can be up to several orders of magnitude faster than with the MLP [16]). These networks also feature good generalization capabilities. However, compared with MLPs, they require much more training data for achieving the same level of accuracy [17].

In fact, it is possible to combine the feature extraction capabilities of an MLP with an RBF, as in the system described in [18] for discrete utterance recognition.

8.2.3.2 Probabilistic Neural Networks

By using nonparametric estimators of probability density functions, it is possible to define a feedforward, multilayer neural network with other training algorithms, by equivalence with parallel analog networks that classify patterns [19, 20]. This so-called Probabilistic Neural Network (PNN), emulates under the form of a network the Parzen window method described in Section 8.2.2. Figure 8.2 shows the organization of such a PNN for the classification of an input pattern $\mathbf{x} = \{x_1, x_2, \ldots, x_D\}$ into one out of I classes (cf. section 8.2.1.).

The first layer is the input layer to which pattern $\mathbf{x}$ is presented. The

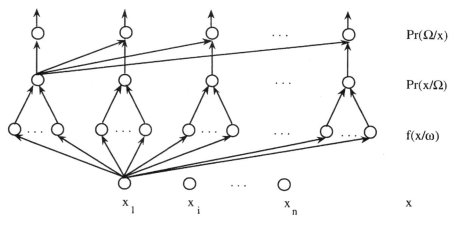

Figure 8.2 Typical architecture of a probabilistic neural network (PNN).

outputs of the second layer compute the mixture probability density functions $f(\mathbf{x}/w_{ij})$. The weights of connections between second and third layers are the mixture coefficients $P(w_{ij})$. Nodes in the third layer sum up the weighted mixture probabilities in order to produce the class-conditional probabilities. The Bayes rule is finally implemented in the fourth layer.

The main advantage of PNNs lies in the simplicity and the speed of learning compared with the back-propagation algorithm of perceptrons. Moreover, the learning process is incremental and easily allows for the inclusion of new pattern examples. Finally, PNNs provide a solution to a classification problem which is asymptotically Bayesian.

However, these networks present two main drawbacks. An important one is the extensive amount of data storage when the training sets are large (all patterns have to be stored). Another difficulty is related to the fact that all patterns in the training set are considered as equally important, and thus have the same contribution to the probability density estimation.

In order to overcome these drawbacks, several solutions have been investigated. The following ones can be cited among others:

- The use of a Kohonen topological map to reduce the number of training patterns by defining a subset of representative patterns through learning vector quantization (LVQ) [21];

- The reduction of the size of the training set and the definition of a weighting coefficient associated with each training pattern [22];

- A semiparametric training method in which the probability density function is considered as a mixture of Gaussian functions with covariance matrices that are not identical for the different classes.

The last idea led to the design of a Gaussian clustering network, GCN [23]. GCNs present some similarities with RBF networks, but they are basically unsupervised networks. They require some kind of lateral interaction between the nodes of the hidden layer, as in competitive, self-organizing models like Kohonen maps. Another model, also based on semiparametric training, is the probabilistic mapping network, PMN [24, 25].

An improvement in PNN, called adaptive PNN, has also been proposed [26]. The idea is to adapt separate smoothing parameters for each measurement dimension. This can often improve the generalization accuracy. Practical results show that the dimensionality of the problem and the complexity of the network can be simultaneously reduced.

Several variants of PNNs have been considered and tested in various areas of pattern recognition. Hierarchical mixtures of experts (HME) [27] are based on a tree-structured architecture for supervised learning. The underlying statistical model is a hierarchical mixture model in which both the mixture coefficients and the mixture components are generalized linear models. The learning model uses maximum likelihood estimation (MLE) in the framework of an estimation-maximization (EM) algorithm (cf. section 8.3.2).

This model has elements in common with the so-called neural tree algorithm [28], which is a decision tree with MLPs at the nonterminals. However, in the neural tree, MLPs are trained by a back-propagation algorithm that has not the same clear link with the overall classification performance as in the probabilistic model of HME.

The generalized Fisher training PNN [24] model is also based on MLE training with the EM algorithm. As in the GCN model, probability density functions are estimated as homoscedastic mixtures of Gaussian functions. The learning algorithm is a generalization of Fisher's linear discrimination algorithm which allows for nonlinear discrimination. The PNN trained with this generalized Fisher algorithm has been shown to require fewer nodes and interconnections than the original PNN in most practical problems.

The Gaussian potential function network (GPFN) [29] is able to generate an arbitrary shape of an input field by synthesizing a set of Gaussian functions provided by hidden units, called Gaussian potential function units. These units are determined by an original learning algorithm, the hierarchically self-organizing learning. However, there are no stochastic constraints imposed on the mixture weights of the Gaussian units and the output units of the model, so that the output of a GPFN cannot be considered as a probability measure.

8.2.4 Summary

PONNs (RBF networks and the various kinds of PNNs) are neural network models that learn statistics about a set of training patterns by techniques based on local basis functions. This point distinguishes them from other neural networks whose training methods are based on global basis functions, such as the well-known error gradient back-propagation.

RBFs can be used instead of MLPs, with a similar performance, when training data is plentiful and cheap, or if online training is necessary. PNNs and RBFs classify an unknown pattern in a similar way. These two models differ mainly in their training methods. RBF training consists of an iterative adaptation of the parameters of the network, whereas PNN training is based on Parzen window estimation of probabilities. Basic PNNs have the important advantage of instantaneous learning. This property makes them very interesting for exploring new sets of data, with frequent retraining phases.

8.3 Learning Methods for PNNs

8.3.1 Position of the Problem

As pointed out in Section 8.2, parametric methods for learning found in statistical classifiers consist in assuming the functional forms of the distribution and of computing the parameters of the function from a set of training data. The search for these parameters is often carried out using maximum likelihood estimation (MLE). Several PNNs have been developed under the framework of MLE learning (as in [24, 25]). MLE aims at maximizing the likelihood of the training observations $\mathbf{x}$ as output of a given correct answer A, while neglecting the output overall likelihood:

$$\lambda = \operatorname{Argmax} p_\lambda(\mathbf{x}/A) \qquad (8.9)$$

It can be demonstrated that, under certain assumptions, the function estimated according to MLE will converge to the true distribution. However, these assumptions are almost never met in practical speech recognition applications. A drawback of the MLE strategy also lies in the fact that the parameters of a pattern class are estimated independently from other classes.

Therefore, a different type of estimation, called maximum mutual information estimation (MMIE), has been more recently proposed [30]. In MMIE, the objective optimization function is the a posteriori probability of the model corresponding to the training data, given the data. This discriminative learning is thus consistent with the Bayes decision rule, and leads to a better classification performance [31–34]. We will now briefly review MLE and MMIE methods in relation to PNN learning.

8.3.2 MLE and EM Algorithms for PNNs

A general technique for maximum likelihood estimation was proposed in [35] under the form of the expectation-maximization (EM) algorithm. One of its most successful applications is the Baum-Welch estimation algorithm used in HMM training [36]. The EM algorithm has also been used in the training of NNs in general, and of PNNs in particular.

EM is an iterative approach to MLE. Each iteration in an EM algorithm is organized into two stages:

1. An expectation (E) phase, during which a likelihood function is estimated;

2. A maximization (M) phase, which involves the maximization of the likelihood computed during the E-phase.

If the algorithm simply increases the likelihood during the M-phase, instead of maximizing it, then the algorithm is called a generalized EM (GEM) algorithm [37]. An example in the NN domain is the Boltzmann learning algorithm.

GEM algorithms usually have a slower convergence than EM algorithms. However, a special GEM algorithm has been proposed, which converges more rapidly than its EM counterpart [37]. The basic idea underlying this algorithm is to only partially update the likelihood at each iteration. This algorithm has been implemented in a special type of PNN, the probabilistic mapping network. This network has produced very good results in a speaker recognition task in an experiment with a 60-speaker base [38].

8.3.3 MMIE for PNNs

As stated previously, the discriminative learning framework of MMIE has led to a substantial increase in the performance of speech recognition systems based on stochastic models. Unfortunately, contrary to MLE, there are no known reestimation formulas with theoretically guaranteed convergence. Therefore, various optimization techniques have to be used. A commonly used technique is the gradient descent.

For instance, an MMIE learning algorithm of this type has been developed for PNNs [38]. The learning strategy used embeds the MMI estimation, based on a gradient descent method, and the Bayes decision rule to give an efficient algorithm. Application to a speaker recognition task has shown that MMIE outperforms MLE on the same training database, with similar algorithm complexity [39] (cf. section 8.4.1).

8.4 Applications to Automatic Speech Recognition

RBFs and PNNs have been applied with some success to various speech recognition tasks, but the most significant results reported so far have been obtained in the area of speaker verification and recognition. It is noticeable that all types of PONNs are basically static networks that cannot directly handle time-varying phenomena. This is a serious limitation, especially for continuous speech recognition. A solution consists of designing neural network architecture that can, to a certain extent, handle the temporal dimension, like time-delay neural networks (TDNNs), or recurrent networks (see for instance [2] or the corresponding chapters in this book). We will see in Section 8.5 that hybrid architectures combining static neural networks (including PONNs) with stochastic models (like HMMs) make it possible to overcome this drawback, while retaining the classification capabilities of neural networks.

8.4.1 Speech Recognition

RBF networks have been used for speech recognition tasks, especially static phoneme classification [40–43], as well as in the framework of hybrid architectures (cf. Section 8.5). Performance figures are not better than those obtained with classical ANNs, but their training complexity is less important. An approach similar to RBFs, called reduced kernel discriminant analysis (RDKA) [44], has also been applied to a phoneme classification task [45]. This approach features good classification capabilities with relatively simple design.

PNNs have been used in several small-scale speech recognition experiments: recognition of isolated words [25], and recognition of Swedish phonemes [23]. However, the limited scope of these experiments makes it difficult to draw general conclusions on the performance of these models.

8.4.2 Speaker Recognition

The recognition of an unknown speaker can be carried out either by using a fixed test utterance and text-dependent techniques, or with varying test utterances. Text-independent and text-dependent experiments have been carried out independently with RBF networks.

In [46], a text-independent speaker recognition system is proposed. This system is made up of a set of RBF networks, one for each person to be recognized. Each RBF model can be used in isolation for verification purposes, or the RBF models can be combined together to form an identification system. The training of the system is twofold. Cluster analysis is first performed to design the nodes of the first layer. The weights of the output nodes are then adapted on the basis of a mean square error criterion. Speech parameterization uses cepstral coefficients and the experiments were carried out with a 40-speaker database. Experimental results indicate that the RBF network outperforms a standard MLP, as well as a vector-quantization-based system. In addition, a substantial reduction in training time over the MLP is obtained.

The system described in [47] bears upon text-dependent speaker identification. The system is made up of a single RBF network in which an output node is assigned to each speaker to be recognized. Speech analysis is performed through cepstral analysis, and two databases of 6 and 104 speakers are used in the experiments. Again, this system outperforms an MLP system in the various experiments that are reported.

In the two previous systems, the authors pointed out an interesting feature of RBF networks, i.e., the fact that the hidden layer is more readily interpreted by human observers [47], and that useful information can be directly incorporated into this layer [46].

It is in fact difficult to use a simple RBF network to recognize many classes, since the training phase becomes tedious and the recognition performance decreases. A compromise between a single, large network (as in [47]) and one network per person to be recognized (as in [46]) has been proposed in [48]. The result is a hybrid system combining (1) a set of RBFs corresponding to groups of similar speakers and used as coarse classifiers and (2) a set of multi-Gaussian classifiers (or Gaussian Mixture models, GMMs [49]) that carry out a fine recognition among the reduced set of candidates selected by the RBFs.

In a text-independent test based on the SPIDRE database, it was found that the use of RBFs made it possible to exclude 40% of the speakers from the second step of recognition without decreasing the performance.

Interesting results have also been obtained in speaker recognition and verification with the use of PNNs. A PNN with a mixture of Gaussian functions and using MLE learning based on a generalized EM algorithm is described in [37]. This system has been tested with a database of 60 speakers. An accuracy close to 99% was obtained in the best case for text-dependent speaker recognition.

A similar system, including discriminative MMI learning, has also been designed [38]. Results obtained on a 100-speaker database (50 female and 50 male) show that MMI learning is slightly, but significantly, better than MLE learning for PNNs.

8.5 Hybrid Connectionist/Stochastic Models

8.5.1 Position of the Problem

HMMs are now widely used in automatic speech processing to recognize both isolated words and continuous sentences, as well as to check the identity of a speaker [50]. These models are optimal in the sense that if the probability distributions of pattern classes are known, the classification of these patterns by some Bayesian decision strategy will yield a minimum error rate. In practice, the probability distributions have to be determined from learning data corpora. If the volume of available data is limited, it can be interesting to consider neural net models that feature some capability of generalizing from incomplete data or that necessitate limited amounts of training data.

HMMs produce very good results but suffer from some limitations that are mainly related to the strong hypotheses on which their optimization algorithms are based. ANNs have also proved useful in the classification of speech patterns, but require improvement in their ability to deal with the temporal and sequential nature of speech.

From what precedes, combining the time-domain modeling capability of HMMs with the discriminative power of ANNs can constitute an attractive solution. However, it is not so easy to design an adequate interface with which the respective qualities of the two models can be fully exploited. The next section presents the main solutions found in the current literature.

8.5.2 Proposed Solutions

ANNs and HMMs can be made to cooperate in a variety of ways in hybrid systems. Three categories of solutions can be defined:

1. ANNs as front-ends for HMMs;
2. ANNs as postprocessors of HMMs;
3. Unified models that try to combine the two theories.

These three categories are detailed next.

8.5.2.1   ANNs as Front-Ends for HMMs

A large number of studies have concentrated on the use of ANNs as front-end for HMMs. It was proved (see for instance [13, 51–53]) that a properly trained MLP for pattern classification is asymptotically equivalent to an estimator of a posteriori class probabilities. Several experiments have confirmed the quality of this estimation. So, from a practical point of view, the outputs of an ANN in a perfectly mastered framework are reliable. A hybrid system of this kind has the general structure presented in Figure 8.3.

Several systems have been developed on the basis of this structure (cf. [54-58]). Some improvements have also been proposed concerning the learning algorithm and the normalization of the MLP outputs by their a priori probabilities. For instance, the use of the REMAP learning algorithm (recursive estimation and maximization of a posteriori probabilities) has significantly reduced the error rate in a test on small vocabulary, isolated word, multispeaker, noisy speech recognition, compared with a classical hybrid system [59].

Estimators other than MLPs, principally recurrent nets and RBF, have also been used in such hybrid systems [60, 61], especially in order to simplify the learning phase. It was shown that the hybrid system's performance on the recognition of E-set letters is better than that of a

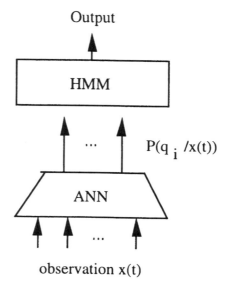

Figure 8.3 Structure of a hybrid ANN-HMM system.

tied-mixture recognizer of comparable complexity and near that of a tied-mixture system of considerably greater complexity [61].

The use of TDNNs has also been investigated. In the system proposed in [62], a TDNN is incorporated into a Viterbi framework for performing time alignment. The outputs of the TDNN can then be normalized so as to deliver a posteriori probabilities that are then fed into an HMM. A hybrid TDNN-HMM model has also been designed that uses the TDNN as a phone labeler for the HMM, instead of a standard vector quantizer [63]. Another idea consists of using a new parameter smoothing technique based on the average values of the activation vectors from the second hidden layer of the TDNN [64].

Hybrid systems like those presented above are also reported to enhance slightly the recognition performances of HMM systems (see for instance [65] for a test on the ARPA Resource Management (RM) task and [66] for the Abbott system of Cambridge University in the SQUALE project). Such systems present several advantages as compared with purely stochastic systems [67]; moreover, they have good recognition performances [68, 69]. However, their implementation and use are not straightforward owing to the large number of parameters involved and to the computation cost. Moreover, the network must be trained with a sufficient amount of data to ensure a convergence to a global minimum. A variety of solutions have been proposed to tackle this problem. One solution uses the ANN to compute an additional set of symbols as transformed observations for the HMM [70]. A further improvement of this method is achieved thanks to a global optimization of both the ANN and the HMM [71]. This approach uses the gradient of the HMM optimization criterion with respect to the transformed observations to estimate the weights of the ANN connections. A unified framework, named hidden neural network (HNN), has also been proposed for the same purpose [72], as well as a minimum classification error (MCE) method for a unified approach to feature extraction and classifier design into a single training process [73].

Another solution consists in using the ANN not as a probability estimator but as a labeler for a discrete parameter HMM [74]. An MLP plays the role of assigning a phonetic label to each speech frame, the label of which is then passed on to the HMM. This amounts to vector quantization involving phonetic information. A related idea is described in [75], but the ANN is trained using an information theory based on an unsupervised training algorithm that tries to achieve maximum mutual information (MMI) between the generated output labels and the underlying phonetic description. This approach has been successfully tested for the Resource Management database [76] and for the *Wall Street Journal* database [77]. The idea of using an ANN as a labeler was first proposed in [78, 79], but with use of an LVQ model, instead of an MLP. Other related experiments have been reported more recently [80, 81]. Similar ANN/HMM hybrids have also been developed with recurrent ANNs instead of classical feedforward networks [82, 83].

8.5.2.2   ANNs as Postprocessors of HMMs

A further approach to hybrid system design considers ANNs as HMM postprocessors. This allows an efficient combination of the time-alignment capability of HMMs with the discriminative power of ANNs, a feature that is of particular relevance to continuous speech recognition. Some systems feed all sentences recognized by the HMM into the ANN [84–86]. This method was also successfully used for spotting words in continuous speech [87], but it works only for applications requiring a limited vocabulary.

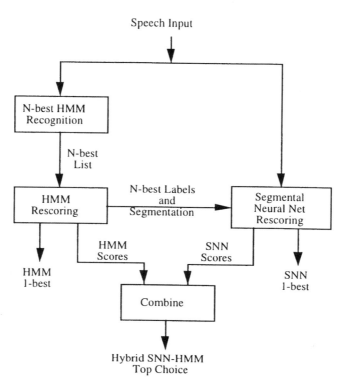

Figure 8.4 Principle of a hybrid SNN/HMM system (after [90]).

Another approach is to limit the number of hypotheses sent to the ANN as candidates elected by an N-best search algorithm [88–91]. The system described in [90], for instance, uses a so-called segmental neural network, which is an ANN that accepts the acoustic frames of a phonetic segment as input and produces an estimation of the probability of a phoneme corresponding to this segment. To be efficient, this network requires a reliable phone segmentation procedure. Figure 8.4 illustrates the principle of the system proposed by the authors.

A similar system was used in [92] to recognize connected spelled letters. This system is made up of two parts, the first being a second-order HMM [93] trained to recognize letters of the alphabet. The second is a selectively trained neural net (STNN) [94], basically an MLP. The latter incorporates acoustic-phonetic knowledge to help separate confusable letter pairs like /p/ and /t/ on the one hand, and /m/ and /n/ on the other hand. The STNN focuses on the discriminative part of words, i.e., the area where the distinctive acoustic information is localized. For instance, in the case of the plosive consonants /p/ and /t/, the discriminative part is made up of the shape of the burst and the vowel transitions. The HMM provides an N-best response string with indications of word boundaries. These boundaries are then used by the STNN to localize the discriminative parts of words. It has been shown that the combination of HMM and STNN substantially improves the identification of confusable letters both in English and in French.

Hybrid HMM-ANN structures have also been used for speaker recognition tasks. For instance, in [95] a two-stage system that uses HMMs for segmenting the input speech signal into phone units was described. Then, phone-dependent RBF networks are used for the verification of identity claims.

8.5.2.3 Unified Models

A third solution used to design hybrid models consists of trying to derive a unifying formalism for the two theories of HMMs and ANNs. Such a formalism is totally absent in the solutions presented in the two preceding sections. Three classes of models have been proposed so far, i.e., the neural network implementation of HMMs, the HMMs with ANNs learning modules, and the alpha-nets. Let's briefly review these three approaches. All these models contribute to a better formalism, even though they have produced only limited increase of the performance so far.

HMM nets are a good example of the neural network implementation of HMMs category [96]. This model is based on the fact that the forward-backward learning algorithm used in HMMs is equivalent to the error gradient back-propagation algorithm of MLPs. An HMM net is thus basically an ANN implementation of an HMM using a recurrent network.

Another model, also focused on learning aspects, is the competitive HMM [97]. It must be recalled that ANNs are trained to discriminate between classes of patterns, whereas HMMs are usually trained to yield estimations of interclass maximum likelihood. Several attempts have been made in order to enhance the recognition performance of an HMM based on the same data training set, with the use of discriminative training algorithms: maximum mutual information or minimum discriminant information [98]. The idea of competitive HMMs consists of using an ANN in order to optimize a discrimination measure based on the a posteriori probability of state occupancy. This method allows for the substantial enhancement of the Baum-Welch algorithm.

A third model in this category is the Viterbi net [1], which emulates the recursion step of the dynamic programming Viterbi algorithm. To each HMM is associated an ANN in which each neuron of the output layer corresponds to a state of the HMM. In the usual case of Gaussian HMMs, the first hidden layer of the ANN computes the set of Gaussian outputs for the input speech frames.

An example of HMMs with ANN learning modules is the connectionist Viterbi learning [54], a variant of the classical Viterbi algorithm that uses an ANN trained by back-propagation in order to represent the output probability distributions associated with the transitions of the HMM. Another representative example of this category is the neuronal predictive HMM [99].

The basic idea of the *alpha-nets* [100] model is to implement the forward part of the Baum-Welch algorithm under the form of a recurrent ANN. Several variants have been proposed [96, 97, 101]. Alpha-nets have been used for speech recognition and also speaker verification [102]. In this latter case, the performance was better than the one obtained with a classical HMM because it allowed discriminative training for each talker pair HMM.

8.5.3 Summary

Hybrid architectures combining ANNs and HMMs allow for the optimal use of the advantages of both models. The use of ANNs (including PONNs) as front-ends of HMMs has proven to be highly efficient, and can even be found in some recent commercial systems. It is also interesting to use ANNs as postprocessors of HMMs, but the training of the overall system is more questionable. Unified models would be very useful, but this is still an open issue for research.

8.6 Conclusion

This chapter has addressed two important topics in the field of neural networks used for classification, i.e., probability-oriented neural networks (PONNs) and hybrid connectionist/stochastic models, which have both been studied in the context of speech and speaker recognition. These two research areas have received particular attention during the past few years, but it would be premature to draw final conclusions.

Compared to classical NN models like multilayer perceptrons, PONNs (PNNs and RBF networks) present the major advantage of an easy and fast training, with the disadvantage of necessitating more training data. Interesting results have been reported concerning the use of such networks in speech recognition and also speaker recognition and verification.

Hybrid connectionist/stochastic models tend to take advantage of the complementary capabilities of connectionist networks (in particular their discriminative power) and of stochastic models (in particular their capacity of handling time sequences) in order to derive new models that are well suited for automatic speech processing tasks. Very good performance has been obtained so far, and this field remains an active research issue. Until now, hybrid models combining a neural network as front-end have given the most interesting results. This is due to the fact that the neural network can compute good approximates of a posteriori probabilities of classes that can be further exploited by the HMM. From a theoretical point of view, hybrid models that tend to embed the two paradigms of neural networks and HMMs in a unified framework also exhibit an undoubted interest in a long-term perspective.

References

[1] Lippmann, R. P., "Review of Neural Networks for Speech Recognition," *Neural Computation*, Vol. 1, no. 1, 1989, pp. 1–38.

[2] Haton, J.-P., "Neural Networks for Speech Recognition," in C. H. Chen, ed., *Fuzzy Logic and Neural Network Handbook*, New York: McGraw-Hill, 1996.

[3] Parzen, E., "On Estimation of Probability Density Function and Mode," *Ann. Math. Statistics*, Vol. 33, 1962, pp. 1065–1076.

[4] Loftsgaarden, D. O., and Quesenbury, G. P., "A Nonparametric Estimate of a Multivariable Density Function," *Ann. Math. Statistics*, Vol. 36, 1965, pp. 1049–1051.

[5] Breiman, L., J. H. Friedman, R. A. Olshen, and C. J. Stone, *Classification and Regression Trees*, Belmont, Calif.: Wadsworth, 1984.

[6] Moody, J., and C. Darken, "Learning with Localized Receptive Fields," *Proc. 1988 Connectionist Models Summer School*, ed. D. S. Touretzky, G. E. Hinton and T. J. Sejnowski, San Mato, CA: Morgan Kaufman, 1989, pp. 133–143.

[7] Reilly, D. L., L. N. Cooper, and C. Erlbaum, "A Neural Model for Category Learning," *Biol. Cybern.*, Vol. 45, 1982, pp. 35–41.

[8] Singer, E., and R. P. Lippmann, "A Speech Recognizer Using Radial Basis Function Neural Networks in a HMM Framework," *Proc. IEEE Int. Conf. Acoustics, Speech, and Signal Processing-ICASSP'92*, San Francisco, Vol. 1, 1992, pp. 629–632.

[9] Lippmann, R. P., and E. Singer, "Hybrid Neural-Network/HMM Approaches to Wordspotting," *Proc. IEEE Int. Conf. Acoustics, Speech, and Signal Processing*, ICASSP'93, Minneapolis, 1993, pp. I.565–I.568.

[10] Poggio, T., and F. Girosi, "Networks for Approximation and Learning," *Proc. IEEE*, Vol. 78, 1990, pp. 1481–1497.

[11] Aizerman, M. A., E. M. Braverman, and L. I. Rozoner, "Theoretical Foundations of the Potential Function Method in Pattern Recognition Learning," *Automation and Remote Control*, Vol. 25, 1964, pp. 821–827.

[12] Broomhead, D. S., and D. Lowe, "Multivariable Function Interpolation and Adaptive Networks," *Complex Systems*, Vol. 2, 1988, pp. 269–303.

[13] Richard, M. D., and R. P. Lippmann, "Neural Network Classifiers Estimate Bayesian *a posteriori* Probabilities," *Neural Computation*, Vol. 3, No. 4, 1991, pp. 461–483.

[14] Musavi, M. T., W. Ahmed, K. H. Chan, K. B. Faris, and D. M. Hummels, "On the Training of Radial Basis Function Classifiers," *IEEE Trans. on Neural Networks*, Vol. 5, 1992, pp. 595–603.

[15] Hush, D. R., and B. G. Horme, "Progress in Supervised Neural Networks," *IEEE Signal Processing Magazine*, 1993. pp. 8–39.

[16] Mazurek, J., A. Krzyzak, and A. Cichocki, "Rates of Convergence of the Recursive Radial Basis Function Networks," *Proc. IEEE Int. Conf Acoustics, Speech, and Signal Processing-ICASSP'97*, Munich, 1997, pp. 3317–3320.

[17] Morgan, D. P., and C. L. Scofield, *Neural Networks and Speech Processing*, Kluwer Academic, Dordrecht, 1992.

[18] Zemany, P. D., et al., "Experiments in Discrete Utterance Recognition Using Neural Networks," *Proc. Boston Miniconference*, 1989.

[19] Specht, D. F., "Probabilistic Neural Networks," *IEEE Transactions on Neural Networks*, Vol. 3, 1990, pp. 109–118.

[20] Specht, D. F., "Probabilistic Neural Networks and General Regression Neural Networks," in C. H. Chen, ed., *Fuzzy Logic and Neural Network Handbook*, New York: McGraw-Hill, 1996.

[21] Burrascano, P., "Learning Vector Quantization for the Probabilistic Neural Network," *IEEE Trans. on Neural Networks*, Vol. 2, 1991, pp. 458–461.

[22] Maloney, P. S., et al., "Successful Applications of Expanded Probabilistic Neural Networks," in C. H. Dagli et al., eds., *Intelligent Engineering Systems Through Artificial Neural Networks*, New York: ASME Press, 1991.

[23] Travén, H. G. C., "A Neural Network Approach to Statistical Pattern Classification by Semiparametric Estimation of Probability Density Functions," *IEEE Trans. on Neural Networks*, Vol. 2, No. 1, 1991, pp. 366–377.

[24] Streit, R. L., and T. E. Luginbuhl, "Maximum Likelihood Training of Probabilistic Neural Networks," *IEEE Trans. on Neural Networks*, Vol. 5, No. 1, 1994, pp. 764–783.

[25] Wu, J., and C. Chan, "Isolated Word Recognition by Neural Network Models with Cross-correlation Coefficient for Speech Recognition," *IEEE Trans. on Pattern Analysis and Machine Intelligence*, Vol. 15, No. 11, 1993, pp. 1174–1185.

[26] Specht, D. F., "Enhancements to Probabilistic Neural Networks," *Proc. IEEE Int. Joint Conf. on Neural Networks*, Baltimore, 1990.

[27] Jordan, M. I., and R. A. Jacobs, "Hierarchical Mixtures of Experts and the EM Algorithm," *Neural Computation*, Vol. 6, 1994, pp. 181–214.

[28] Strömberg, J.-E., J. Zrida, and A. Isaksson, "Neural Trees-Using Neural Nets in a Tree Classifier Structure," *Proc. IEEE Int. Conf. on Acoustics, Speech, and Signal Processing*, ICASSP'91, Toronto, 1991, pp. 137–140.

[29] Lee, A. S., and R. M. Kil, "A Gaussian Potential Function Network with Hierarchically Self-organizing Learning," *IEEE Trans. on Neural Networks*, Vol. 4, No. 1, 1991, pp. 207–224.

[30] Bahl, L. R., P. F. Brown, P. V. de Souza, and R. L. Mercer, "Maximum Mutual Information Estimation of Hidden Markov Model Parameters for Speech Recognition," *Proc. IEEE Int. Conf. Acoustics, Speech, and Signal Processing*, ICASSP'86, Tokyo, 1986, pp. 49–52.

[31] Cardin, R., Y. Normandin, and R. de Mori, "High Performance Connected Digit Recognition Using Maximum Mutual Information Estimation," *Proc. IEEE Int. Conf. Acoustics, Speech, and Signal Processing*, ICASSP'91, Toronto, 1991, pp. 533–536.

[32] Cardin, R., Y. Normandin, and E. Millien, "Inter-word Coarticulation Modelling and MMIE Training for Improved Connected Digit Recognition," *Proc. IEEE Int. Conf. Acoustics, Speech, and Signal Processing*, ICASSP'93, Minneapolis, 1993, pp. II.243–II.246.

[33] Willett, D., C. Neukirchen, and J. Rottland, "Dictionary-Based Discriminative HMM Parameter Estimation for Continuous Speech Recognition Systems," *Proc. IEEE Int. Conf. Acoustics, Speech, and Signal Processing*, ICASSP'97, Munich, 1997, pp. 1515–1518.

[34] Merialdo, B., "Phonetic Recognition Using Hidden Markov Models and Maximum Mutual Information Training," *Proc. IEEE Int. Conf. Acoustics, Speech, and Signal Processing*, ICASSP'88, New York, 1988, pp. 111–114.

[35] Dempster, A. P., N. M. Laird, and D. B. Rubin, "Maximum Likelihood from

Incomplete data via the EM Algorithm," *Journal of Royal Statistical Society*, Vol. 39, 1977, pp. 1–38.

[36] Baum, L. E., "An Inequality and Associated Maximization Technique in Statistical Estimation of Probabilistic Functions of Markov Processes," *Inequalities*, Vol. 3, 1972, pp. 1–8.

[37] Li, H., Y. Gong, and J.-P. Haton, "EM Algorithms for Probabilistic Mapping Networks," *Rapport INRIA*, No. 2614, 1995.

[38] Li, H., J.-P. Haton, and Y. Gong, "The MMI Learning of Probabilistic Neural Networks," *Proc. Int. Conf. Artificial Neural Networks*, Paris, Vol. 2, 1995, pp. 123–128.

[39] Li, H., J.-P. Haton, and Y. Gong, "On MMI Learning of Gaussian Mixture for Speaker Models," *Proc. 4th European Conf. Speech Communication and Technology*, Madrid, 1995, pp. 363–366.

[40] Renals, R., and R. Rohwer, "Phoneme Classification Experiments using Radial Basis Functions," *Proc. Int. Joint Conf. Neural Networks*, 1989, pp. 1461–1467.

[41] Niranjan, M., and F. Fallside, "NNs and RBFs in Classifying Static Speech Patterns," *Techn. Report CUEDIF-INFENG 17R22*, Eng. Dept, Cambridge University, 1988.

[42] Moody, J., and C. Darken, "Fast Learning in Networks of Locally-tuned Processing Unit," *Neural Computation*, Vol. 1, No. 2, 1989, pp. 281–294.

[43] Nowlan, S. J., "Maximum Likelihood Competitive Learning," in D. Touretzky, ed., *Advanced Neural Information Processing 2*, San Mateo, CA: Morgan Kaufmann, 1990.

[44] Holmström, L., P. Koistinen, J. Laaksonen, and E. Oja, "Neural and Statistical Classifiers: Taxonomy and Two Case Studies," *IEEE Trans. on Neural Networks*, Vol. 8, No. 1, 1997, pp. 5–17.

[45] Holmström, L., and A. Hämäläinen, "The Self-Organizing Reduced Kernel Density Estimator," *Proc. Int. Conf. Neural Networks*, San Francisco, Vol. 1, 1993, pp. 417–421.

[46] Oglesby, J., and J. S. Mason, "Radial Basis Function Networks for Speaker Recognition," *Proc. IEEE Int. Conf. Acoustics, Speech, and Signal Processing*, ICASSP'91, Toronto, 1991, pp. 393–400.

[47] Fredrickson, S. E., and L. Tarassenko, "Radial Basis Functions for Speaker Identification," *Proc. ESCA Workshop Automatic Speaker Recognition, Identification and Verification*, Martigny, 1994, pp. 107–110.

[48] Li, W. Y., and D. O'Shaughnessy, "Hybrid Network Based on RBFN and GMM for Speaker Recognition," *Proc. Eurospeech'97*, Rhodes, 1997, pp. 955–958.

[49] Reynolds, D. A., and R. C. Rose, "Robust Text-Independent Speaker Identification Using Gaussian Mixture Speaker Models," *IEEE Trans. Speech and Audio Processing*, Vol. 3, No. 1, 1995, pp. 72–83.

[50] Rabiner, L. R., "A Tutorial on Hidden Markov Models and Selected Applications in Speech Recognition," *Proc. IEEE*, Vol. 77, No. 2, 1989, pp. 257–285.

[51] Hopfield, J. J., "Learning Algorithms and Probability Distributions in Feed-forward Networks," *Proc. Nat. Acad. Sci.*, 1987, pp. 8429–8433.

[52] Bourlard, H., and C. J. Wellekens, "Links Between Markov Models and Multilayer Perceptrons," *IEEE Trans. on Pattern Analysis and Machine Intelligence*, Vol. 12, 1990, pp. 1167–1178.

[53] Gish, H., "A Probabilistic Approach to the Understanding and Training of Neural Network Classifiers," *Proc. IEEE Int. Conf. Acoustics, Speech, and Signal Processing*, Albuquerque, 1990, pp. 1361–1364.

[54] Franzini, M., K. F. Lee, and A. Waibel, "Connectionist Viterbi Training: a New Hybrid Method for Continuous Speech Recognition," *Proc. IEEE Int. Conf. Acoustics, Speech, and Signal Processing*, Albuquerque, 1990, pp. 425–428.

[55] Morgan, N., and H. Bourlard, "Continuous Speech Recognition using Multilayer Perceptrons with Hidden Markov Models," *Proc. IEEE Int. Conf. Acoustics, Speech, and Signal Processing*, Albuquerque, 1990, pp. 26–30.

[56] Cohen, M., H. Franco, N. Morgan, D. Rumelhart, and V. Abrash, "Hybrid Neural Network/Hidden Markov Model Continuous-Speech Recognition," *Proc. Int. Conf. Spoken Language Processing*, ICSLP'92, Banff, 1992, pp. 915–918.

[57] Fanty, M., P. Schmid, and R. Cole, "City Name Recognition over the Telephone," *Proc. IEEE Int. Conf. Acoustics, Speech, and Signal Processing*, ICASSP'93, Minneapolis, 1993, pp. 549–552.

[58] Lubensky, D. M., A. O. Asadi, and J. M. Naik, "Connected Digit Recognition Using Connectionist Probability Estimators and Mixture-Gaussian Densities," *Proc. Int. Conf. Spoken Language Processing*, ICSLP'94, Yokohama, 1994, pp. 295–298.

[59] Bourlard, H., Y. Konig, and N. Morgan, "REMAP: Recursive Estimation and Maximization of A Posteriori Probabilities in Connectionist Speech Recognition," *Proc. Eurospeech'95*, Madrid, 1995, pp. 1663–1666.

[60] Renals, S., N. Morgan, M. Cohen, and H. Franco, "Connectionist Probability Estimation in the Decipher Speech Recognition System," *Proc. IEEE Int. Conf. Acoustics, Speech, and Signal Processing*, San Francisco, Vol. 1, 1992, pp. 601–604.

[61] Singer, E., and R. P. Lippmann, "A Speech Recognizer using Radial Basis Function Neural Networks in a HMM Framework," *Proc. IEEE Int. Conf. Acoustics, Speech, and Signal Processing*, San Francisco, Vol. 1, 1992, pp. 629–632.

[62] Dugast, C., L. Devillers, and X. Aubert, "Combining TDNN and HMM on a Hybrid System for Improved Continuous-speech Recognition," *IEEE Trans. on Speech and Audio Processing*, Vol. 2, No. 1, 1994, pp. 217–223.

[63] Ma, W. and D. Van Compernolle, "TDNN Labeling for a HMM Recognizer," *Proc. IEEE Int. Conf. Acoustics, Speech, and Signal Processing*, Albuquerque, 1990, pp. 421–423.

[64] Jang, C. S., and C. K. Un, "A New Parameter Smoothing Method in the Hybrid TDNN/HMM Architecture for Speech Recognition," *Speech Communication*, Vol. 19, No. 4, 1996, pp. 317–324.

[65] Renals, S., N. Morgan, H. Bourlard, M. Cohen, and H. Franco, "Connectionist Probability Estimators in HMM Speech Recognition," *IEEE Trans. on Speech and Audio Processing*, Vol. 2, No. 1, 1994, pp. 161–174.

[66] Steeneken, J. M., and D. A. van Leeuwen, "Multi–lingual Assessment of Speaker Independent Large Vocabulary Speech Recognition Systems: the SQUALE Project," *Proc. Eurospeech'95*, Madrid, 1995, pp. 1271–1274.

[67] Bourlard, H. "Toward Increasing Speech Recognition Error Rates," *Proc. Eurospeech'95*, Madrid, 1995, pp. 883–894.

[68] Konig, Y., H. Bourlard, and N. Morgan, "Remap: Experiments with Speech Recognition," *Proc. IEEE Int. Conf. Acoustics, Speech, and Signal Processing*, Atlanta, 1996, pp. 3350–3353.

[69] Dupont, S., H. Bourlard, O. Deroo, V. Fontaine, and J.-M. Boite, "Hybrid HMM/ANN Systems for Training Independent Tasks: Experiments on Phonebook and Related Improvements," *Proc. IEEE Int. Conf. Acoustics, Speech, and Signal Processing*, Munich, 1997, pp. 1767–1770.

[70] Bengio, Y., R. Cardin, R. de Mori, and Y. Normandin, "A Hybrid Coder for Hidden Markov Models Using a Recurrent Neural Network," *Proc. IEEE Int. Conf. Acoustics, Speech, and Signal Processing*, Albuquerque, 1990, pp. 537–540.

[71] Bengio, Y., R. de Mori, G. Flammia, and R. Kompe, "Global Optimization of a Neural Network-Hidden Markov Model Hybrid," *IEEE Trans. on Neural Networks*, Vol. 3, No. 2, 1992, pp. 252–259.

[72] Riis, S. K., and A. Krogh, "Hidden Neural Networks: a Framework for HMM/NN Hybrids," *Proc. IEEE Int. Conf. Acoustics, Speech, and Signal Processing*, Munich, 1997, pp. 3233–3236.

[73] Rahim, M. G., and C. H. Lee, "Simultaneous ANN Feature and HMM Recognizer Design Using String-based Minimum Classification Error (MCE) Training," *Proc. Int. Conf. Speech and Language Processing*, Yokohama, 1996, pp. 1824–1827.

[74] Le Cerf, P., W. Ma, and D. Compernolle, "Multilayer Perceptrons as Labelers for Hidden Markov Models," *IEEE Trans. on Speech and Audio Processing*, Vol. 2, No. 1, 1994, pp. 185–193.

[75] Rigoll, G., "Maximum Mutual Information Neural Networks for Hybrid Connectionist-HMM Speech Recognition Systems," *IEEE Trans. on Speech and Audio Processing*, Vol. 2, No. 1, 1994, pp. 175–184.

[76] Rigoll, G., Ch. Neukirchen, and J. Rottland, "A New Hybrid System Based on MMI-neural Networks for the RM Speech Recognition Task," *Proc. IEEE Int. Conf. Acoustics, Speech, and Signal Processing*, Atlanta, 1996, pp. 865–868.

[77] Rottland, J., Ch. Neukirchen, and D. Willett, "Performance of Hybrid MMI-connectionist/HMM Systems on the WSJ Speech Database," *Proc. IEEE Int. Conf. Acoustics, Speech, and Signal Processing*, Munich, 1997, pp. 1747–1750.

[78] Kimber, D. G., M. A. Bush, and G. N. Tajchman, "Speaker-Independent Vowel Classification Using Hidden Markov Models and LVQ2," *Proc. IEEE Int. Conf. Acoustics, Speech, and Signal Processing*, Albuquerque, 1990, pp. 497–500.

[79] Iwamida, H., S. Katagiri, E. McDermott, and Y. Tohkura, "A Hybrid Speech Recognition System Using HMMs with an LVQ-trained Codebook," *Proc. IEEE Int. Conf. Acoustics, Speech, and Signal Processing*, Albuquerque, 1991, pp. 489–492.

[80] Schmidbauer, O., and J. Tebelskis, "An LVQ Based Reference Model for Speaker-Adaptive Speech Recognition," *Proc. IEEE Int. Conf. Acoustics, Speech, and Signal Processing*, San Francisco, Vol. 1, 1992, pp. 629–632.

[81] Kapadia, S., V. Valtchev, and S. Young, "MMI Training for Continuous Phoneme

Recognition on the TIMIT Database," *Proc. IEEE Int. Conf. Acoustics, Speech, and Signal Processing*, Minneapolis, 1993, pp. 491–494.

[82] Robinson, A., "An Application for Recurrent Nets to Phone Probability Estimation," *IEEE Trans. on Neural Networks*, Vol. 5, 1994, pp. 298–305.

[83] Valtchev, V., S. Kapadia, and S. Young, "Recurrent Input Transformations for Hidden Markov Models," *Proc. IEEE Int. Conf. Acoustics, Speech, and Signal Processing*, Minneapolis, 1993, pp. 287–290.

[84] Howell, D. N. L., "The Multi–layer Perceptron as a Discriminating Post Processor for Hidden Markov Networks," *Proc. of Speech, 7th FASE Symposium*, 1988.

[85] Huang, W. Y., and R. P. Lippmann, "HMM Speech Recognition with Neural Net Discrimination," in *Advances in Neural Information Processing System* 2, Morgan Kaufmann, 1990, pp. 194–202.

[86] Guo, J., and H. C. Lui, "A Multilayer Perceptron Postprocessor to Hidden Markov Modeling for Speech Recognition," *Proc. IEEE Int. Conf. Acoustics, Speech, and Signal Processing*, Minneapolis, 1993, Vol. 2, pp. 263–266.

[87] Lippmann, R. P., and E. Singer, "Hybrid Neural-Network/HMM Approaches to Wordspotting," *Proc. IEEE Int. Conf. Acoustics, Speech, and Signal Processing*, ICASSP'93, Minneapolis, 1993, Vol. 1, pp. 565–568.

[88] Ostendorf, M., A. Kannan, S. Austin, and O. Kimball, "Integration of Diverse Recognition Methodologies through Reevaluation of the N-best Sentence Hypotheses," *Proc. of DARPA Speech and Natural Language Workshop*, San Mateo, CA: Morgan Kaufman, 1991.

[89] Austin, S., G. Zavaliagkos, J. Makhoul, and R. Schwartz, "Speech Recognition Using Segmental Neural Nets," *Proc. IEEE Int. Conf. Acoustics, Speech, and Signal Processing*, San Francisco, 1992, pp. 625–628.

[90] Zavaliagkos, G., Y. Zhao, R. Schwartz, and J. Makhoul, "A Hybrid Segmental Neural Net/Hidden Markov Model System for Continuous Speech Recognition," *IEEE Trans. on Speech and Audio Processing*, Vol. 2, No. 1, 1994, pp. 151–160.

[91] Boiteau, D., and P. Haffner, "Connectionist Segmental Post-processing of the N-best Solutions in Isolated and Connected Word Recognition Task," *Proc. Eurospeech'93*, Berlin, 1993, pp. 1933–1936.

[92] Mari, J.-F., D. Fohr, Y. Anglade, and J.-C. Junqua, "Hidden Markov Models and Selectively Trained Neural Networks for Connected Confusable Word Recognition," *Proc. Int. Conf. Spoken Language Processing*, Yokohama, 1994, pp. 1519–1522.

[93] Mari, J.-F., and J.-P. Haton, "Automatic Word Recognition Based on Second-order Hidden Markov Models," *Proc. Int. Conf. Spoken Language Processing*, Yokohama, 1994, pp. 247–250.

[94] Anglade, Y., D. Fohr, and J.-C. Junqua, "Speech Discrimination in Adverse Conditions Using Acoustic Knowledge and Selectively Trained Neural Networks", *Proc. IEEE Int. Conf. Acoustics, Speech, and Signal Processing*, Minneapolis, 1993, pp. 279–282.

[95] Olsen, J. O., "A Two Stage Procedure for Phone Based Speaker Verification," in J. Bigün, G. Chollet and G. Borgefors, eds., *Audio- and Video-based Biometric Person Authentication*, Berlin: Springer, 1997, pp. 219–226.

[96] Niles, L. T., and H. F. Silverman, "Combining Hidden Markov Models and Neural Network Classifiers," *Proc. IEEE Int. Conf. Acoustics, Speech, and Signal Processing*, ICASSP'90, Albuquerque, 1990, pp. 417–420.

[97] Young, S. J., "Competitive Training in Hidden Markov Models," *Proc. IEEE Int. Conf. Acoustics, Speech, and Signal Processing*, ICASSP'90, Albuquerque, 1990, pp. 681–684.

[98] Ephraim, Y. and Rabiner, "On the Relations Between Modeling Approaches for Information Sources," *Proc. IEEE Int. Conf. Acoustics, Speech, and Signal Processing*, New York, 1988, pp. 24–27.

[99] Hassanein, K., L. Deng, and M. I. Elmasry, "A Neural Predictive Hidden Markov Model for Speaker Recognition," *Proc. ESCA Workshop on Automatic Speaker Recognition, Identification and Verification*, Martigny, 1994, pp. 115–118.

[100] Bridle, J. S., "Training Stochastic Model Recognition Algorithms as Networks Can Lead to Maximum Information Estimation of Parameters," in D. S. Touretzky, ed., *Advances in Neural Information Processing Systems* 2, San Mateo, CA: Morgan Kaufman, 1990, pp. 211–217.

[101] Hwang, J. N., J. A. Vlontzos, and S. Y. Kung, "A Systolic Neural Network Architecture for Hidden Markov Models," *IEEE Trans. on Acoustics, Speech and Audio Processing*, Vol. 37, No. 12, 1989, pp. 1967–1979.

[102] Carey, M. J., E. S. Parris, and J. S. Bridle, "A Speaker Verification System Using Alpha-nets," *Proc. IEEE Int. Conf. Acoustics, Speech, and Signal Processing*, Toronto, 1991, pp. 397–400.

9

Minimum Classification Error Networks

Shigeru Katagiri

9.1 Introduction

9.1.1 Speech Pattern Recognition Using Modular Systems

Speech (pattern) recognition is a process of mapping a dynamic (variable-durational) instantiation, which belongs to one of a given set of M speech classes, to a class index; C_j ($j = 1, \cdots, M$). We especially consider the design problem of training the adjustable parameter set Ψ of a recognizer, aiming to achieve the optimal (best in recognition accuracy for all future instantiations) recognition decision status.

One of the fundamental approaches to this problem is the Bayes approach using the following Bayes decision rule, rigorous execution of which is well known to lead to this optimal status:

$$C(u_1^{T_0}) = C_i, \quad \text{iff} \quad i = \arg\max_j p(C_j \mid u_1^{T_0}) \tag{9.1}$$

where $u_1^{T_0}$ is a dynamic speech instantiation with the length T_o, $C(\cdot)$ represents the recognition operation, and it is assumed that an a posteriori probability for the dynamic instantiation, $p(C_j \mid u_1^{T_0})$, exists. A training goal in this approach is to find a state of Ψ that enables the corresponding estimate $p_\Psi(C_j \mid u_1^{T_0})$, which is a function of Ψ, to precisely approximate the true a posteriori probability (density) $p(C_j \mid u_1^{T_0})$, or in other words, to adjust Ψ so that $p_\Psi(C_j \mid u_1^{T_0})$ can approximate $p(C_j \mid u_1^{T_0})$ as precisely as possible. For example, one

307

could attempt a direct estimate of this a posteriori probability by using a sufficiently large artificial neural network (ANN), which is known to have considerable potentialities of approximation. However, to the best of this author's knowledge, there has not been a successful design example based on such an optimistic strategy. It is reckless to replace the opaque (meaning that an internal process is not explicitly described) human auditory and cognitive organs, which must actually be highly sophisticated, with an artificial system made up of only limited resources. (It can probably be assumed that the human brain contains infinite resources.) Also, it is inherently difficult to directly compute the a posteriori probability.

Actually, most speech recognizers are transparent (meaning that an internal process is explicitly described) modular systems. Such a recognizer consists of several observable modules, each carefully designed based on scientific experiences. As illustrated in Figure 9.1, a typical recognizer consists of (1) a feature extractor (feature extraction module) and (2) a classifier (classification module) that is further divided into a language model and an acoustic model. Let us represent the adjustable parameter sets of the feature extractor, the acoustic model, and the language model by Φ, Λ, and Ξ, respectively: $\Psi = \Phi \cup \Lambda \cup \Xi$.

The feature extraction module converts a speech wave sample $u_1^{T_0}$ to a dynamic pattern $x_1^T = (z_1, z_2, \cdots, z_t, \cdots, z_T)$ that is a sequence of static (fixed-dimensional) F-dimensional acoustic feature vectors, where T is the duration of the dynamic pattern and z_t is the t-th feature vector of the sequence. The feature vector is generally made up of cepstrum or bank-of-filters output coefficients, of which scientific validity is supported by psychoacoustic findings about the auditory system. The

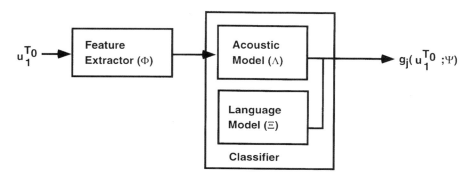

Figure 9.1 Typical structure of a modular speech recognizer.

designable parameter set, such as lifter and bank-of-filters functions, that controls the nature of the feature vectors is then Φ.

Next, the classification module assigns a class index to this converted feature pattern. This assignment is generally performed by using the classification rule

$$C(x_1^T) = C_i, \quad \text{iff} \quad i = \arg\max_j p(x_1^T \mid C_j)p(C_j) \tag{9.2}$$

which is conceptually equivalent to (9.1) but, precisely, $C(\cdot)$ is the classification operation. Note in (9.2) that in accordance with the Bayes rule of probability, the a posteriori probability is replaced by the conditional probability (density) and the a priori probability, which are both suited for the estimation based on the well-analyzed maximum likelihood (ML) method. In fact, in most cases of speech recognizer design, a target has been to achieve the accurate estimation of the probability functions, and considerable research efforts have been made to develop useful training procedures for this achievement in the ML framework. Especially, studies have focused on the probability estimation using hidden Markov model (HMM) for $p(x_1^T \mid C_j)$ and the estimation of probabilistic language models such as bigram and trigram for $p(C_j)$; the estimates of $p(x_1^T \mid C_j)$ and $p(C_j)$ are computed by $p_\Lambda(x_1^T \mid C_j)$ and $p_\Xi(C_j)$, respectively.

9.1.2 Classifier Design

In the modular recognition strategy, the quality of the feature extraction or, in other words, the suitability for the successive classification, is obviously crucial: in principle, features suited for classification can reduce the load of the post-end classifier. In fact, many studies have been done with the aim of discovering useful speech features, resulting in a noticeable family of the linear-predictive-coding-based modeling methods (e.g., [1]), the cepstrum computation, and the filter-bank spectrum computation (e.g., [2]). However, since the advent of HMM application to speech recognition and computational language modeling in the 1970's, the research focus in speech recognition has shifted to classifier design.

In the same sense as (9.1), substituting accurate estimates for the probabilities in (9.2) enables one to fundamentally achieve the optimal, minimum classification error status. However, this conventional ML-based approach actually has a basic problem in that the functional form of the class distribution (the conditional probability density) function to be estimated is in practice rarely known, and the likelihood maximization of these estimated functions, performed to model each entire class distribution individually, is not direct with regard to the minimization of classification errors.

To alleviate the difficulties of ML-based design, discriminative training has gradually come to attract research concern, specially aiming to improve the acoustic model parameter Λ.

In the discriminative training, a discriminant function $g_j(x_1^T; \Lambda)$ is introduced to measure the class membership of the input x_1^T (the degree to which x_1^T belongs to one class), where one should note that the discriminant function is a function of the classifier parameters Λ. This discriminant function does not need to be a probability function; it can be any reasonable type of measure, such as distance or similarity. In the approach, the following decision rule is used in place of (9.2);

$$C(x_1^T) = C_i, \quad \text{iff} \quad i = \arg \max_j g_j(x_1^T; \Lambda) \tag{9.3}$$

In this approach, Λ is trained in order to reduce a loss that reflects a classification error in a certain manner. Since the classification result is evaluated in the design stage, this approach is fundamentally more direct with regard to the minimization of classification errors than the ML-based approach. In fact, designs using this approach have successfully improved the classification accuracy of ML-based baseline systems in various speech classification tasks. However, there was plenty of room left for improvement in this apparently powerful discriminative training: most discriminative training methods did not prove their training optimality (being the minimum classification error status) with appropriate mathematical rigor, and they did not sufficiently cope with the variety of speech recognition task settings, such as isolated word recognition, connected word recognition, and open-vocabulary speech recognition [2].

9.1.3   What Is an Artificial Neural Network?

In the speech recognition field, HMM has been, as cited above, a main system structure of classifier, and most speech recognition researchers probably categorize HMM in a traditional system class instead of the recent class of ANN. However, as in [3, 4], HMM can naturally be considered as an ANN system. In fact, the term "ANN" can be confusing in its application, and solving this confusion is clearly recommended for clarity of discussion in this chapter as well as for further advancement of research.

Let us consider the confusion surrounding the multilayered perceptron (MLP) network, which has been a central class of network since the memorable readvent of ANNs in the 1980's [5]. Originally, the term "MLP" simply referred to the network structure, in that the network connection was based on the classical perceptron network [6]. However, due to the significant impact of the training method known as error back-propagation with the minimum squared error (MSE) objective (loss), which was tightly incorporated into the design of "MLP", the term MLP has often been used to refer only to the MLP network trained with the MSE objective. Obviously, this confusion makes it easy to forget that the fundamentals of the MSE-based discriminative training methods were fully analyzed and that the issue of training method is basically separate from the issue of system structure selection. A way to clear up the confusion is to return to the fundamentals.

Taking into account the above consideration, we define, in the following paragraphs of the chapter, ANN to be a class of system structure that consists of many simple computation units and the trainable (adjustable) interaction mechanism among the units in the parallel distributed form. Accordingly, we shall refer to the structure of most classifiers, such as the system using prototype-based distance computation [7], HMM, as well as so-called artificial neural networks (MLP and recurrent networks, etc.), as ANNs, and discuss the technical issues of speech recognition, especially addressing problems in achieving a minimum recognition error network. It may be worth while noting that the prototype-based system and the HMM belong to kernel-function-based networks such as the radial basis function network.

9.1.4 Minimum Recognition Error Network

As summarized above, most conventional ways of designing speech recognizers, based on either the ML-based classifier design or the discriminative-training-based classifier design, were unsatisfactory. The resultant classifiers could not guarantee to classify feature patterns, computed in the feature extractor, in the optimal (minimum classification error, or MCE) condition. Also, because there was no interaction between the feature extractor module and the classifier module, the features were not necessarily sufficiently useful for classification. Obviously, a novel design method was desired in order to alleviate the limitations of the conventional ways and enable one to pursue the overall optimality of a recognizer.

One solution would be to train the entire recognizer in a manner consistent with the minimization of classification errors, retaining the modular system structure. The transparent modular structure and the global adjustment aimed at the MCE condition can be practical and well-balanced complements of each other. Actually, with the aim of achieving this design strategy, a family of discriminative training called the generalized probabilistic descent method (GPD) [8] has been extensively investigated. A modular recognizer of which the trainable feature extractor and the classifier are globally trained with GPD would deserve a privileged name, i.e., minimum recognition error network. Note here that the term "minimum recognition error" refers to the condition in which the entire recognizer is trained in the sense of MCE. Actually, in the light of the noteworthy results of GPD study, we dedicate this chapter to the realization of the minimum recognition error network.

9.1.5   Chapter Organization

The chapter is organized as follows. In Section 9.2 we provide the fundamentals of the pattern classification based on discriminative training, and discuss the background of GPD development. In Section 9.3 we introduce GPD. Its formalization and mathematical nature, such as the training optimality, are described in detail. Also in Section 9.3, we summarize the concept of minimum classification error (MCE) learning, which is closely related to the GPD formalization and has a significant theoretical contribution to pattern classification study [9]. In Section 9.4 we introduce derivatives of GPD or, in other words, algorithms based on extensions of the GPD concept. The chapter is then concluded in Section 9.5. Several additional issues related to (GPD-based) minimum recognition error network implementation are finally provided in the appendixes.

9.2   Discriminative Pattern Classification

9.2.1   Bayes Decision Theory

We first summarize the Bayes decision theory, which underlies most approaches to pattern classification, including the discriminative training approach. We assume for simplicity's sake that given an observed feature pattern sample, we aim to classify it accurately. What is "accuracy" here? The meaning of this common term is not necessarily clear for a particular operation of classification. A significant contribution of the Bayes decision theory is to give accuracy a general statistics-based definition, i.e., the minimum expected loss (risk) situation, and to show that this situation can be achieved by observing the Bayes decision rule.

For the general task of classifying M-class patterns, the Bayes theory is formulated as follows (see [10] for details). According to the conventional presentation of this theoretical framework, we here assume a sample to be static (fixed-dimensional). A static feature pattern $\mathbf{x}(\in C_k)$ is given. To measure the accuracy, the theory first introduces an (individual) loss $\ell_k(C(\mathbf{x}))$ that is incurred by judging the given pattern's class C_k to be one of the M possible classes, where $C(\cdot)$ denotes the classification operation as in (9.2) and (9.3). It is obvious that the accuracy of the task should be evaluated over all of the possible samples. Thus, using statistical theory, the theory next introduces an expected loss incurred by classifying $\mathbf{x}$, called the conditional risk, as

$$L(C(\mathbf{x}) \mid \mathbf{x}) = \sum_k \ell_k(C(\mathbf{x}))1(\mathbf{x} \in C_k)P(C_k \mid \mathbf{x}) \tag{9.4}$$

and also introduces an expected loss associated with $C(\cdot)$, called the overall risk, as

$$\mathscr{L} = \int L(C(\mathbf{x}) \mid \mathbf{x})p(\mathbf{x})d\mathbf{x} \tag{9.5}$$

where $1(\mathscr{A})$ is the following indicator function

$$1(\mathscr{A}) = \begin{cases} 1 & \text{(if } \mathscr{A} \text{ is true)} \\ 0 & \text{(otherwise)} \end{cases} \tag{9.6}$$

The accuracy is accordingly defined by the overall risk. The smaller the overall risk, the better its corresponding classification result. A desirable decision is thus the one that minimizes the overall risk. Consequently, the theory leads to the following well-known Bayes decision rule: *To minimize the overall risk, compute all of the (M) possible conditional risks $L(C_i \mid \mathbf{x})$ ($i = 1, \cdots, M$) and select the C_j for which $L(C_j \mid \mathbf{x})$ is minimum.*

9.2.2 Minimum Error Rate Classification

Using the general concept of loss enables one to evaluate classification results flexibly. This is one of the big attractions of the Bayes decision theory. However, in practice, a loss that evaluates results evenly for all of the possible classes has been used widely. The most natural example of this type of loss is the following error count loss

$$\ell_k(C(\mathbf{x})) = \begin{cases} 0 & (C(\mathbf{x}) = k) \\ 1 & (\text{otherwise}) \end{cases} \tag{9.7}$$

and then its corresponding conditional risk becomes

$$L(C(\mathbf{x}) \mid \mathbf{x}) = 1 - P(C(\mathbf{x}) \mid \mathbf{x}) \tag{9.8}$$

and the overall risk becomes the average probability of error. Thus, the minimization of the overall risk based on (9.7) leads to the minimization of the average probability of error, in other words, the minimum error rate classification (e.g., [10]).

According to the above Bayes decision rule, it is obvious that the desirable classification decision is the one that minimizes the conditional risk (9.8), or maximizes the a posteriori probability $P(C(\mathbf{x}) \mid \mathbf{x})$. As far as the minimum error classification is concerned, the best classifier design is theoretically achieved by simulating (9.8) as accurately as possible. Hence this way of thinking justifies the ML-based approach of aiming at a correct estimation of the a posteriori probability, or its corresponding a priori probability and conditional probability. However, in realistic situations where only limited resources are available and the forms of the probability functions are rarely known, an accurate estimation of these probability functions is difficult and generally this approach does not achieve satisfactory design results.

9.2.3 Discriminative Training

Originally, the probability functions are given for the task at hand. In the ML-based approach, the estimation of these functions constitutes the classifier design phase. Taking into account the fact that these given, true functions are unobservable and essentially difficult to estimate accurately, one may find an alternative approach to the design; that is, one can consider that the design should involve an accurate evaluation based on (9.7) for every sample. This is the very concept of the discriminative training design. However, unfortunately, this point of view has not been explicitly indicated in most conventional discriminative training design attempts. Usually, the design formalization has started with the following decision rule, selecting arbitrary types of discriminant functions, which are not necessarily the probability functions, and also selecting some optimization criteria which are used to design these discriminant function but do not necessarily have a direct relationship to (9.7),

$$C(\mathbf{x}) = C_i, \quad \text{iff} \quad i = \arg\max_j g_j(\mathbf{x}; \Lambda) \tag{9.9}$$

The implementation of discriminative training is basically characterized by the following factors:

- Functional form of the discriminant function (classifier structure or measure);
- Design (training) objective;
- Optimization method;
- Consistency with unknown samples (robustness, generalization capability, or estimation sensitivity).

Classifier structure, which determines the type of Λ, is generally selected based on the nature of the patterns, and this selection often determines the selection of the measure or the selection of $g_j(\mathbf{x}; \Lambda)$. The linear discriminant function is a typical example of a classical classifier structure, and the measure used therein is a linearly weighted sum of input vector components. A distance classifier (distance network) that uses the distance between an input and a prototype vector as the measure is another widely used example.

The design objective is a function used for evaluating a classification result in the design stage and it corresponds to the concept of loss. Usually, an individual loss that is a design criterion for an individual design sample is first introduced; the individual loss for $\mathbf{x}$ ($\in C_k$) is denoted by $\ell_k(\mathbf{x}; \Lambda)$. It should be noted here that the individual loss used for the design phase is a function of Λ. As discussed in the previous section, a natural form of the individual loss is the classification error count as:

$$\ell_k(\mathbf{x}; \Lambda) = \begin{cases} 0 & (C(\mathbf{x}) = k) \\ 1 & (\text{otherwise}) \end{cases} \tag{9.10}$$

Next, similar to the overall risk, an ideal overall loss, i.e., the expected loss, is defined by using the individual loss as

$$L(\Lambda) = \sum_k P(C_k) \int \ell_k(\mathbf{x}; \Lambda) p(\mathbf{x}|C_k) d\mathbf{x} \qquad (9.11)$$

However, since sample distributions are essentially unknown, it is impossible to use this expected loss in practice. For this reason, an empirical average loss, defined in the following, is usually used;

$$L_0(\Lambda) = \frac{1}{N} \sum_k \sum_n \ell_k(\mathbf{x}_n; \Lambda) 1(\mathbf{x}_n \in C_k) \qquad (9.12)$$

where n of $\mathbf{x}_n$ explicitly means that the sample $\mathbf{x}_n$ is the n-th sample of the finite (consisting of N samples) design sample set $X = \{\mathbf{x}_1, \ldots, \mathbf{x}_n, \ldots, \mathbf{x}_N\}$. In the case of using the error count loss of (9.10), this empirical average loss becomes the total count of classification errors measured over X.

   In addition to (9.10), several different types of loss forms have been used; e.g., perceptron loss, squared error loss (between a discriminant function and its corresponding supervising signal), and mutual information. However, these are suboptimal selections, and it is known that the design results obtained by using them are not necessarily consistent with the minimum classification error probability situation (e.g., see [10]).

   Optimization is the process of finding in a competitive manner with regard to classes the state of Λ that minimizes the loss over X, and its embodiments can be grouped into heuristic methods and mathematically proven algorithms. They are also categorized from another point of view; i.e., the batch type search versus the adaptive (sequential) search.[1] Among many approaches such as error correction training, stochastic approximation, and simulated annealing, the methods based on gradient search over the average empirical loss surface have been widely used due to their practicality and mathematical validity [6–7, 10–11].

   Originally, the purpose of design is to realize the state of Λ that leads to accurate classification over all of the samples of the task at hand, not just the given design samples. It should be recalled that the expected

---

1. The term "adaptive" means a procedure in which training adjustment is performed every time one design sample is presented.

loss was defined as an ideal overall loss. Thus it is desirable that the design result obtained by using design samples be consistent with unknown samples too. Obviously, a pursuit of this nature requires some assumptions concerning unobservable unknown samples, or the entire sample distribution of the task. Some additional information is needed that brings (9.12) closer to (9.11). In contrast with ML-based modeling, which introduces a parametric probability function in order to estimate the overall sample distribution, measures to increase the consistency in the discriminative training are generally moderate: the loss is individually defined for every design sample, and the design is fundamentally performed over the given design samples, as shown in the use of the average empirical loss (9.12). In many cases of discriminative training, consistency is merely a result of the selection of the discriminant function or the measure.

We have considered the classification of static patterns in this section. The issues discussed above also hold true in the classification of dynamic patterns.

9.3 Generalized Probabilistic Descent Method

9.3.1 Overview

The recent discriminative training methods led to a number of successful results [12–14]. However, the resulting recognizers were not necessarily satisfactory. In particular, they sometimes resulted in less robust recognition; i.e., they generalized poorly to testing data, while achieving rather high accuracy over training data. Main possible causes of this are summarized as follows:

- Empirical rules such as corrective training [6, 15] and learning vector quantization (LVQ) [16] did not have enough of a mathematical basis to guarantee design optimality.

- As is well known, the MSE-based discriminative training, which is generally used for the MLP-based classifiers, is not necessarily equivalent to the minimization of misclassifications [10].

- The maximization-of-mutual-information training does not necessarily imply the minimization of misclassifications [17].

- Efforts were much too limited to acoustic modeling, often to the modeling of static acoustic feature vectors, and therefore lacked the global scope of designing an overall process of recognizing dynamic speech wave patterns.

Therefore, there was a clear need to alleviate these problems in the discriminative training approach.

GPD was originally developed as a renewed version of the classical, adaptive discriminant function design method called the probabilistic descent method (PDM) [18]. A key point in this revision is to overcome the unsmoothness problem, or the lack of smoothness of the original PDM formalization ("smoothness" will be defined in Section 9.2.2) [8, 19–21]. To achieve this, GPD uses the L_p norm form and a sigmoidal function specifically for approximating the classification error count loss of (9.10). In the beginning, the GPD research focus was on analyzing implementational possibilities and investigating the relationship between GPD and LVQ, which was developed for the self-organizing feature map network [4]. After [9], however, the focus was moved to investigating the significance of using the proposed smooth GPD formalization with the smooth error count loss. It was shown that in principle, the GPD design using the smoothed error count loss enabled one to arbitrarily approximate the optimal, minimum classification error probability situation [9]. Then, in deference to this important finding, the GPD came to be sometimes called the minimum classification error/generalized probabilistic descent (MCE/GPD) method [22].

The GPD has been extensively applied to various design experiments for speech pattern classification (e.g., [21, 23–26]). The superiority of this method, which guarantees the theoretical validity of the training target, is being proven experimentally too. Recent success in large-scale, real-world applications provides clear evidence of the promising nature of this rather new design method [27].

Since the GPD originates in PDM, most of these implementations adaptively update the classifier parameters. However, the formalization concept of GPD can apply to batch-type updating, such as the steepest descent method, without any loss of its mathematical rigor. Actually, several examples such as [26] fully achieved the design goal of the GPD by using the standard steepest descent optimization method.

As said before, a recognizer usually consists of a feature extractor module and a classifier module, and this classifier is further divided into an acoustic model module and a language model module. Obviously, all of these modules are jointly concerned in the final classification decision. Thus, in principle, all of these modules should be jointly optimized so as to increase classification accuracy, or in other words, to minimize classification errors. According to this understanding, the GPD was extended as discriminative feature extraction (DFE) which trains both the front-end feature extractor and the post-end classifier consistently under a single design objective of minimizing classification errors [22, 28, 29]. A key point in this extension is to use the chain rule of differential calculus. The trainable parameters of the feature extraction module are updated by using the classification result information back-propagated from the classifier module. In the development of DFE, the design focus was extended to the feature extractor, which affects the difficulty (ease) of the post-end classification decision. One may easily note here that the same extension can be directed to the language model and other modules, if there are any. Actually, such extension for the language module was proposed in parallel with development of DFE [30, 31].

In addition to many experimental evaluation reports in the literature, both the GPD and the DFE are being further evolved. For example, the GPD has been successfully reformulated for the special recognition environment called keyword spotting [32, 33], and the DFE has been studied from the viewpoint of the subspace method [34]; also, the DFE has been further revised as the discriminative metric design (DMD) [35]. Therefore, the GPD, which was originally developed for classifier design, is now forming a new, large family of discriminative design methods for pattern recognition.

In the successive subsections, we present the essense of the GPD-based training methodology by selectively introducing implementation examples and several important topics.

9.3.2 Formalization Fundamentals

9.3.2.1 Distance Classifier for Classifying Dynamic Patterns: Preparation

GPD can basically be implemented for any reasonable system structure. To introduce the formalization fundamentals of GPD, we use a distance classifier according to the original presentation in [8, 19, 20, 36].

Let us assume that an input speech wave sample is converted to its corresponding feature vector sequence pattern;

$$x_1^T = f(u_1^{T_0}) = (\mathbf{z}_1, \ldots, \mathbf{z}_\tau, \ldots, \mathbf{z}_T) \in \underbrace{\Re^F \times \cdots \times \Re^F}_{T} \tag{9.13}$$

where $f(\cdot)$ is a feature extraction function. A distance classifier consists of trainable prototypes as

$$\Lambda = \{\lambda_j\} = \{\{r_j^b\}_{b=1}^{b=B_j}\}_{j=1}^{j=M}$$

(9.14)

where r_j^b is C_j's b-th finite-length dynamic prototype, which is defined in the same sample space with that of x_1^T. A classification decision is made by comparing an input (feature vector sequence) pattern and each prototype pattern in some distance measurement. The most natural intuition here is to use the decision rule as

$$C(x_1^T) = C_i, \quad \text{iff} \quad i = \arg\min_j g_j(x_1^T; \Lambda)$$

(9.15)

which is essentially the same as (9.9), assuming that the pattern is dynamic and also the discriminant function $g_j(x_1^T; \Lambda)$ is a class-oriented distance measure between x_1^T and the set of C_j's prototypes $\{r_j^b\}_{b=1}^{b=B_j}$. There are many possibilities for defining the class-oriented distance measure. A typical case is based on the nearest-neighbor concept and given as follows. First, a dynamic programming (DP) matching path and its corresponding path distance are introduced between an input pattern and each prototype. The path distance is given as

$$D_\theta(x_1^T, r_j^b) = \sum_{\tau=1}^{T} w_{j,\tau}^b \delta_{\varpi(j,b,\tau,\theta)}$$

$$= \|\mathbf{z}_{\varpi(j,b,\tau,\theta)} - \mathbf{r}_{j,\tau}^b\|^2$$

(9.16)

where $\mathbf{z}_{\varpi(j,b,\tau,\theta)}$ is the $\varpi(j,b,\tau,\theta)$-th component vector of x_1^T, to which the τ-th component vector of r_j^b, $\mathbf{r}_{j,\tau}^b$, corresponds along the θ-th matching path of r_j^b; $\delta_{\varpi(j,b,\tau,\theta)}$ is a local distance, defined by the squared Euclidean distance between these corresponding component vectors; $w_{j,\tau}^b$ is a weight coefficient. Next, using the path distances, the distance between an input and each prototype, i.e., a prototype distance, is defined as the smallest (best) path distance for each prototype:

$$d_A(x_1^T, r_j^b) = \min_\theta \{D_\theta(x_1^T, r_j^b)\}$$

(9.17)

where operation "min" is executed by DP matching. Lastly, the discriminant function that represents the degree to which an input belongs to each class is defined as the prototype distance of the closest (in the sense of (9.17)) prototype to the input:

$$g_j(x_1^T; \Lambda) = d_A(x_1^T, r_j^{c_j}), \quad \text{where } c_j = \arg\min_b d_A(x_1^T, r_j^b) \qquad (9.18)$$

Accordingly, based on the rule (9.15), the classifier classifies an input to the class having the smallest prototype pattern distance among all the $(B_1 + B_2 + \cdots + B_M)$ possible prototype distances.

9.3.2.2 Emulation of Decision Process

The design target for the above classifier is to find the optimal status of Λ. According to the discriminative training concept, one may attempt to achieve the optimal situation by evaluating the loss for each training input pattern. Given a training sample, its corresponding loss, such as the perceptron loss and the squared error loss, would be computed and the resulting adjustment amount would be fed back to the updating of the prototypes. Apparently, such an approach works in practice. However, it actually suffers from a serious mathematical problem; i.e., the computation of the adjustment amount is indispensably based on the gradient calculation of the loss, and the loss for the above classifier is not differentiable in terms of the trainable parameters, the prototypes: the "min" operation included in the discriminant function (see (9.17) and (9.18)) is not smooth, or in other words, at least first-differentiable with respect to the prototypes, and consequently the loss is not smooth either. From the viewpoint of mathematical rigor, a desirable formalization of the training method should obviously overcome this unsmoothness problem.

To solve the above problem, GPD makes use of the smooth L_p norm function and the smooth sigmoidal function in its formulation. The method first replaces the discriminant function (9.18) by

$$g_j(x_1^T; \Lambda) = \left[\sum_{b=1}^{B_j} \left\{ D(x_1^T, r_j^b) \right\}^{-\zeta} \right]^{-1/\zeta} \qquad (9.19)$$

where ζ is a positive constant, and $D(x_1^T, r_j^b)$ is a generalized prototype distance between x_1^T and r_j^b, defined as

$$D(x_1^T, r_j^b) = \left[\sum_{\theta=1}^{\Theta_j^b} \left\{ D_\theta(x_1^T, r_j^b) \right\}^{-\xi} \right]^{-1/\xi} \qquad (9.20)$$

where ξ is also a positive constant and Θ_j^b is the number of all the possible DP-matching paths due to r_j^b.

The most important development concept of GPD is to embed the entire classification decision process in a functional-form training procedure. To do this, GPD introduces a smooth misclassification measure as follows: For $x_1^T \in C_k$,

$$d_k(x_1^T; \Lambda) = g_k(x_1^T; \Lambda) - \left[\frac{1}{M-1} \sum_{j, j \neq k} \{ g_j(x_1^T; \Lambda) \}^{-\mu} \right]^{-1/\mu} \quad (9.21)$$

where μ is a positive constant. Clearly, this misclassification measure represents the decision operation in scalar value computation; i.e., $d_k(\ ) > 0$ emulates a misclassification, and $d_k(\ ) < 0$ emulates a correct classification. Then, the decision result can be directly evaluated by embedding the misclassification measure in a loss as

$$\ell_k(x_1^T; \Lambda) = l_k\big(d_k(x_1^T; \Lambda)\big) \quad (9.22)$$

where $l_k(\cdot)$ is a smooth, monotonically-increasing function of the misclassification. The individual losses, each for one design sample, should be reduced with some optimization method. To do this, GPD uses the probabilistic descent theorem [9, 18].

Theorem 1. (Probabilistic descent theorem)

Assume that a given design sample $x_1^T(t)$ belongs to C_k. If the classifier parameter adjustment

$\delta\Lambda(x_1^T(t), C_k, \Lambda)$ is specified by

$$\delta\Lambda(x_1^T(t), C_k, \Lambda) = -\epsilon U \nabla \ell_k(x_1^T(t); \Lambda) \quad (9.23)$$

where U is a positive-definite matrix and ϵ is a small positive real number, then

$$E[\delta L(\Lambda)] \leq 0 \quad (9.24)$$

Furthermore, if an infinite sequence of randomly selected samples $\mathbf{x}_t$ is used for learning (designing) and the adjustment rule of (9.23) is utilized with a corresponding (learning) weight sequence $\epsilon(t)$ which satisfies

$$\sum_{t=1}^{\infty} \epsilon(t) \to \infty, \quad \text{and} \quad \sum_{t=1}^{\infty} \epsilon(t)^2 < \infty, \quad (9.25)$$

then the parameter sequence $\Lambda(t)$ (the state of Λ at t) according to

$$\Lambda(t+1) = \Lambda(t) + \delta\Lambda(x_1^T(t), C_k, \Lambda(t)) \qquad (9.26)$$

converges with probability one (1) at least to Λ^* which results in a local minimum of $L(\Lambda)$.

Since all of the elemental functions, such as the discriminant function and the misclassification measure, are smooth, the adjustment based on (9.23) can be rigorously applied to the trainable parameters, i.e., the prototypes. The resulting adjustment rule is accordingly given as

$$r_{j,\tau}^b(t+1) = \begin{cases} r_{j,\tau}^b(t) + 2\epsilon(t)\nu_k w_{j,\tau}^b \phi_j \rho_j \varphi_j & \text{(for } j = k) \\[2mm] r_{j,\tau}^b(t) - \dfrac{2}{M-1}\epsilon(t)\nu_k w_{j,\tau}^b \sigma_j \phi_j \rho_j \varphi_j & \text{(for } j \neq k) \end{cases} \qquad (9.27)$$

where

$$\nu_k = \alpha l_k(d(_k(x_1^T(t); \Lambda))\{1 - l_k(d(_k(x_1^T(t); \Lambda))\} \qquad (9.28)$$

$$\phi_j = \left[\sum_{b'=1}^{B_j} \left\{ \frac{D(x_1^T(t), r_j^b)}{D(x_1^T(t), r_j^{b'})} \right\}^\zeta \right]^{-(1+\zeta)/\zeta} \qquad (9.29)$$

$$\rho_j = \left[\sum_{\theta=1}^{\Theta_1^b} \{D_\theta(x_1^T(t), r_j^b)\}^{-\xi} \right]^{-(1+\xi)/\xi} \qquad (9.30)$$

$$\varphi = \sum_{\theta=1}^{\Theta_j^b} \frac{z_{\varpi(j,b,\tau,\theta)}(t) - r_{j,\tau}^b}{\{D_\theta(x_1^T(t), r_j^b)\}^{\xi+1}} \qquad (9.31)$$

$$\sigma_j = \left[\frac{1}{M-1} \sum_{j,j\neq k}^{M} \left\{ \frac{g_k(x_1^T(t); \Lambda)}{g_j(x_1^T(t); \Lambda)} \right\}^\mu \right]^{-(1+\mu)/\mu} \qquad (9.32)$$

and $\varpi(j, b, \tau, \theta)$ indicates the component vector index of $x_1^T(t)$, to which the τ-th component vector of r_j^b, $r_{j,\tau}^b$, corresponds along the θ-th matching path of r_j^b. In (9.27), the adjustment is done for all of the possible paths and all of the possible prototypes. This is markedly distinct from the conventional distance classifier in which the adjustment is selectively done with the "min" operations.

Treating $w_{j,\tau}^b$'s as adjustable parameters, one can achieve an adjustment rule similar to (9.27), though we omit the result. See this point in [20].

The use of the L_p norm form affords one an interesting flexibility in the implementation. Let ζ and ξ approach ∞ in (9.19) and (9.20). Then, clearly, (9.19) approximates the operation of searching for the closest prototype, and (9.20) for the operation of searching for the best matching path. Also, controlling μ of (9.21) enables one to simulate various decision rules. In particular, when μ approaches ∞, (9.21) comes to resemble rule (9.15).

9.3.2.3   Selection of Loss Functions

The formalization described so far has not specified the type of loss function, such as the classification error count loss or the squared error loss. Among the many possible selections of loss forms, GPD often uses a smooth classification error count loss as

$$\ell_k(x_1^T, \Lambda) = l_k(d_k(x_1^T; \Lambda)) = \frac{1}{1 + e^{-(\alpha d_k(x_1^T;\Lambda)+\beta)}} \quad (\alpha > 0) \qquad (9.33)$$

where α and β are constants, and this special selection plays a significant role in the analytic study of the discriminative training and also in the practical use of GPD. Actually, one may note that the infinite repetition of GPD adjustment using (9.33) leads at least to the local minimum status of the expected smooth error count loss in the probabilistic descent sense and this status is linked to the ideal, minimum error probability situation. In this section, we briefly describe this point, according to the result in [9].

Let us assume that the probability measure of a dynamic (feature vector sequence) pattern can be computed appropriately through the probability computation using an HMM network and also that the related conditional probability and the joint probability are properly defined. Then, the discriminant function is given as

$$g_j(x_1^T; \Lambda) = p_\Lambda(C_j \mid x_1^T) \qquad (9.34)$$

where $p_\Lambda(C_j \mid x_1^T)$ is an estimate of the a posteriori probability.

For simplicity, we first assume that the true functional form, determined by $\dot{\Lambda}$, of the probability is known. Then, defining the misclassification measure as

$$d_k(x_1^T; \Lambda) = -g_k(x_1^T; \Lambda) + \left[\frac{1}{M-1} \sum_{j, j \neq k} \{g_j(x_1^T; \Lambda)\}^\mu \right]^{1/\mu} \quad (9.35)$$

for example, we can rewrite the expected loss, defined by using the smooth classification error count loss (9.33) in the GPD formalization, as follows:

$$L(\dot{\Lambda}) = \sum_k \int_\Omega p(x_1^T, C_k) \ell_k(x_1^T; \dot{\Lambda}) 1(x_1^T \in C_k) dx_1^T$$

$$\simeq \sum_k \int_\Omega p(x_1^T, C_k) 1(x_1^T \in C_k)$$

$$1\big(p_{\dot{\Lambda}}(C_k \mid x_1^T) \neq \max_j p_{\dot{\Lambda}}(C_k \mid x_1^T)\big) dx_1^T, \quad (9.36)$$

where Ω is the entire sample space of the dynamic patterns x_1^T's ($T < T_{\max}$), and it is assumed that $dp(x_1^T, C_k) = p(x_1^T, C_k) dx_1^T$. Controlling the smoothness of functions such as L_p norm and the sigmoidal function, one can arbitrarily increase the approximation accuracy of the last equation to $L(\dot{\Lambda})$ in (9.36). Note here that we use $\dot{\Lambda}$. Based on this fact, the status of $\dot{\Lambda}$ that corresponds to the minimum situation of $L(\dot{\Lambda})$ in (9.36), which is achieved by adjusting $\dot{\Lambda}$, is clearly equal to the Λ^* that corresponds to a true probability, or in other words, achieves the maximum a posteriori probability situation. Accordingly, it turns out that the minimum situation of $L(\dot{\Lambda})$ can get indefinitely close to the minimum classification error probability

$$\mathcal{E} = \sum_k \int_{\Omega_k} p_{\dot{\Lambda}^*}(x_1^T, C_k) 1(x_1^T \in C_k) dx_1^T \quad (9.37)$$

where Ω_k is a partial space of Ω that causes a classification error according to the maximum a posteriori probability rule, i.e.,

$$\Omega_k = \{x_1^T \in \Omega \mid p_{\dot{\Lambda}^*}(C_k \mid x_1^T) \neq \max_j p_{\dot{\Lambda}^*}(C_k \mid x_1^T)\} \quad (9.38)$$

This result may sound quite natural and trivial. However it seems, the result was the first message that showed that one could achieve the minimum classification error probability situation through discriminative training, which had long been considered a deterministic and empirical training framework.

The assumption that $\dot{\Lambda}$ is known is obviously impractical. However, recent results concerning the approximation capability of ANN and Gaussian kernel function have provided useful suggestions for studying this inadequacy. Actually, based on the recent results such as [37, 38], one would be able to argue that an HMM with sufficient adjustable parameters has the fundamental capability of modeling a (unknown) true probability function. If $L(\Lambda)$ (and $\mathcal{E}$) has a unique minimum situation, and if Λ and $L(\Lambda)$ (and $\mathcal{E}$) are monotonic to each other, then the minimum situation corresponds to the case in which $g_j(x_1^T; \Lambda)$ is equal to the true probability function. It thus turns out that under this assumption, which is softer (more realistic) than that in the above paragraph, even if the true parametric form of the probability function is unknown, the GPD-based discriminative training enables one to fundamentally achieve the minimum classification error probability situation.

As above, the GPD training using the smooth classification error count loss, i.e., MCE/GPD, successfully provided the fundamental link between discriminative training and the optimal Bayes situation for classification. The effect of the MCE/GPD training will be given in the next subsection.

9.3.2.4 Design Optimality in Practical Situations

In realistic situations in which only finite design samples are available, the state of Λ that can be achieved by the probabilistic descent training is at most a locally optimal situation over a set of design samples. Moreover, in the case of finite training repetitions, Λ does not necessarily achieve even the local optimum solution. However, the training algorithms of GPD, such as (9.27), have been shown to be quite useful even in these realistic settings.

The discussions in the previous subsection assumed the impractical condition that design samples and classifier parameters are sufficiently available. However, it has also been clear that MCE/GPD possesses a high degree of utility and originality in the practical circumstances in which only finite resources are available. In fact, the more limited the design resources, the more distinctive from other methods is the MCE/GPD result. A reason for this practical utility is that MCE/GPD always directly pursues the minimum classification error situation, conditioned by given circumstances, while the ML-based approach aims at estimating the entire probability function; moreover, the conventional, discriminative training methods aim at minimizing the average empirical loss that is not necessarily consistent with the classification error count.

The smoothness incorporated in the GPD formalization also contributes toward increasing the practical utility by substantially increasing the number of training samples and making softer the empirical average loss. This point is related to the global search problem of optimal loss status and the increase of training robustness. Details about the point are described in [29].

9.3.3 GPD-Based Classifier Design

We report on the utility of GPD by summarizing the experimental results of training speech pattern classifiers with MCE/GPD in two tasks: (1) classifying 9-class spoken American E-rhyme letters (*E*-set task) and (2) classifying 41-class spoken Japanese phonemes (*P*-set task). In the experiments, the multiprototype distance classifier, represented in (9.14), was used. The prototypes were first initialized by using the conventional, modified *k*-means clustering, and then were adjusted based on the rules that were basically the same as (9.27).

9.3.3.1 *E*-Set Task

The *E*-set data consisted of the 9-class *E*-rhyme letter syllables: {b, c, d, e, g, p, t, v, z}. Each sample was recorded over dialup telephone lines from one hundred (50 male and 50 female) untrained speakers. Speaking was done in the isolated-word mode. Since all of the samples included the common phoneme {e} and they were recorded over telephone lines, this task was intrinsically rather difficult. Actually it has been reported that the achievable conventional recognition rate in this task was usually slightly higher than 60% and at most 70% by using a larger-size HMM recognizer. Recognition experiments were done in the multispeaker mode; i.e., each speaker uttered each of the *E*-set syllables twice, once for designing and once for testing. Thus, for every class, the design and testing data sets consisted of 100 samples, respectively.

The recognizer possesses several factors, such as the number of prototypes, that may affect its recognition capability. To investigate these factors thoroughly, this task was carefully tested by several separate research groups. We summarize here the results of these separately conducted experiments [19–21].

In [21], the modified k-means clustering results ranged from 55.0% to 59.8% in the case of using one prototype for every class. For this small-size classifier, MCE/GPD successfully achieved rates ranging from 74.2% to 75.4%. In the case of using three prototyes for every class, the modified k-means clustering resulted in the range of 64.1% to 64.9%, with 74.0% to 77.2% for MCE/GPD. Again, the superiority of MCE/GPD is clearly demonstrated in this large-size case.

Some studies [19] and [20] investigated an implementation somewhat different from (9.27), using an exponential-form distance measure and also considering the weights, $w_{j,\tau}^b$'s, to be adjustable. Consequently, in this case too, MCE/GPD increased about 60% of the modified k-means clustering with the use of only one prototype per class to 79.4%, and also, remarkably, reached 84.4% by using four prototypes for every class.

9.3.3.2   *P*-Set Task

The data of this task consisted of 41-class phoneme segments included in the ATR speech database, which has been widely used as a standard large-scale Japanese speech database. Each phoneme sample was extracted from 5,240 common words, spoken by a male speaker, by using manually-set acoustic-phonetic labels (about 26,000 samples in total). This sample set was split into two independent sets of roughly equal sizes: one for designing and one for testing. The experiments were conducted in the speaker-dependent mode.

According to [21], the MCE/GPD-trained classifier that used five prototypes per class (especially ten for every vowel class) achieved the accuracy of 96.2%. Here too, compared with the 86.8% for the segmental k-means clustering that was used for system initialization, we can clearly see the utility of MCE/GPD.

9.4   Derivatives of GPD

9.4.1   Overview

The key concept of GPD formalization is to emulate the overall process of a task at hand in a tractable functional form. Actually, the implementation of MCE/GPD presented in Section 9.3 properly realized this concept for the task of classifying a dynamic speech pattern. However, in the example task, the pattern to be classified was a priori extracted from its surrounding input signal and was represented in the preset form of an acoustic feature vector sequence. The task defined therein is one simplified and limited case among many. Clearly, the design concept of GPD should be applied directly to more complicated and realistic task situations. For example, the concept of GPD development should be applied to either a connected word recognition task or an open-vocabulary speech recognition task. It should also be applied to the entire process of recognition, which includes the spotting (detection) of target speech sounds, the design of the feature extraction parameters (Φ), and the design of the language model (Ξ). In this light GPD has actually been quite extensively studied, and its original formalization for classification has been dramatically extended to a new family of discriminative design methods, which are more suitable for handling complex real-world tasks. In the following pages of this section, we shall summarize several important members of the GPD family, such as segmental GPD, minimum spotting error learning (MSPE) [32], discriminative utterance verification [39], discriminative feature extraction (DFE), e.g., [22, 40], and also introduce application examples of GPD to speaker recognition [41].

9.4.2 Segmental GPD for Continuous Speech Recognition

Most definitions used in the simple example task of classifying spoken syllables, in Sections 9.3, can be applied to a more realistic task of classifying continuous speech utterances. The most important issue in this advanced application is still to embed the entire process of classification in an appropriate GPD-based functional form. For example, for a connected word recognition task, a discriminant function should be defined so as to directly measure the degree to which a connected word sample belongs to its corresponding class. Due to resource restrictions, most continuous-speech recognizers employ sub-word models, such as phonemes, as the basic unit of acoustic modeling, and represent words and connected words by concatenating the sub-word models. The classification process in this task consequently includes the mechanism of dividing a long input of continuous-speech into short subword segments. Also, due to the need for tackling the statistical variation of speech sounds, most present continuous speech recognizers use HMM-based acoustic models. Efforts of applying GPD to continuous speech recognition must appropriately take into account these two requirements, i.e., the segmentation mechanism and the probabilistic modeling. Indeed, the development of segmental GPD was motivated by these concerns [23, 42].

Assume the use of a modular connected word recognizer that

consists of a feature extraction module and a classification module. Then, consider $x_1^T = (z_1, z_2, \cdots, z_T)$ to be speech input to the classifier, where z_t is an acoustic feature vector (observation) at time index t. Also let $W = (w_1, w_2, \cdots, w_S)$ be a word sequence that usually constitutes a sentence. Generally, an HMM classifier for x_1^T is defined by using a first-order S-state Markov chain governed by the following manifolds:

1. A state transition probability matrix $A = [a_{\iota\kappa}]$, where $a_{\iota\kappa}$ is the probability of making a transition from state ι to state κ;

2. An initial state probability $\pi_\iota = P(q_0 = \iota)$, which specifies the state of the system q_0 at time index $t = 0$;

3. An observation emission probability according to a distribution $b_{q_t}(z_t) = P(z_t \mid q_t), q_t = 1, 2, \cdots, S$, which is usually defined with a mixtured Gaussian distribution.

Accordingly, assuming that Λ is a set of the above probability parameters, i.e., the state transition probabilities, the initial state probabilities, and the observation emission probabilities, the discriminant function for x_1^T is given as the following likelihood measure

$$g_{W_r}(x_1^T; \Lambda) = \log P(x_1^T, q_{W_r}, W_r \mid \Lambda) \qquad (9.39)$$

where

$$W_r = \arg\max_{W \neq W_1, \cdots, W_{r-1}} P(x_1^T, q_{W_r}, W_r \mid \Lambda)$$

$$= r\text{th best word sequence,}$$

$$q_{W_r} = \text{best state sequence corresponding to Wr} \qquad (9.40)$$

and $P(x_1^T, q_{W_r}, W_r \mid \Lambda)$ is the joint state-word sequence likelihood.

The goal of training is to find an optimal Λ that leads to accurate classification of word sequences. Given $x_1^T \in W_k$ for training, segmental GPD defines the misclassification measure as

$$d_{W_r}(x_1^T; \Lambda) = -g_{W_k}(x_1^T; \Lambda) + \log\left\{\frac{1}{r_m}\sum_{r=1}^{r_m} e^{g_{W_r}(x_1^T; \Lambda)\cdot\eta}\right\}^{1/\eta} \qquad (9.41)$$

where r_m is the total number of the competing word sequences, different from W_k, that will be taken into consideration in training. The training procedure of segmental GPD is then completed by embedding this misclassification measure in the smooth error count loss, as in Section 9.3.

Clearly, the Markov modeling satisfies the above two requirements for continuous speech recognition: it performs the segmentation process with its state transition mechanism and it provides a probabilistic representation framework. An important point to note here is that an implementation with probabilistic models should maintain the probabilistic constraint of the model parameters, i.e., the sum of the probabilistic parameters, such as the state transition probability, over all the possible cases should be equal to one. Segmental GPD thus uses parameter transformation as

$$a_{ij} \rightarrow \ddot{a}_{ij} \quad \text{where } a_{ij} = \frac{e^{\ddot{a}_{ij}}}{\sum_{m} e^{\ddot{a}_{im}}} \tag{9.42}$$

in the parameter adaptation.

The power of segmental GPD has been demonstrated in several experimental tasks. For example, [23] reported quite successful recognition results in the 9-class American English E-rhyme task, used in Section 9.3; i.e., an HMM recognizer with 10-state, 5-component mixtured Gaussian(/state) models scored 99% over the training data and 88% over the testing data. For the connected word case, [43] and [44] reported results in the TI connected-digit recognition task. In particular, in [44] the best results reported so far on the database (a string error rate of 0.72% and a word error rate of 0.24% on testing data) were successfully achieved by a segmental-GPD-trained context-dependent subword model system.

The same design concept employed in segmental GPD, i.e., optimizing to increase recognition accuracy for concatenated word inputs, has been tested in several slightly different ways and has further demonstrated its high utility [25–26].

9.4.3 Minimum Error Training for Open-Vocabulary Speech Recognition

9.4.3.1 Open-Vocabulary Speech Recognition

Usually, spontaneous conversation utterances contain speech segments that are not directly relevant to tasks at hand, such as interjections and repairs. Also, it is not very realistic to attempt to fully model large-vocabulary words and recognize a tremendous number of sentences that are generated by concatenating these words. Therefore,

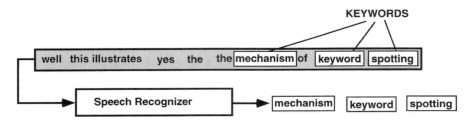

Figure 9.2 Mechanism of open-vocabulary speech recognition.

it is recommendable that a practical speech recognizer employ an open-vocabulary paradigm in which only selected keywords are to be recognized. Figure 9.2 illustrates the mechanism of open-vocabulary speech recognition.

There are two main approaches to open-vocabulary recognition [32, 45]. The first is keyword spotting based on a threshold comparison. A system prepares keyword models, computes the similarity between a segment of input utterance and each keyword model, and decides whether the segment contains the keyword or not by comparing a similarity value and a preset threshold. The second is to use a continuous speech recognizer having a "filler" model. In this case, the recognizer attempts to segment and classify an input utterance in the same manner as conventional speech recognizers for isolated or concatenated spoken words do, but here a nonkeyword segment is classified as the class corresponding to the filler model. There are clear differences between the two approaches, especially in terms of segmentation strategy and implementation approach. However, the approaches are still closely linked, because the similarity between the filler class and keyword classes works as a kind of thresholding operation. Evaluation of these approaches is an important ongoing research issue in the speech recognition field.

9.4.3.2 Mimimum Spotting Error Learning

Depending on the decision strategy used for spotting, one can clearly note that spotting performance relies heavily upon the adequacy of the model and the threshold. Conventionally, the model has been designed using ML training, as is done for standard continuous speech recognition, and the threshold has been simply determined in a trial-and-error fashion. Obviously, such an unintegrated design method does not guarantee optimal spotting accuracy. One natural solution to this problem is to design both the model and the threshold jointly in the sense of minimizing spotting errors. In this light, GPD was reformulated as minimum spotting error learning (MSPE) for keyword spotting in [32].

Since spotting can be done in keyword-by-keyword mode, or in other words independently for each keyword class, in the algorithm described here we will consider a simple task with only one keyword class and thus a spotter (spotting system) that contains a single keyword model λ (such as prototypes or HMMs) and its threshold b. The goal of MSPE design is to optimize $\Lambda = \{\lambda, b\}$. For clarity of description, we assume λ to be a prototype-based keyword model, i.e., a sequence of prototypes, each defined in the acoustic feature vector space. A spotting decision is fundamentally done at every time index. Therefore, a discriminant function $g_t(x_1^T, S_t; \lambda)$ is defined as a function that measures a distance between a selected segment S_t of input utterance x_1^T and the model λ, and the spotting decision rule is formalized as follows: If the discriminant function meets

$$g_t(x_1^T, S_t; \lambda) < b \tag{9.43}$$

then the spotter judges at "t" that a keyword exists in the segment S_t; no keyword is spotted otherwise. Importantly here, this type of decision could produce in principle two types of spotting errors: false detection (the spotter decides that S_t does not include the keyword when S_t actually includes it) and false alarm (the spotter decides that S_t includes the keyword even when S_t does not include it).

Similar to original GPD, the MSPE formalization aims to embed the above cited spotting decision process in an optimizable functional form and provide a concrete algorithm of optimization so that one can consequently reduce the spotting errors. The spotting decision process is then emulated as spotting measure $d_t(X; \Lambda)$, which is defined as

$$d_t(X; \Lambda) = b - \ln\left\{\frac{1}{|I_t|} \sum_{s \in I_t} \exp(-\xi g_s(X; \Lambda))\right\}^{-1/\xi} \tag{9.44}$$

where I_t is a short segment, which is set for increasing the reliability and stability of the decision, around the time position t, $|I_t|$ is the size of I_t (the number of acoustic feature frames in I_t), and ξ is a positive constant. A positive value of $d_t(X; \Lambda)$ implies that at least one keyword exists in I_t and a negative value implies that no keyword exists in I_t. Decision results are evaluated by using a loss function that is defined by using two types of smoothed 0−1 functions as:

$$\ell_t(X;\Lambda) = \begin{cases} \acute{\ell}(d_t(X;\Lambda)) & \text{(if } I_t \text{ includes a training keyword)} \\ \gamma\hat{\ell}(d_t(X;\Lambda)) & \text{(if } I_t \text{ includes no training keyword)} \end{cases}$$

(9.45)

$\ell_t(\cdot)$ approximates (1) one for one false detection, (2) γ for one false alarm, and (3) zero for correct spotting, where γ is a parameter controlling the characteristics of the spotting decision. By setting γ to one, one can treat all of the errors evenly. By letting γ be less than one, one can tune the system toward reduction of false-detection errors, which are often more troublesome than false-alarm errors in many application situations.

The training procedure of MSPE is accordingly obtained by applying the adjustment rule (9.26) of the probabilistic descent theorem to the above defined functions, i.e., the loss, the spotting measure, and the discriminant function. In [32], readers can find its promising aspects, demonstrated in a task of spotting Japanese consonants.

Even in the open-vocabulary situation, a system is often required to spot a sequence (or set) of keywords instead of only a single keyword. In order to meet this requirement, MSPE was reformulated as the minimum error classification of keyword-sequences method (MECK). See [33] for details.

9.4.3.3 Discriminative Utterance Verification

The keyword spotting procedure is considered a hypothesis testing problem, based especially on Neyman-Pearson testing and a likelihood ratio test (LRT). The design concept of GPD was also used to increase the LRT-based keyword spotting capability of a continuous speech recognizer employing a filler model.

Figure 9.3 illustrates a recognizer used for the above cited spotting purpose. The system consists of two subclassifiers: (1) the target classifier and (2) the alternate classifier. The target classifier labels an input as a sequence of the hypothesized keyword and the out-of-vocabulary filler segments. The alternate classifier then generates an alternate hypothesis for the LRT test at the segment where the keyword is hypothesized by the target classifier. An LRT tests the hypothesis that an input x_1^T is generated by the model λ_C corresponding to the keyword W_C versus x_1^T having been generated by an imposter model

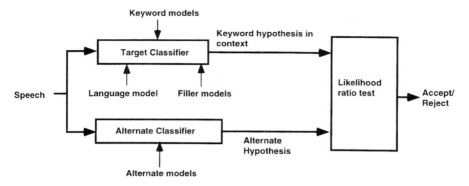

Figure 9.3 A speech recognizer capable of verifying the word hypotheses produced by a continuous speech recognizer (after [45] with some simplification).

λ_I corresponding to the alternate hypothesis W_I, according to the likelihood ratio $p(x_1^T \mid \lambda_C)/p(x_1^T \mid \lambda_I)$. Accordingly, if the ratio exceeds a preset threshold, the system accepts the hypothesized keyword; otherwise it rejects the hypothesis.

Obviously, the quality of hypothesis testing relies on the adequacy of the estimated likelihood values. However, similar to most cases of the ML-based speech recognition, it is quite difficult to achieve accurate estimates due to the fact that both the probability density functions and their parameterization forms are often unknown. To alleviate this problem, GPD was used as a design algorithm for directly minimizing the false-detection errors and the false-alarm errors [39]. Figure 9.4 illustrates the GPD-based training scheme of the keyword hypothesis verification, i.e., discriminative utterance verification (DUV). The training first defines a distance based of the likelihood ratio as:

$$d_C(x_1^T; \Lambda) = \log p(x_1^T \mid \lambda_I) - \log p(x_1^T \mid \lambda_C) \qquad (9.46)$$

One should notice here that a positive value of the distance implies hypothesis rejection and a negative value implies hypothesis acceptance. Then, a loss is defined so as to embed the two types of errors, i.e., the false detection and the false alarm, directly as

$$\ell(x_1^T; \Lambda) = \ell^*(d_C(x_1^T; \Lambda))1(x_1^T \in W_C) + \ell^*(-d_C(x_1^T; \Lambda))1(x_1^T \in W_I) \qquad (9.47)$$

where ℓ^* is a smooth monotonic $0-1$ function. Clearly, a training target here is to minimize the loss over possible design samples in the same

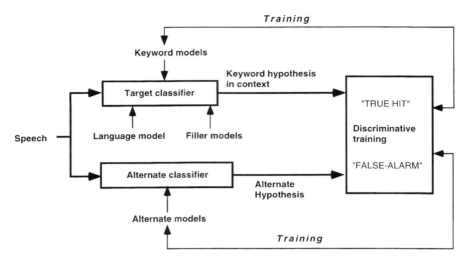

Figure 9.4 Block diagram illustrating a GPD-based discriminative training for a speech recognizer verifying hypothesized vocabulary words (after [45] with some simplification).

manner as the other GPD training cases. Refer to [39] for details.

9.4.4   Discriminative Feature Extraction

9.4.4.1   Fundamentals

In the previous sections, we have considered only the GPD applications to the post-end decision modules, such as the classification (or identification) module, the spotting decision module, and the verification module. Actually, several more advanced applications of GPD have been reported where the GPD training was applied to both the acoustic modeling and the language modeling [30, 31], and the training scope was extended to the front-end feature extraction stage [28, 40, 46]. In particular, in recent years, GPD-based overall design of the feature extraction and classification modules has attracted the attention of speech researchers who recognize the importance of feature space design (on which the quality of the post-end classification decision indispensably relies), and who also seek solutions to unsatisfactory aspects of current feature extractor design techniques. DFE, which is one of the most straightforward embodiments of GPD's concept of global optimization, is becoming a novel paradigm for design algorithm study [47–53].

The formalization of DFE is provided by simply replacing the classification rule (9.3) with the following recognition rule for an input instantiation $u_1^{T_0}$ in the GPD paradigm:

$$C(u_1^{T_0}) = C_i, \quad \text{iff} \quad i = \arg\max_j g_j(\mathcal{T}_\Phi(u_1^{T_0}); \Lambda) \quad (9.48)$$

where $\mathcal{T}_\Phi( )$ is a feature extraction operation with the designable parameters Φ. Note that the feature extraction process is embedded in the discriminant function, and that both the feature extractor parameters Φ and the classifier parameters Λ now become the targets of the GPD optimization. Λ is trained in the very same manner as that of GPD training for classification, and Φ is trained by using the chain rule of differential calculus, which is additionally used for back-propagating the derivative of the loss to the feature extraction module. Figure 9.5 illustrates the DFE training strategy. One should clearly notice here that the DFE design jointly optimizes the overall recognizer with the single objective of minimizing recognition errors.

In (9.48), we used the speech wave sample $u_1^{T_0}$. However, one can use any reasonable form of input, which may be interpreted as an intermediate feature representation of the original utterance input, such as a sequence of FFT-computed power spectrum vectors. In such cases, the DFE training scope is slightly limited but the training can be performed without any degration of

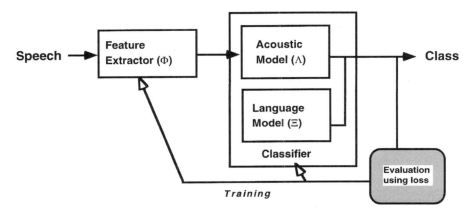

Figure 9.5 Training strategy of discriminative feature extraction.

formalization rigor.

In the rest of this subsection we focus on three topics related to DFE training.

9.4.4.2 An Example Implementation for Cepstrum-Based Speech Recognition

Among many implementation possibilities, [40] studies DFE applied to the design of a cepstrum-based speech recognizer. Cepstrum is one of the most widely used parameter selections for the acoustic feature vector. It has been shown that its low-quefrency components contain salient information for speech recognition and various lifter shapes have been investigated with the aim of more accurate recognition. However, since there were no methods to design the lifter shape under the criterion of minimizing recognition errors, the shape has usually been determined in a trial-and-error fashion. Obviously, such conventional lifter shapes are not guaranteed to be optimal. An expectation of DFE here is therefore that DFE training can lead to more accurate recognition through the realization of an optimal design for both the lifter feature extractor and the post-end classifier.

Figure 9.6 shows a typical DFE-trained lifter shape, which was

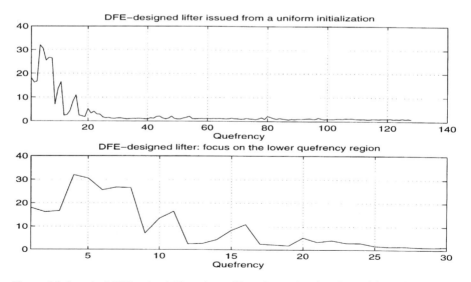

Figure 9.6 A typical DFE-trained lifter shape (Top: the entire view issued from the uniform initialization that set the lifter values at a constant before DFE training. Bottom: the same lifter with a focus on the lower frequency region.) (After [40].)

obtained in the Japanese five-vowel pattern recognition task [40]. This lifter deemphasizes (1) the high quefrency region that corresponds to pitch harmonics and minute spectral structure and (2) the lower quefrency region (0–2 quefrency region) that is dominated by the bias and slant of the overall spectrum while enhancing the region of 3–20 quefrency that corresponds primarily to the formant structure. The observed shape suggests that the DFE training successfully extracted the salient features for vowel pattern recognition. The training actually achieved a clear error reduction, from 14.5% (best by a baseline system) to 11.3% (DFE-trained system), over unknown testing data using the very same size of recognizer.

9.4.4.3 Discriminative Metric Design

Feature extraction can be viewed as a process that forms a framework for measuring the class membership of an input pattern. In general, however, the definition of class identity can be different from class to class. Thus, the feature representation framework, or metric, does not have to be common to all of the classes; i.e., each class could have its own metric that is suitable for representing its identity as effectively as possible. From this point of view, DFE, whose formalization originally assumed a feature extractor common to all sample classes, is not quite sufficient. Motivated by this point, DFE was extended to the discriminative metric design (DMD) method by using the subspace method (SM) paradigm [35].

A recognition decision rule of the DMD framework is as follows:

$$C(u_1^{T_0}) = C_i, \quad \text{iff} \quad i = \arg\max_j g_j(\mathscr{T}_\Phi^j(u_1^{T_0}); \Lambda) \qquad (9.49)$$

where $\mathscr{T}_\Phi^j(\ )$ is a class-specific feature extractor for C_j. Note that for each class the feature extractor is embedded in its corresponding discriminant function, i.e., $g_j(\cdot)$. Similar to the DFE formalization, DMD has many implementation possibilities. Among them, [35] in particular studied implementation using a quadratic discriminant function, which is shown to be closely related to MCE/GPD training for Gaussian kernel functions and accordingly to recent MCE/GPD applications in HMM speech recognizers.

Readers are referred to [35] and [54] for details.

9.4.4.4 Minimum Error Learning Subspace Method

Recognition-oriented feature design has been most extensively studied in the subspace method (SM) paradigm, especially for problems in character and image recognition [55]. SM originated from the similarity method, in which the recognition (classification) decision was made by measuring the angle between a pattern (vector) to be recognized and a class model for every class. In particular, a recognizer based on this method is defined by (lower-dimensional) class subspaces, each assumed to represent the salient features of its corresponding class. There is no distinction between the feature extraction process and the classification process in this system framework. Given an input sample, the recognizer computes the orthogonal projection of the input onto each class subspace and then classifies the pattern to the class giving the maximum projection value.

The performance of SM-based recognizers relies on the quality of the class subspaces. The most fundamental algorithms for designing the subspaces were the CLAFIC [56] and the multiple similarity method (MSM) [57], where each class subspace was designed by running the Karhunen-Loève transformation or principal component analysis over the design data of its corresponding class. Obviously, this class-by-class design does not directly guarantee recognition error reduction: the recognition error of design samples is not reflected in the subspace design. These methods were later improved, aiming at increasing recognition accuracy. The CLAFIC was extended to the learning subspace method (LSM) (e.g., [55]); the MSM was reformed as the compound similarity method [57]. In these new versions, subspaces are trained iteratively (adaptively) depending upon the recognition result of each design sample; i.e., when an input design pattern is misrecognized, the subspace of the true class and that of the most likely but incorrect class are adjusted so that projection onto the true class subspace increases and projection onto the competing incorrect class subspace decreases. This discriminative iteration actually contributed to improvements in accuracy. However, the training mechanism had only an intuitive validity, and its mathematical optimality in terms of error minimization has long remained unclear.

To alleviate the above problem, [34] studied a DFE formalization for the SM framework and proposed the minimum error learning subspace (MELS) method, which guarantees the optimal subspace design in the MCE/GPD sense. Detailed discussions including the training convergence proof is included in [34] and [58].

9.4.5 Speaker Recognition Using GPD

Speaker recognition is usually classified into two main categories, speaker identification, which is the process of identifying an unknown speaker from a known population, and speaker verification, which is the process of verifying the identity of a claimed speaker from a known population. Given a test utterance x_1^T, which is assumed to be represented as a sequence of feature vectors, the likelihood of observing x_1^T being generated by speaker k is denoted as $g_k(x_1^T; \Lambda)$ where Λ is a set of trainable recognizer parameters. Then, the decision operation of speaker identification is formalized as

$$\hat{k} = \arg\max_k g_k(x_1^T; \Lambda) \qquad (9.50)$$

with $\hat{k}$ being the identified speaker, attaining the highest likelihood score among all competing speakers; the decision operation of speaker verification is formalized as follows: If the likelihood meets

$$g_k(x_1^T; \Lambda) > b_k \qquad (9.51)$$

then the recognizer accepts the claimed speaker identity k; the recognizer rejects it otherwise. One can easily notice here that the formalization for speaker identification is equivalent to that for speech recognition, and also that the formalization for speaker verification is essentially the same as that for keyword spotting. One may also notice that the recognizer training here incurs the same difficulty in the likelihood estimation as does the conventional training of continuous speech recognizers and keyword spotters. To alleviate the problem, in [41], GPD training was successfully applied to the design of both an HMM-based speaker identification system and an HMM-based speaker verification system. The derivation of GPD-based training is fundamentally the same as that for speech recognition. Note that whereas GPD training was applied to threshold training in MECK [32], the threshold b_k was determined experimentally in [41]. Refer to [41] for details of the implementation and experimental results.

For speaker verification, [41] also proposed the use of a GPD-trained normalized likelihood score. The decision rule in this case is formalized as follows: If the likelihood meets

$$\frac{\log g_k(x_1^T; \Lambda)}{\log g_{k'}(x_1^T; \Lambda)} > b_k \qquad (9.52)$$

then the recognizer accepts the claimed speaker identity k; the recognizer rejects it otherwise, where k is the claimed speaker and k' is the antispeaker (the set of speakers other than k). Clearly, the likelihood ratio of (9.52) is quite similar to the misclassification measure of GPD, and one can naturally expect that the discriminative power of the normalized score is increased by GPD training. In [41], a remarkable improvement due to GPD training was actually demonstrated.

To see other GPD applications to the task of speaker recognition, readers are referred to recent literature such as [59], [60], [61], and [62].

9.5 Summary

In this chapter, we have reviewed a recent approach to minimum recognition error network design, based on the generalized probabilistic descent method. We have elaborated the characteristics of GPD, focusing on the problems of classifying dynamic speech patterns. The most important point of the GPD concept is to embed the entire process of a given recognition task in a smooth functional form so that one can optimize all of the adjustable system parameters in a manner consistent with the design objective of minimizing recognition errors.

Generally, discriminative design methods such as GPD can achieve higher recognition accuracy using fewer trainable system parameters than ones based on the ML principle. This is a natural feature of the discriminative design method, which focuses on the classification results near the class borders. However, this advantage has often been criticized: the high discriminative power is considered to lead to a less robust design result, i.e., an overlearning of given design samples. However, such criticism is beside the point. The overlearning phenomenon is commonly observed in ML design using many adjustable parameters; on the other hand, it is rarely observed in discriminative design using fewer parameters. The overlearning of a discriminative design has its basis in the directness by which such designs are linked to a given classification task, and thus is clearly an advantage; in fact, many experimental results in the literature demonstrate that descriminative designs using fewer system parameters (than those of the ML case) achieve higher recognition accuracy over unknown samples.

The significance of GPD is that it has both mathematical rigor and high practicality. GPD provided attractive solutions to the three major issues of discriminative training, i.e., (1) the design objective, (2) the optimization method, and (3) the design consistency with unknown samples. In addition, the high utility of GPD has been demonstrated in various experiments, as introduced here. The remaining one of the four discriminative training issues, i.e., the selection of the discriminant function forms, has not been fully studied so far. This point should be investigated in a task-by-task fashion, and GPD, which gives a sound mathematical framework for the other design issues, could be quite useful in this future study.

As the original form of GPD for classification problems has been greatly extended so as to cope with various existing paradigms of pattern recognition such as the open-vocabulary problem and speaker verification, we could further extend the family to other types of tasks. The simple but significant concept of GPD, i.e., formalizing the overall procedure of the task at hand into an optimizable design process, holds great potential in the fields of system design and signal and information processing technology.

Acknowledgments

We thank our colleagues Eric Woudenberg and Erik McDermott for their valuable assistance during the preparation of materials for this chapter.

References

[1]   Karkel, J. D., and A. H. Gray, Jr., *Linear Prediction of Speech*, Berlin: Springer-Verlag, 1978.

[2]   Rabiner, L., and B.-H. Juang, *Fundamentals of Speech Recognition*, Englewood Cliffs, NJ: Prentice Hall, 1993.

[3]   Kung, S.-Y., *Digital Neural Networks*, Englewood Cliffs, NJ: Prentice Hall, 1993.

[4]   Katagiri, S., C.-H. Lee, and B.-H. Juang, "Discriminative Multi-Layer Feed-Forward Networks," *IEEE Neural Networks for Signal Processing*, 1991, pp. 11–20.

[5]   Rumelhart, D. E., J. L. McClelland, and the PDP Research Group, *Parallel Distributed Processing*, MIT Press, 1986.

[6]   Nilsson, N., *The Mathematical Foundations of Learning Machines*, Morgan Kaufmann Publishers, 1990.

[7]   Fukunaga, K., *Introduction to Statistical Pattern Recognition*, Academic Press, 1972.

[8]   Katagiri, S., C.-H. Lee, and B.-H. Juang, "A Generalized Probabilistic Descent Method," *ASJ, Proc. Conf.*, 1990, pp. 141–142.

[9]   Juang, B.-H., and S. Katagiri, "Discriminative Learning for Minimum Error Classification," *IEEE Trans. SP.*, Vol. 40, No. 12, 1992, pp. 3043–3054.

[10] Duda, R., and P. Hart, *Pattern Classification and Scene Analysis*, New York: Wiley, 1973.

[11] Geman, S., and D. Geman, "Stochastic Relaxation, Gibbs Distributions, and the Bayesian Restoration of Images," *IEEE Trans. on PAMI*, Vol. 6, No. 6, 1984, pp. 721–741.

[12] Hampshire II, J., and A. Waibel, "A Novel Objective Function for Improved Phoneme Recognition Using Time-Delay Neural Networks," *IEEE Trans. on Neural Networks*, Vol. 1, No. 2, 1990, pp. 216–228.

[13] McDermott, E., and S. Katagiri, "LVQ-Based Shift-Tolerant Phoneme Recognition," *IEEE Trans. SP*, Vol. 39, No. 6, 1991, pp. 1398–1411.

[14] Waibel, A., T. Hanazawa, G. Hinton, K. Shikano and K. Lang, "Phoneme Recognition Using Time-Delay Neural Networks," *IEEE Trans. ASSP*, Vol. 37, No. 3, 1989, pp. 328–339.

[15] Bahl, L., P. Brown, P. de Souza, and R. Mercer, "A New Algorithm for the Estimation of Hidden Markov Model Parameters," *IEEE Proc. ICASSP88*, Vol. 1, 1988, pp. 493–496.

[16] Kohonen, T., "The Self-Organizing Map," *IEEE Proc. IEEE*, Vol. 78, No. 9, 1990, pp. 1464–1480.

[17] Bahl, L., P. Brown, P. de Souza, and R. Mercer, "Maximum Mutual Information Estimation of Hidden Markov Model Parameters for Speech Recognition," *IEEE Proc. ICASSP86*, Vol. 1, 1986, pp. 49–52.

[18] Amari, S., "A Theory of Adaptive Pattern Classifiers," *IEEE Trans. on Electronic Computers*, Vol. EC-16, No. 3, 1967, pp. 299–307.

[19] Chang, P.-C., and B.-H. Juang, "Discriminative Template Training for Dynamic Programming Speech Recognition," *IEEE Proc. ICASSP92*, Vol. 1, 1992, pp. 493–496.

[20] Chang, P.-C., and B.-H. Juang, "Discriminative Training of Dynamic Programming Based Speech Recognizers," *IEEE Trans. SAP*, Vol. 1, No. 2, 1993, pp. 135–143.

[21] Komori, T., and S. Katagiri, "GPD Training of Dynamic Programming-Based Speech Recognizers," *J. Acoust. Soc. Jpn.* (E), Vol. 13, No. 6, 1992, pp. 341–349.

[22] Katagiri, S., B.-H. Juang, and A. Biem, "Discriminative Feature Extraction," in R. Mammoner, ed., *Artificial Neural Networks for Speech and Vision*, London: Chapman and Hall, 1994, pp. 278–293.

[23] Chou, W., B.-H. Juang, and C.-H. Lee, Segmental GPD Training of HMM Based Speech Recognition," *IEEE Proc. ICASSP92*, Vol. 1, 1992, pp. 473–476.

[24] McDermott, E., and S. Katagiri, "Prototype-Based Discriminative Training for Various Speech Units," *IEEE Proc. ICASSP92*, Vol. 1, 1992, pp. 417–420.

[25] McDermott, E., and S. Katagiri, "Prototype-Based MCE/GPD Training for Various Speech Units," *Computer Speech and Language*, Vol. 8, 1994, pp. 351–368.

[26] Rainton, D., and S. Sagayama, "Minimum Error Classification Training of HMMs—

Implementation Details and Experimental Results," *J. Acoust. Soc. Jpn.* (E), Vol. 13, No. 6, 1992, pp. 379–387.

[27] Juang, B.-H., "Automatic Speech Recognition: Problems, Progress & Prospects," Handout at the 1996 IEEE Signal Processing Society Workshop on Neural Networks for Signal Processing, 1996.

[28] Biem, A., and S. Katagiri, "Feature Extraction Based on Minimum Classification Error/Generalized Probabilistic Descent Method," *IEEE Proc. ICASSP93*, Vol. 2, 1993, pp. 275–278.

[29] Katagiri, S., "A Unified Approach to Pattern Recognition," *Proc. ISANN94*, 1994, pp. 561–570.

[30] Huang, X., M. Belin, F. Alleva, and M. Hwang, "Unified Stochastic Engine (USE) for Speech Recognition," *IEEE Proc. ICASSP93*, Vol. 2, 1993, pp. 636–639.

[31] Su, K.-Y., T.-H. Chiang, and Y.-C. Lin, "A Unified Framework to Incorporate Speech and Language Information in Spoken Language Processing," *Proc. ICASSP92*, Vol. 1, 1992, pp. 185–188.

[32] Komori, T., and S. Katagiri, "A Minimum Error Approach to Spotting-Based Pattern Recognition," *IEICE Trans. Inf. & Syst.*, Vol. E78-D, No. 8, 1995, pp. 1032–1043.

[33] Komori, T., and S. Katagiri, "A Novel Spotting-Based Approach to Continuous Speech Recognition: Minimum Error Classification of Keyword-Sequences," *J. Acous. Soc. Jpn.* (E), Vol. 16, No. 3, 1995, pp. 147–157.

[34] Watanabe, H., and S. Katagiri, "Discriminative Subspace Method for Minimum Error Pattern Recognition," *IEEE Neural Networks for Signal Processing* V, 1995, pp. 77–86.

[35] Watanabe, H., T. Yamaguchi, and S. Katagiri, "A Novel Approach to Pattern Recognition Based on Discriminative Metric Design," *IEEE Neural Networks for Signal Processing* V, 1995, pp. 48–57.

[36] Katagiri, S., C.-H. Lee, and B.-H. Juang, "New Discriminative Training Algorithms Based on the Generalized Probabilistic Descent Method," *IEEE Neural Networks for Signal Processing*, 1991, pp. 299–308.

[37] Funahashi, K., "On the Approximate Realization of Continuous Mappings by Neural Networks," *Neural Networks*, Vol. 2, No. 3, 1989, pp. 183–191.

[38] Sorenson, H., and D. Alspach, "Recursive Bayesian Estimation Using Gaussian Sums," *Automatica*, Vol. 7, 1971, pp. 465–479.

[39] Rose, R. C., B.-H. Juang, and C.-H. Lee, "A Training Procedure for Verifying String Hypothesis in Continuous Speech Recognition," *Proc. ICASSP95*, 1995, pp. 281–284.

[40] Biem, A., S. Katagiri, and B.-H. Juang, "Pattern Recognition Using Discriminative Feature Extraction," *IEEE Trans. on Signal Processing*, Vol. 45, No. 2, 1997, pp. 500–504.

[41] Liu, C.-S., C.-H. Lee, W. Chou, B.-H. Juang, and A. Rosenberg, "A Study on Minimum Error Discriminative Training for Speaker Recognition," *J. Acoust. Soc. Am.*, Vol. 97, No. 1, 1995, pp. 637–648.

[42] Juang, B.-H., W. Chou, and C.-H. Lee, "Minimum Classification Error Rate Methods

for Speech Recognition," *IEEE Trans. SAP*, Vol. 5, No. 3, 1997, pp. 257–265.

[43] Chou, W., C.-H. Lee, and B.-H. Juang, "Minimum Error Rate Training Based on N-Best String Models," *IEEE Proc. ICASSP93*, Vol. 2, 1993, pp. 652–655.

[44] Chou, W., C.-H. Lee, and B.-H. Juang, "Minimum Error Rate Training of Inter-Word Context Dependent Acoustic Model Units in Speech Recognition," *Proc. ICSLP94*, 1994, pp. 439–442.

[45] Rose, R. C., "Word Spotting from Continuous Speech Utterances," in C.-H. Lee, F. K. Soong, and K. K. Paliwal, eds. *Automatic Speech and Speaker Recognition: Advanced Topics*, Kluwer Academic Publishers, 1996, pp. 303–329.

[46] Biem, A., S. Katagiri, and B.-H. Juang, "Discriminative Feature Extraction for Speech Recognition," *Neural Networks for Signal Processing III—Proc. 1993 IEEE Workshop*, 1993, pp. 392–401.

[47] Chengalvarayan, R., and L. Deng, "Use of Generalized Dynamic Feature Parameters for Speech Recognition," *IEEE Trans. SAP*, Vol. 5, No. 3, 1997, pp. 232–242.

[48] Chengalvarayan, R., and L. Deng, "HMM-Based Speech Recognition Using State-Dependent, Discriminatively Derived Transforms on Mel-Warped DFT Features," *IEEE Trans. SAP*, Vol. 5, No. 3, 1997, pp. 243–256.

[49] Chou, W., M. G. Rahim, and E. Buhrke, "Signal Conditioned Minimum Error Rate Training," *Proc. EUROSPEECH95*, 1995, pp. 495–498.

[50] Euler, S., "Integrated Optimization of Feature Transformation for Speech Recognition," *Proc. EUROSPEECH95*, 1995, pp. 109–112.

[51] Hernando, J., J. Ayarte, and E. Monte, "Optimization of Speech Parameter Weighting for CDHMM Word Recognition," *Proc. EUROSPEECH95*, 1995, pp. 105–108.

[52] Paliwal, K. K., M. Bacchiani, and Y. Sagisaka, "Minimum Classification Error Training Algorithm for Feature Extractor and Pattern Classifier in Speech Recognition," *Proc. EUROSPEECH95*, 1995, pp. 541–544.

[53] Rahim, M. G., and C.-H. Lee, "Simultaneous ANN Feature and HMM Recognizer Design Using String-Based Minimum Classification Error (MCE) Training," *Proc. ICSLP96*, 1995, pp. 1824–1827.

[54] Watanabe, H., T. Yamaguchi, and S. Katagiri, "Discriminative Metric Design for Robust Pattern Recognition," *IEEE Trans. Signal Processing*, Vol. 45, No. 11, 1997.

[55] Oja, E., *Subspace Methods of Pattern Recognition*, Research Studies Press, 1983.

[56] Watanabe, S., P. F. Lambert, C. A. Kulikowski, J. L. Buxton, and R. Walker, "Evaluation and Selection of Variables in Pattern Recognition," in J. T. Tou, ed., *Computer and Information Sciences II*, Academic Press, 1967, pp. 91–122.

[57] Iijima, T., *Pattern Recognition Theory*, Morikita, 1989 (in Japanese).

[58] Watanabe, H., and S. Katagiri, "Subspace Method for Minimum Error Pattern Recognition," *IEICE Trans. Information and Systems* (forthcoming).

[59] Farrell, K. R., R. J. Mammone, and K. T. Assaleh, "Speaker Recognition Using Neural Networks and Conventional Classifiers," *IEEE Trans. SAP*, Vol. 2, No. 1, Part II, 1994, pp. 194–205.

[60] Matsui, T., and S. Furui, "A Study of Speaker Adaptation Based on Minimum Classification Error Training," *Proc. EUROSPEECH95*, 1995, pp. 81–84.

[61] Setlur, A., and T. Jacobs, "Results of a Speaker Verification Service Trial Using HMM Models," *Proc. EUROSPEECH95*, 1995, pp. 639–642.

[62] Sugiyama, M., and K. Kurinami, "Minimal Classification Error Optimization for a Speaker Mapping Neural Network," *IEEE Neural Networks for Signal Processing II*, 1992, pp. 233–242.

[63] Makino, S., M. Endo, T. Sone, and K. Kido, "Recognition of Phonemes in Continuous Speech Using a Modified LVQ2 Method," *J. Acoust. Soc. Jpn.* (E), Vol. 13, No. 6, 1992, pp. 351–360.

[64] McDermott, E., "LVQ3 for Phoneme Recognition," *ASJ, Proc. Spring Conf.*, 1990, pp. 151–152.

[65] Ando, A., and K. Ozeki, "A Clustering Algorithm to Minimize Recognition Error Function," *IEICE Trans. A*, Vol. J74-A, No. 3, 1991, pp. 360–367.

[66] Katagiri, S., and C.-H. Lee, "A New Hybrid Algorithm for Speech Recognition Based on HMM Segmentation and Learning Vector Quantization," *IEEE Trans. SAP*, Vol. 1, No. 4, 1993, pp. 421–430.

[67] Su, K.-Y., and C.-H. Lee, "Speech Recognition Using Weighted HMM and Subspace Projection Approaches," *IEEE Trans. SAP*, Vol. 2, No. 1, 1994, pp. 69–79.

[68] Iwamida, H., S. Katagiri, and E. McDermott, "Re-Evaluation of LVQ-HMM Hybrid Algorithm," *ASJ, J. Acoust. Soc. Jpn.* (E), Vol. 14, No. 4, 1993, pp. 267–274.

[69] Asou, H., and N. Otsu, "Nonlinear Data Analysis and Multilayer Perceptrons," *IEEE Proc. IJCNN*, Vol. 2, 1989, pp. 411–415.

[70] Gish, H., "A Probabilistic Approach to the Understanding and Training of Neural Network Classifiers," *IEEE Proc. ICASSP90*, Vol. 3, 1990, pp. 1361–1364.

Appendix 1

Probabilistic Descent Theorem for Probability-Based Discriminant Functions

In the case of using a probability function as the discriminant function, the adjustment convergence of (9.26) no longer holds true: the probabilistic descent theorem assumed the classifier parameter Λ to be arbitrary scalar variables and/or arbitrary vector variables, and did not anticipate that each class parameter vector $\lambda_j = (\lambda_j[1], \cdots, \lambda_j[D])^T$ had to meet the following probability constraint:

$$\sum_d \lambda_j[d] = 1 \quad \text{and} \quad \lambda j[d] \geqslant 0 \text{ (for } \forall d) \qquad (9.53)$$

where we assume that λ_j is of D-dimension. The HMM state transition probability and the mixture coefficients of the mixtured Gaussian HMM are typical examples of this type of constrained parameter vector.

One solution to this problem is to use the transformation of (9.42). In addition, [8] and [36] provide a more general solution, i.e., the following, constrained probabilistic descent theorem for applying the GPD-based adjustment to the constrained parameters.

Theorem 1 (Constrained probabilistic descent theorem)

Assume that a design sample $x_1^T(t)(\in C_k)$ is given and a classifier parameter set λ_j for C_j includes the parameter vector set $\{\rho_j\}$ that satisfies the constraint of (9.53). Then, the adjustment using $\big(\delta\Lambda(x_1^T(t), C_k, \Lambda)\big)_\Phi$, which is obtained by projecting $\delta\Lambda(x_1^T(t), C_k, \Lambda)$ of (9.26) to the ρ_j's subspace that is spanned according to the constraint, reduces $\delta L(\Lambda)$ in the sense of expectation; here $(\ )_\Phi$ is the orthogonal projection of the parenthesized parameter vector onto the (9.53)-based subspace of the parameter vector space.

Proof

Considering that the gradient computation is independently done for every dimensional element of the parameter vector, we can prove the theorem by showing that the adjustment in terms of one parameter vector ρ_j, which satisfies the constraint, reduces $\delta L(\Lambda)$ in the expectation sense. In fact, the following inequality clearly proves that the adjustment can reduce $\delta L(\Lambda)$ in the expectation sense:

$$E[\delta L(\Lambda)] = E[(\delta\Lambda(x_n, C_k, \Lambda))_\Phi \cdot \nabla L(\Lambda)]$$

$$= -\epsilon\{UE[\nabla \ell_k(x_n; \Lambda)]\} \cdot (\nabla L(\Lambda))_\Phi$$

$$= -\epsilon\{U\nabla L(\Lambda)\} \cdot (\nabla L(\Lambda))_\Phi$$

$$= -\epsilon U\|\nabla L(\Lambda)\|^2 \cos\vartheta \leqslant 0 \qquad (9.54)$$

where ϑ is the angle between $\nabla L(\Lambda)$ and $(\nabla L(\Lambda))_\Phi$, and $\vartheta \leqslant \pi/2$ always holds true.

Now it is clear that the characteristics of the original theorem, such as the stochastic approximation-based convergence to a locally optimal situation, hold true.

Let us point out that even in the case of using the probability-based discriminant function, all the parameters do not need to observe the constrained probabilistic descent adjustment. For example, the mean vector of the component Gaussian distribution of the mixtured Gaussian HMM system can be updated according to (9.26). Moreover, the covariance matrix of the component distribution can be updated in the same way. The adjustment of the diagonal covariance matrix, which is widely used in speech pattern recognition, is actually easy. The adjustment of the full covariance matrix is somewhat complicated and difficult: It should satisfy the positive-definiteness constraint of the matrix. However, it can be done fundamentally based on (9.26). A GPD implementation for this case is presented in detail in [35].

Appendix 2

Relationships Between MCE/GPD and Others

- **Relation with LVQ**

As seen in [16] and [63], several implementations of LVQ have been reported. However, the following fundamental concept of LVQ underlies all of the implementations; that is, *only if a given design sample is misclassified and is located near the actual class boundary are the prototypes of the correct class (of the given sample) and those of some competing classes adjusted so as to increase the possibility of classifying the sample correctly.* Moreover, all of them have in common a heuristic window function for restricting the adjustment to the class boundary region.

According to [8], let ζ, ξ, and μ approach ∞ in (9.27). Then, we obtain the following simple adjustment rule:

$$\mathbf{r}^b_{j,\tau}(t+1) = \begin{cases} \mathbf{r}^b_{j,\tau}(t) + 2\epsilon(t)\nu_k w^b_{j,\tau}(\mathbf{z}_{1(j,b,\tau)}(t) - \mathbf{r}^b_{j,\tau}(t)) \\ \qquad \text{(for } j = k \text{ and } b = 1) \\ \mathbf{r}^b_{j,\tau}(t) - 2\epsilon(t)\nu_k w^b_{j,\tau}(\mathbf{z}_{1(j,b,\tau)}(t) - \mathbf{r}^b_{j,\tau}(t)) \\ \qquad \text{(for } j = i \text{ and } b = 1) \\ \mathbf{r}^b_{j,\tau}(t) \qquad \text{(otherwise)} \end{cases} \qquad (9.55)$$

where $1(j, b, \tau) = \theta(j, b, \tau)|_{\theta=1}$, and C_i is the class having the smallest discriminant function value among classes other than C_k, i.e., the most likely competing class. One can easily see that in the circumstances of handling static patterns, this rule is quite similar to an improved version of LVQ, which is characterized as follows [64]: *If and only if a given sample is misclassified and is located in a small region (window) near the actual class boundary, then (1) the closest prototype (to the sample) of the correct class is pushed closer to the sample, and (2) the closest prototype of the best (most probable) competing class is pulled farther from the sample; otherwise no adjustment is incurred.* Clearly, the adjustment mechanism of this LVQ training is realized in (9.55). In particular, it is worthwhile noting that the window set in LVQ corresponds to a derivative function of the MCE/GPD's smooth classification error count loss. Consequently, this correspondence proves that in substance, LVQ uses the design objective of classification error count.

In LVQ, the adjustment was generally performed only when a sample was misclassified. In contrast, the MCE/GPD training based on (9.55) performs the adjustment when a sample is correctly classified but the corresponding misclassification measure is small (close to zero); in other words, when the certainty of the corresponding classification decision is small. This is a natural result of the MCE/GPD formalization using the smooth decision process. Consequently, this "learning-in-correct" contributes to increasing design robustness [21]. The effect of "learning-in-correct" is also reported in [18], [65], [66], and [67].

The close relation between MCE/GPD and LVQ enables us to use many application results of LVQ as circumstantial evidence of the effectiveness of MCE/GPD [13, 16, 66]. The inadequacy of LVQ, which was reported in the literature, e.g., [68], may be due to a lack of consistency with regard to the design objective (loss) formulation between a given task and the LVQ training. In most cases, LVQ has been used for reducing the misclassification of static patterns, such as the elemental acoustic feature vectors of the dynamic speech pattern to be classified originally. Solving this inconsistency was one of the motivations for the development of MCE/GPD.

- **Relation with Maximization of Mutual Information**
As befits the Bayes decision theory, we have considered classification error count loss in this paper. On the other hand, various other objective functions have also been extensively studied for the sake of increasing classification accuracy. Among them, the use of mutual information has attracted much research interest in speech pattern classification [17].

This approach aims to select the state of Λ so as to increase as much as possible the mutual information between class C_j and a sample x_1^T ($\in C_k$), which is defined in the following:

$$I_k(x_1^T; \Lambda) = \ln \frac{p_\Lambda(x_1^T \mid C_k)}{\sum\limits_{j}^{M} p_\Lambda(x_1^T \mid C_j)P(C_j)} \tag{9.56}$$

Let us consider the effect of maximizing the mutual information in the GPD (instead of MCE/GPD) framework. For convenience, we use a negative value of mutual information. Then, the goal of GPD design is to minimize this negative measure. The negative mutual information is rewritten as:

$$-I_k(x_1^T; \Lambda) = \ln \frac{\sum_j^M p_\Lambda(x_1^T \mid C_j)P(C_j)}{p_\Lambda(x_1^T \mid C_k)}$$

$$= \ln \left\{ P(C_k) + \frac{\sum_{j,j\neq k}^M p_\Lambda(x_1^T \mid C_j)P(C_j)}{p_\Lambda(x_1^T \mid C_k)} \right\}$$

$$\geqslant \ln \frac{\sum_{j,j\neq k}^M p_\Lambda(x_1^T \mid C_j)P(C_j)}{p_\Lambda(x_1^T \mid C_k)}$$

$$= -\ln p_\Lambda(x_1^T \mid C_k)$$

$$+ \ln \left\{ \sum_{j,j\neq k}^M P(C_j)e^{\ln p_\Lambda(x_1^T \mid C_j)} \right\} \tag{9.57}$$

Here, defining the logarithmic likelihood, $\ln p_\Lambda(x_1^T \mid C_k)$, as the discriminant function, one can treat the bottom-line expression of (9.57) as a kind of misclassification measure:

$$d_k(x_1^T; \Lambda) = -g_k(x_1^T; \Lambda) + \ln \left\{ \sum_{j,j\neq k}^M P(C_j)e^{g_j(x_1^T; \Lambda)} \right\} \tag{9.58}$$

Then, the inequality,

$$-I_k(x_1^T; \Lambda) \geqslant d_k(x_1^T; \Lambda) \tag{9.59}$$

holds true, and therefore, maximizing the mutual information leads at least to minimizing the misclassification measure (9.58). Consequently, it turns out that a classifier design based on maximizing mutual information has the same effect as a GPD design that uses the misclassification measure (9.58) and the linear loss function. Note here that the loss used is not a smoothed error count but a simple linear function of the misclassification measure. Obviously, this method is not guaranteed to be consistent with the minimum classification error condition unless a $0-1$ nonlinear function is imposed.

• Relation with Minimization of Squared Error Loss

Most ANN classifiers have used the squared error between the teaching signal and the classifier output (the value of the discriminant function) as the loss; the design employed therein aims to minimize the expected loss or the average empirical loss of this individual squared error loss. Similar to the mutual information maximization method, let us consider the nature of this squared error loss minimization method in the GPD framework. First, the squared error loss is represented as

$$\ell_k(x_1^T;\Lambda) = \frac{1}{2}\sum_j^M \{g_j(x_1^T;\Lambda) - \varepsilon_j\}^2 \tag{9.60}$$

where $\{\varepsilon_j\}$ is a teaching signal which is usually set, for a design sample $x_1^T \ (\in C_k)$, to

$$\varepsilon_j = \begin{cases} 1 & (j=k) \\ 0 & \text{(otherwise)} \end{cases} \tag{9.61}$$

The loss is thus rewritten as

$$\ell_k(x_1^T;\Lambda) = -g_k(x_1^T;\Lambda) + \frac{1}{2}\{(g_k(x_1^T;\Lambda))\}^2$$

$$+ \frac{1}{2} + \frac{1}{2}\sum_{j,j\neq k}^M \{(g_j(x_1^T;\Lambda))\}^2$$

$$> -g_k(x_1^T;\Lambda) + \frac{1}{2}\sum_{j,j\neq k}^M \{(g_j(x_1^T;\Lambda))\}^2 \tag{9.62}$$

Then, we can treat the bottom line expression of (9.76) as a kind of misclassification measure:

$$d_k(x_1^T;\Lambda) = -g_k(x_1^T;\Lambda) + \frac{1}{2}\sum_{j,j\neq k}^M \{(g_j(x_1^T;\Lambda))\}^2 \tag{9.63}$$

It turns out that the reduction of the squared error loss results in the reduction of this misclassification measure. Hence, we can conclude that the squared error loss minimization leads at least to an optimal situation, in the GPD sense, which is based on the linear loss using the misclassification measure (9.63). Note here again that the loss used is not a smoothed classification error count but a simple linear function of the misclassification measure. Obviously, it is not guaranteed that the resulting status of this method is consistent with that of the minimum classification error condition.

It is is difficult to use (9.63) as the misclassification measure. Since the second term (of the right-hand side) of (9.63), i.e., the average discriminant function of the competing classes, is not normalized by the number of possible classes (M), the design result using this measure is fundamentally affected by the number of classes. Furthermore, since the discriminant function of C_k is compared with the squared values of the other competing class discriminant functions, the scalar decision based on (9.63) does not emulate the classification appropriately. In the literature, it is shown that a discriminant function based on this minimization method can be a good estimate of the corresponding class a posteriori probability function in the statistical sense (e.g., see [69] and [70]). However, it is also pointed out that fundamentally, such an estimate is not accurate in the class boundary region, which is important for classification but is usually represented by a small value of $p(x_1^T)$ [10].

Part III:
Current Issues in
Speech Signal Processing

10

Networks for Speaker Recognition
Kevin Farrell

10.1 Introduction

Speaker recognition refers to the ability to recognize a speaker based on his or her voice. The two main applications within speaker recognition are speaker identification and speaker verification. Speaker identification addresses the problem of identifying a person among a population when given an unknown voice sample. Speaker verification consists of determining whether or not a voice sample provides sufficient match to a claimed identity.

There are many commercial applications for speaker recognition. For example, when initiating a bank account transaction over the phone or at an automatic teller machine (ATM), speaker verification can provide an additional level of security over personal identification numbers (PINs). For similar applications in which a PIN is not available, speaker identification may be used. Speaker recognition has the advantage over other forms of biometric authentication, such as fingerprint, retinal scan, etc., in that it can be applied over the telephone network.

Speaker recognition applications are generally text independent or text dependent. Text-independent speaker verification systems do not require that the text used for training and testing be the same. Text-dependent speaker verification systems require that the same text be used during both training and testing. Text-independent systems are convenient from a user standpoint in that the person does not have to remember a specific password and will be recognized regardless of what he or she says. The enrollment procedure, however, is more involved. Here, the user will have to provide sufficient input to the enrollment such that the collection of sounds in the user's vocabulary is adequately modeled. Text-dependent systems are password specific and can perform well even with few repetitions. Also, the use of a password provides an additional level of security. Now, an imposter will not only have to have vocal tract characteristics similar to the victim, but will additionally need to guess the correct password.

One key aspect of automatic speaker recognition is how to develop a model that will generate a high response when given an observation from the true speaker, and correspondingly, a low response when given an observation from an imposter. One important consideration in selecting a modeling approach is determined by whether the application is text dependent or text independent. Text-dependent applications should use modeling techniques that incorporate temporal information, since the same text used for training will be used for testing. Examples of models that have been used for text-dependent speaker verification include dynamic time warping (DTW) [1] and hidden Markov models (HMMs) with both whole-word [2, 3] and subword [4, 5] implementations. Text-independent applications, on the other hand, do not require models that maintain temporal information, as the user is not constrained to test the system with the same text used during training. The popular approaches for text-independent speaker recognition applications include vector quantization (VQ) [6] and Gaussian mixture models (GMMs) [7].

Recently, neural networks have been considered for speaker recognition tasks. In many cases they have shown comparable if not better performance than the traditional approaches described. One of the main differences between neural networks and their statistical or distortion-measure-based counterparts is the fact that they use discriminative information. The discriminative information is obtained by training the network to have a high response when given data from the target speaker and a low response when given data from a nontarget speaker. By using the information of different speakers to construct the model for a target speaker, a new dimension of information for speaker recognition becomes available. This information can be used by itself for speaker recognition applications or can be combined with the results of statistical models to capitalize on the diversity of the two measures.

In this chapter, we provide a discussion of neural networks and their applications to speaker recognition. The following section provides an overview of speaker recognition. This is followed by a description of discriminative information with respect to speaker recognition. Next, a discussion is given for some of the neural network solutions that have been evaluated for speaker recognition applications. The concepts of diversity and data fusion as pertaining to the combination of neural networks and statistical methods are discussed. A summary of the chapter is then provided.

10.2 Speaker Recognition Overview

Speaker recognition is a topic within the field of pattern recognition. The two main phases of any pattern recognition problem, including speaker recognition, are feature extraction and classification. Feature extraction consists of extracting characteristics of a signal that are

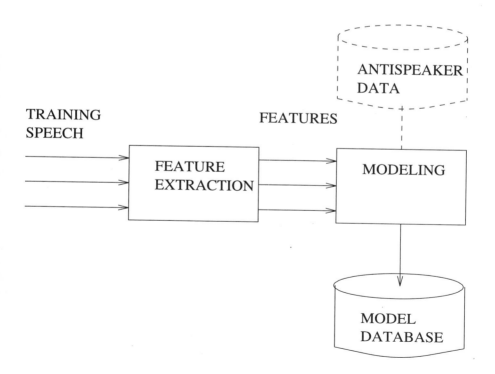

Figure 10.1 Speaker recognition training.

indicative of the class to which that signal belongs. The classification phase uses these features to make a decision about class membership.

Speaker recognition has two modes of operation, namely training and testing. The purpose of training is to produce a model for a specific speaker, or target speaker. The data used for training generally consists of either multiple repetitions of the same text for text-dependent applications, or a passage of phonetically balanced text for text-independent applications. The training process for speaker recognition is shown in Figure 10.1.

The training speech from the target speaker first undergoes feature extraction. The predominant characteristic that causes people's voices to be different from one another is the shape of the vocal tract. The difference in the length and cross-sectional areas in the vocal tract from person to person results in different resonant frequencies and bandwidths. The vocal tract source, or pitch, is another feature that is characteristic of a user. However, this is only a single dimension of information that by itself is insufficient to allow for delineation between a large population of speakers. For capturing the information regarding resonant frequencies and their bandwidths, most feature extraction routines for speaker recognition utilize some type of spectral analysis. A common feature set used for speaker recognition is the cepstrum [9].

The next step for training, as illustrated in Figure 10.1, is modeling. The purpose of the modeling phase is to capture a representation of the feature data for a given user. This representation, or model, will be used in the future when the system is used to either authenticate a claimed identity of a user or identify the user within a population. Hence, it is necessary to store the trained model to a model database. Some modeling approaches require auxiliary data from other users to perform their training. We refer to this data here as antispeaker data. Antispeaker data consists of data from a population of users which is used to construct the data set for training algorithms that require labeled data, such as neural networks.

There are numerous methods for training models for speaker verification that each represent different characteristics of the feature data. Some models, such as vector quantization and dynamic time warping, provide a distortion measure between the feature data observed during training and testing. Other models, such as Gaussian mixture models and hidden Markov models, estimate parameters for an assumed probabilistic model. Neural networks use yet another criterion. They learn how to discriminate between feature data for a target speaker

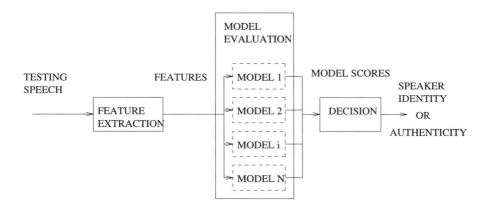

Figure 10.2 Speaker recognition testing.

and feature data from other speakers. Whereas the traditional methods for developing speaker recognition models, such as the distortion-based and statistical models described above, can construct a model from only the target speaker's data, neural networks require data from other speakers.

The testing mode for speaker recognition is illustrated in Figure 10.2. The front-end processing, namely feature extraction, is identical to that used for training. For testing, the models are evaluated for the given features and the decision component outputs a result. For speaker verification, this result corresponds to a decision to accept or reject the claimed speaker identity. For identification, this result corresponds to the identity of the best match within the enrolled user population. This is known as closed-set speaker identification. A more difficult problem is known as open-set speaker identification. Here, the system must decide not only the identity of the speaker but whether the speaker exists within the enrolled population. For the case of open-set speaker identification, the decision component must also provide for a none-of-the-above category to accommodate a case in which the speaker has not enrolled.

The performance of speaker recognition systems is generally expressed in terms of an error rate. For speaker identification, this is simply computed as the number of mistakes made by the system divided by the total number of trials. Speaker verification systems have two types of errors, namely false-reject and false-accept errors. False-reject errors represent the case in which the true user is rejected by the system. False-accept errors represent the case in which an imposter is accepted when performing a verification on a different user's model. Speaker verification performance can be quantified with a single number that can be used as a metric for comparison between different systems. This metric is known as an equal error rate. To compute the equal error rate, the threshold is first shifted through the scores and the percent of false-reject and false-accept errors is recorded for each threshold position. These points can be used to obtain a receiver operating characteristic (ROC) curve, which illustrates the performance tradeoffs between false-accept and false-reject. An example of an ROC curve is shown in Figure 10.3. The point at which false-accept and false-reject are equal represents the equal error rate (EER). In the ROC curve illustrated here, the EER is roughly 2.8%.

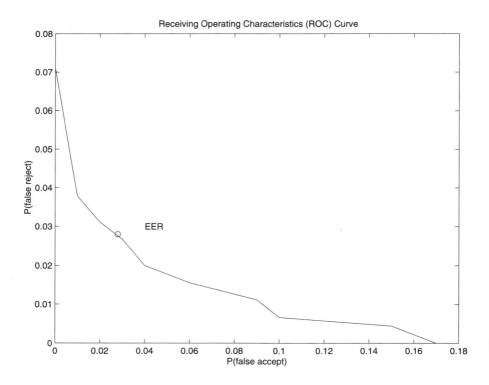

Figure 10.3 Receiver operating characteristic (ROC) curve.

10.3 Discriminative Information

Discriminative information is a useful, if not necessary, component of speaker recognition systems. Discriminative information within speaker recognition consists of data from speakers other than the one currently being enrolled or authenticated. Discriminative information can be incorporated during the training phase by using a supervised training method to develop a model. Discriminative information can be used during testing by using cohort normalization. These techniques are described in more detail in the following sections.

10.3.1 Supervised Training

Supervised training algorithms for speaker recognition learn how to discriminate between feature data for a target speaker and feature data from other speakers. For example, consider a population of size *M* for which a set of feature vectors has been derived for each speaker. A training pool can then be formed as illustrated in Figure 10.4. Here, a training data set for a given speaker is generated by labeling the feature vectors for the target speaker as one and the vectors for everyone else as

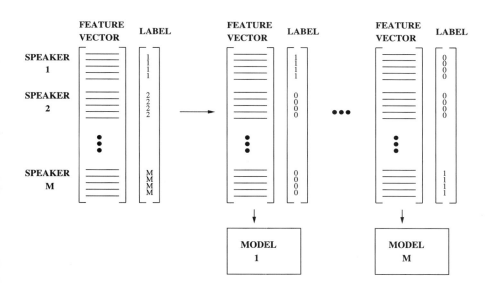

Figure 10.4 Supervised training for speaker recognition.

zero. Most neural network models are trained from such a data-partitioning scheme. This method will create a separate model for each enrolled user. An alternate approach for using this training data is to train one model for all classes. This, however, poses a much more complex data set for the training algorithm and additionally is inconvenient for situations where users must be added to the system afterward.

Neural networks will position their discriminant boundaries to distinguish observations of the target class from those of other classes as illustrated in Figure 10.5. As can also be seen in Figure 10.5, models that are based on distortion criteria, such as VQ or DTW, or probabilistic measures, such as HMM or GMM, use only the in-class observations (shown as the Xs) to determine parameters. Supervised models use the data from out-of-class observations in addition to the in-class data such that discriminant boundaries can be positioned. This technique is advantageous for speaker verification in that the models are being trained specifically to reject nontarget speakers.

A special case of supervised training known as minimum classification error (MCE) has also been considered for speaker recognition. The MCE training algorithm uses a cost function that minimizes the classification error [10, 11]. This is contrasted to the more typical learning algorithms that minimize a norm of the approximation error. Using the MCE approach, speaker recognition models that use statistical

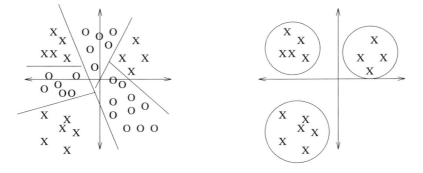

DISCRIMINANT BOUNDARIES   CENTROID OR PROBABILITY ESTIMATION

Figure 10.5 Supervised training versus probabilistic modeling.

and distortion-based algorithms can acquire knowledge about other speakers during their training.

The MCE algorithm has been evaluated for speaker identification and verification [12]. A comparison was made for HMMs that were trained with maximum likelihood (ML) methods and with discriminative learning (MCE). The MCE approach demonstrated a 25% reduction in error rate over the maximum likelihood approach.

10.3.2 Cohort Normalization

Cohort normalization provides a method for incorporating discriminant information during the model testing phase in a speaker verification task. A typical speaker verification decision is implemented by the following equation:

$$\hat{p}(X|S = S_t) > T \qquad (10.1)$$

Here, $\hat{p}$ is a probability estimate, $X = \{x_1, x_2, \ldots, x_N\}$ is a sequence of feature vectors, S_t is a target speaker within the set of speakers S, and T is the threshold. For cohort normalization [13–14], the decision is implemented with a likelihood ratio:

$$\frac{\hat{p}(X|S = S_t)}{\hat{p}(X|S \neq S_t)} > T \qquad (10.2)$$

In (10.2), the denominator represents the probability that the observation is not from the target speaker. This term is typically estimated by applying the observation to a set of models for speakers that are known beforehand to be "close" to the target speaker. These speakers are known as cohorts. By normalizing a speaker's score with scores of models from other speakers that are known to be close, a relative measurement is obtained. If the cohort scores for a given speaker are known to be high, then the normalized score should be close to one. Correspondingly, if the cohort scores for a speaker are known to be low, the normalized score may be much greater than one. The use of the cohort score in the denominator allows a system to set some expectations on the score of a speaker's model with respect to the scores from other speakers' models.

Some common methods for estimating the denominator term are to obtain the average of the cohort scores

$$\hat{p}(X|S \neq S_t) = \frac{1}{N} \sum_{i=1}^{N} p(X|S_i) \qquad (10.3)$$

or to take the maximum

$$\hat{p}(X|S \neq S_t) = \max_{i \neq t} p(X|S_i) \qquad (10.4)$$

Cohort normalization can provide substantial performance improvements to modeling techniques that do not use discriminative information during their training, such as nearest-neighbor approaches [13], HMMs [14], and GMMs [7]. Cohort normalization has been evaluated for neural tree networks [15], but the improvements here were minimal due to the fact that discriminative information was already used during training.

10.4 Speaker Recognition Networks

Neural networks have been considered for speaker recognition applications for the past ten years. For speaker verification, the neural-network-based modeling approaches include multilayer perceptrons [16, 17], radial basis functions [18], predictive neural networks [19], neural tree networks [15], and recurrent neural networks [20]. Several neural network approaches have also been evaluated for speaker identification, including multilayer perceptrons [21], learning vector quantization [22], time-delay neural networks [23], and neural tree networks [24]. A synopsis of these neural-network-based approaches is provided in the following sections.

10.4.1 Multilayer Perceptron

The multilayer perceptron (MLP) is a popular form of neural network that has been considered for various speech processing tasks [25]. For speaker recognition applications, the MLPs are trained with the distributed labeling scheme described in the supervised training section. MLPs by themselves do not incorporate temporal information into their classification procedure. Hence, they are more appropriate for text-independent speaker recognition applications unless they utilize temporal data or segmentation information from other models.

One possible architecture for an MLP-based model is illustrated in Figure 10.6. Here, input feature vectors of dimension D are applied to the input layer. The training algorithm will then learn the weights for the hidden layer w_{ij}^h and output layer w_j^o. The supervision is set up such that training vectors from the speaker have a label equal to one and training vectors from antispeakers have labels equal to zero. Hence, when evaluating this speaker recognition system, test vectors for the target speaker should have a response close to one whereas test vectors from other speakers should have a response near zero. Alternative architectures have also been considered in which the model has two nodes in the output layer as opposed to the model shown in Figure 10.6. In this case, fuzzy logic was applied toward the final decision [8].

A speaker identification system will have an MLP as illustrated in Figure 10.6 for each person that is enrolled in the system. The system will appear as illustrated in Figure 10.7. A speech signal will first

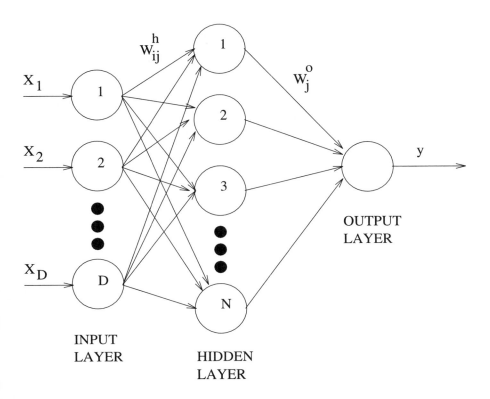

Figure 10.6 Multilayer perceptron.

undergo feature extraction and the features will be applied to the MLP for each user enrolled in the system. The scores of each feature vector for each MLP will be accumulated and the decision unit will output the speaker identity corresponding to the maximum accumulated score.

MLPs have been reported to perform comparable to VQ for speaker identification [21]. An MLP-based classifier for speaker identification has been presented in [21], where each speaker is represented by an MLP as illustrated in Figure 10.7. The features consisted of 10th order linear-prediction-derived cepstral coefficients. The MLP used the back-propagation algorithm for training [26]. The back-propagation algorithm will iteratively adjust the hyperplanes in feature space to best separate the classes. An MLP with one hidden layer and 128 hidden nodes achieved a 92% identification rate for this experiment, which was just slightly worse than the performance obtained with a VQ classifier with 64 codebook entries per speaker. Performance improved as the number of hidden nodes increased. However, it was observed that increasing the number of hidden layers did not improve generalization (performance on the training data improved but not on the testing data). It was also noted that the performance of MLPs degrades rapidly as the speaker population increases.

Another problem that is apparent for MLP evaluations with large antispeaker databases is convergence of the back-propagation algorithm. It is well known that the back-propagation algorithm is susceptible to getting trapped in a local minimum, in which case there will be misclassified training patterns. One method for assisting the

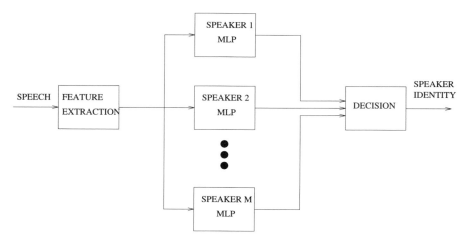

Figure 10.7 Speaker identification using MLPs.

convergence of the MLP is to preprocess the input feature space to improve the separability. This can be accomplished by using a variant of the MLP known as a functional link neural network. The functional link neural network was evaluated for a speaker verification task in [27]. Another method for alleviating the MLP convergence problem was evaluated in [15]. Here, the training data for antispeakers was first compressed using a VQ algorithm. Hence, if there were 20 antispeakers each with 1,000 training vectors, the VQ algorithm would be used to compress each set of 1,000 to a smaller set of vectors such as 128. By reducing the amount of training data, the convergence properties of the MLPs improved, such that larger populations could be evaluated. However, the performance of the MLP was still not as good as the VQ-based modeling approach.

Some MLP experiments for speaker verification were done in [20] where a 6% equal error rate was observed for an application using four speakers. The MLP has also been used to replace the observation probability estimates for an HMM [16] where it was shown that the resulting performance was better than that obtained by using the HMM observation probabilities. The MLP has also been used to model phonemes following segmentation for text-prompted speaker verification [17].

It has been noted [28] that since the MLP boundaries are unbounded at the edges of feature space, there may be problems with the rejection capabilities of the model. This problem is shown in Figure 10.8. Here it can be seen that the three discriminant boundaries successfully partition the training data, shown as zeros and ones. However, consider the case in which several class 0 observations, shown as X's, are encountered during testing. These cases will be classified as one, regardless of the distance that they are from the training data due to the open boundary. One solution to this problem is to use a bounded modeling approach, such as a radial basis function [29].

10.4.2 Radial Basis Functions

Radial basis function (RBF) networks [29] can be viewed as clustering followed by a perceptron. Consider the RBF in Figure 10.9. The training phase consists of first clustering the training data into M clusters. The centroids of these M clusters are used within the kernel functions, which are typically Gaussian kernels or sigmoids. The outputs of the kernel functions are used to train a single layer perceptron. When selecting the

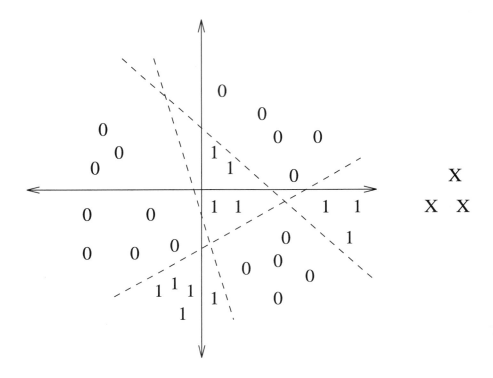

Figure 10.8 Discriminant boundaries.

kernel function and a Gaussian and the perceptron weights as mixture probabilities, the RBF network becomes equivalent to the GMM, with the exception that supervision is available.

Radial basis function networks have been considered for text-independent speaker verification [18]. The experiment was performed for a population of 40 speakers, for which 10 speakers were used as the enrolled speakers and the remaining 30 were used as imposters. The experiment evaluated the false-reject error for a fixed false-accept error of 1%. The RBF-based approach demonstrated a false-reject error of 8%, which was better than that of both the MLP (17%) and VQ (22%).

10.4.3   Time-Delay Neural Networks

Time-delay neural networks (TDNNs) [30] are layered feedforward neural networks that incorporate delayed versions of the inputs, so as to learn the temporal correlations of the input data. TDNNs are the

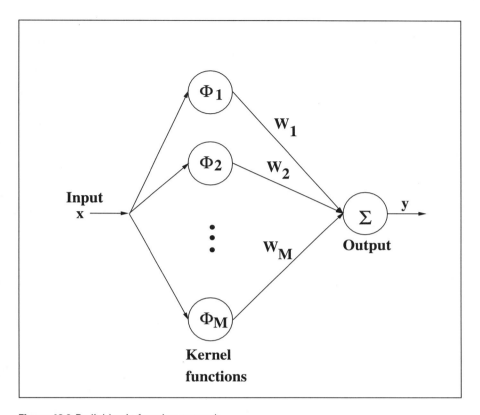

Figure 10.9 Radial basis function network.

supervised counterpart to the HMM in that they attempt to capitalize on the temporal information. TDNNs have been considered for text-independent speaker identification [23]. Here, TDNNs were evaluated for a population of 20 speakers (10 males and 10 females) and demonstrated a 98% identification rate.

10.4.4 Recurrent Neural Networks

A recurrent neural network is similar to the MLP with the exception that it uses feedback connections. One can make the analogy to filter theory, where a TDNN has a finite impulse response (FIR) structure and an RNN has an infinite impulse response (IIR) structure. Hence, RNNs represent another neural network modeling approach that incorporates temporal data.

Recurrent neural networks have been evaluated for text-dependent speaker verification [20]. The vocabulary consisted of digits and was collected for a population having four target speakers and 34 imposters. The equal error rate observed for this experiment was 7.5%. An MLP evaluated on the same test data provided an equal error rate of 6.0%.

10.4.5 Learning Vector Quantization

Learning vector quantization (LVQ) is similar to vector quantization, however, supervision is used in determining the centroids [31]. LVQ has been evaluated for speaker identification in [22]. The database consisted of 10 speakers (5 males and 5 females) and an identification rate of 97% was achieved.

10.4.6 Decision Trees

A decision tree represents a collection of rules, organized in a hierarchical fashion, that implement a decision structure. Each nonterminal

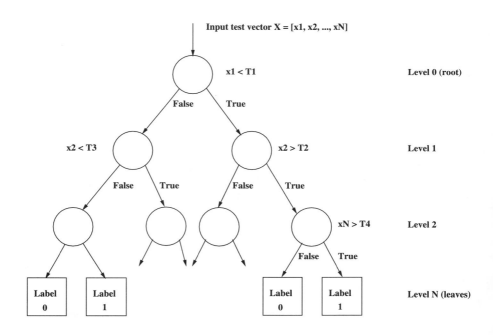

Figure 10.10 Decision tree.

(nonleaf) node of the tree represents a decision, and each terminal (leaf) node corresponds to a class. The leaves represent exclusive partitions of the input data. The concept of a decision tree is illustrated in Figure 10.10.

The decision tree shown here is a binary decision tree. A test vector will be evaluated at each node and directed to one of two subsequent nodes based on a decision regarding a selected element within a feature vector. This decision process continues until the test vector comes to a leaf, at which point it will be assigned the class label of that leaf.

The architecture of a decision tree is found by rule induction. The training algorithm for the ID3 [32] decision tree is briefly reviewed here. Given a set of training patterns $\{x_i; t_i\}$ an ID3 tree can be grown as follows.

Consider a collection C of objects or, in other words, input vectors. If $C = \{\emptyset\}$ or all $x_i \in C$ are of the same class, then make C a leaf. Otherwise, let T be a test on any object with w possible outcomes:

$$T(C_i) \in \{O_1, O_2, \ldots, O_w\} \tag{10.5}$$

where the outcome O_i represents membership to class i. The test T operates on an attribute, i.e., an element of a feature vector, and for a given node will select the attribute that will gain the most information. This is accomplished as follows.

Assume that C is comprised of p objects in class P and n objects in class N. It can then be said that an arbitrary vector x will belong to class P with a probability:

$$p(x \in P) = \frac{p}{p+n} \tag{10.6}$$

and likewise to class N with a probability:

$$p(x \in N) = \frac{n}{p+n} \tag{10.7}$$

If the decision tree is viewed as the source of a message (that consists of a P or a N), the expected information can be expressed as:

$$I(p, n) = -\frac{p}{p+n}\log_2\frac{p}{p+n} - \frac{n}{p+n}\log_2\frac{n}{p+n} \tag{10.8}$$

Now assume that each partition C_i has p_i objects within class P and n_i objects within class N. The expected information for attribute A is evaluated as the sum of the expected information for each class C_i:

$$E(A) = \sum_{i=1}^{v} \frac{p_i + n_i}{p + n} I(p_i, n_i) \qquad (10.9)$$

The information gained by branching on attribute A is defined as:

$$gain(A) = I(p, n) - E(A) \qquad (10.10)$$

The information gain represented by (10.10) is computed for all attributes at a given node. The attribute that maximizes this quantity will be used to partition the data at the node.

Given the attribute A which maximizes the information gain, use the test T to partition C into w mutually exclusive sets $\{C_1, C_2, \ldots, C_w\}$. A node is then grown for each of the C_w partitions and the algorithm is repeated. Decision trees can always be trained to classify all training patterns correctly assuming that there are no conflicting training patterns where two identical observations each have different class labels. Hence, decision trees are not susceptible to convergence problems as are MLPs. However, when a decision tree is trained to classify all training patterns, it is very possible that there will be problems with generalization. This corresponds to the case in which the decision tree has overtrained in order to classify outliers, which is likely to have a negative impact on the performance with test data. A solution to this problem is to remove branches of the tree that may have been grown to classify outliers. This process is known as pruning [33] and is effective in improving the generalization of decision trees. The C4 decision tree [34] is another decision tree that is similar to the ID3 decision tree with some modifications to the pruning algorithm.

Two examples of how a decision tree partitions feature space are shown in Figure 10.11. Consider the leftmost diagram in the figure. The root node of the decision tree will determine if the feature element x_1 is below a threshold T_1. If this condition is found to be true, then class x will be decided. If this condition is false, then the next node of the decision tree will determine if the feature element x_2 is above a threshold T_2. If this condition is true, then class o will be decided, otherwise class x.

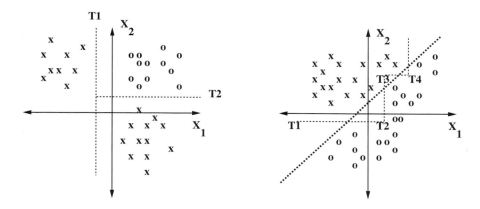

Figure 10.11 Decision tree partitioning of feature space.

A disadvantage of decision trees results from the fact that discriminant locations are determined by evaluating one feature at a time as opposed to using the composite feature vector. This restricts the discriminants to being perpendicular to the feature axes. This problem is illustrated in the right-side diagram of Figure 10.11. Numerous decisions must be made to separate the two classes, whereas a single diagonal discriminant could perform the same task. Each decision requires a node, hence, a four-node tree is necessary to solve a problem that is linearly separable. This same problem could be solved with a single neuron, which consists of one node.

Decision trees were evaluated for speaker identification in [15]. The specific decision tree algorithms included ID3 [32], C4 [34], and CART [33]. For a closed-set speaker identification experiment with 10 male speakers, the performance for the ID3, C4, and CART classifiers was 88%, 84%, and 76%, respectively. The same experiment was also evaluated with 20 male speakers and the resulting performance was 79% and 73% for the ID3 and C4 classifiers, respectively. Though the decision trees trained much faster than MLPs, their performance was not quite as good. One of the limitations of decision trees that may be attributed to this is the fact that they use only one feature element at a time when making their decision. This constrains the discriminant boundaries to be parallel to the feature axes and could require many decisions (or levels of the tree) to solve a problem that may even be separable. These problems are alleviated with the neural tree network described below.

10.4.7   Neural Tree Network

The neural tree network (NTN) [35] is a hierarchical classifier that uses a tree architecture to implement a sequential linear decision strategy. The NTN is a type of decision tree that uses a perceptron to split the data at each node. This overcomes the limitation of standard decision tree algorithms in that the discriminant boundaries are not constrained to being perpendicular to the feature axes. For speaker recognition, a binary NTN is used to solve a two-class problem, namely, whether or not the observation vector belongs to the target speaker. The binary NTN is recursively trained as follows. Given a set of training data at a particular node, if all data within that node belongs to the same class, the node becomes a leaf. Otherwise, the data is split into two subsets, which become the children of this node. This procedure is repeated on the children of the current node until all the data is completely separated and each leaf node contains data from only one class.

In order to prevent overtraining, an additional stopping criterion is used to halt the growth of the tree. This stopping criterion basically consists of stopping the growth of the tree beyond a prespecified level [15]. A relative frequency measurement can then be used at the leaves to estimate a class probability. This additional stopping criterion in essence is pruning the tree. This concept is illustrated in Figure 10.12. By stopping the growth of the NTN beyond a certain level, a measure at the new leaves can be made to determine the confusability of the data at that point.

The neural tree network was evaluated for both text-independent speaker identification and speaker verification in [15]. For the speaker identification experiment with 10 male speakers, the performance was 98%. For the speaker verification experiment, the equal error rate was 2.4%, which was also evaluated with 10 male speakers. These experiments were also evaluated for a data set with 20 male speakers and the corresponding performance was 96% for speaker identification and 2.8% equal error rate for speaker verification. The identification performance was equivalent to that obtained for a VQ-based approach and the verification performance was better than that of VQ.

The neural tree network has also been evaluated for several text-dependent speaker verification applications. A whole-word model implementation [36] demonstrated an equal error rate of 5.2%. Two subword implementations have also been investigated; one that used a transcription from an HMM segmentor had an equal error rate of 0.8%

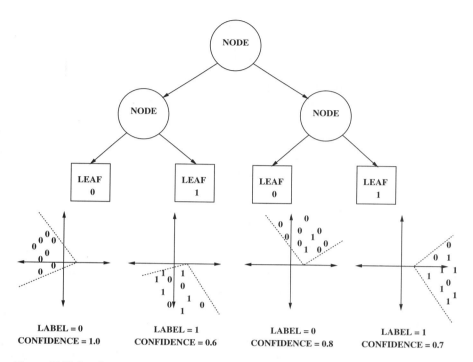

Figure 10.12 Pruning algorithm for the NTN.

[37] and another that used a blind segmentor to partition the utterance had an equal error rate of 2.4% [38].

10.4.8 Performance Summary

The results for the various neural network approaches described as applied to speaker identification are summarized in Table 10.1. The results observed for speaker verification are summarized in Table 10.2. Note that these results apply for different databases and different scenarios. Hence, one should not try to make a relative comparison of performance based on the results in this table.

10.5 Model Combination

The combination of different models has been found to be effective for reducing system error rates. Model combination can be performed to take advantage of information diversity and redundancy. Diversity can be used as a means for improving system performance through the

Table 10.1

Performance summary for speaker identification networks

Neural Network	Database	Error Rate
MLP [21]	10 speakers	8%
LVQ [22]	10 speakers	7%
TDNN [23]	20 speakers	2%
ID3 [15]	20 speakers	21%
C4 [15]	20 speakers	27%
MLP [15]	20 speakers	10%
NTN [15]	20 speakers	4%

incorporation of different information. Similarly, redundancy can achieve the same goals through the reuse of data. These concepts have been thoroughly explored in the field of communications and have also been applied to pattern recognition problems. The basic idea is that if several models can be constructed, whose errors are mutually uncorrelated, then performance advantages can be obtained through the proper combination of the model scores. For example, consider the scatter plot shown in Figure 10.13. This plot shows scores obtained from Gaussian mixture modeling (GMM) and NTN modeling approaches on the x-axis and y-axis, respectively, for a particular speaker. If either of these methods were used by itself, there would be errors. However,

Table 10.2

Performance summary for speaker verification networks

Neural Network	Database	False Accept	False Reject
RBF [18]	10 speakers	1%	8%
RNN [20]	4 speakers	7.5%	7.5%
MLP [20]	4 speakers	6.0%	6.0%
HMM-MLP [16]	100 speakers	1.0%	9.1%
NTN [15]	20 speakers	2.8%	2.8%
NTN [36]	20 speakers	5.2%	5.2%
NTN [37]	138 speakers	0.8%	0.8%
NTN [38]	30 speakers	2.4%	2.4%

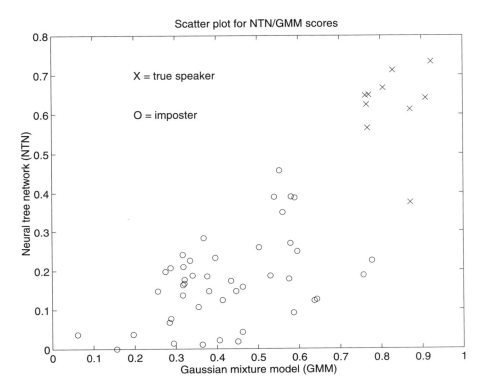

Figure 10.13 Scatter plot for model diversity.

when considering the two-dimensional problem in Figure 10.13, the two classes are now separable. A simple method of combining these scores is to project them onto the diagonal, which would result in the one-dimensional data set shown in Figure 10.14. The transformed data set shown here is a one-dimensional, separable data set.

Several diversity schemes are applicable for speaker recognition. These are data diversity, feature diversity [39], and model diversity [36, 38, 40, 41]. The data illustrated in Figure 10.13 is an example of model diversity. These different forms of diversity may be combined as long as they meet the criterion that the errors are uncorrelated. However, one must accept a tradeoff in memory and speed when using diversity. Each form of diversity that is used will result in an additional model, or models, which will require additional memory to store and processing time to evaluate. The merits of the performance gains must be weighed against these considerations.

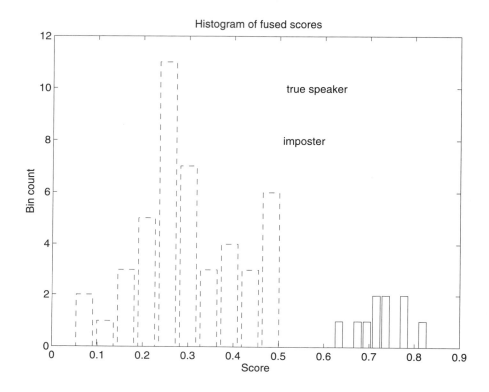

Figure 10.14 Histogram of scores projected onto diagonal.

Redundancy has also been considered for speaker verification through use of the "leave-one-out" method [42]. The leave-one-out method and other resampling techniques were originally proposed for estimating a statistic when only a small number of observations were available. The speaker recognition application is analogous in that the goal is to estimate a speaker model and threshold with a limited number of training repetitions. Resampling techniques have been recently applied to training neural networks [43], and specifically, the leave-one-out method was applied to text-dependent speaker verification [36]. Here, the four training repetitions were partitioned into four sets of three repetitions, where each set had a different repetition left out. The left-out repetition was used as an independent observation for that set so that data for estimating a threshold position could be obtained.

The combination of different sources of information has been explored within a field known by various names including data fusion, consensus building, team decision theory, combination of multiple experts, and a number of other titles. We will refer to the combination of different sources of information here as data fusion. The selection of data fusion techniques can be subdivided based on the type of information that will be combined. For example, if the model outputs are probabilities, then methods such as linear or log opinion pools can be used [44]. If the model outputs are actually class labels, then methods such as voting [45] or ranking [46] can be used. These methods will be described in more detail below.

10.5.1 Model Combination Approaches

10.5.1.1 Linear Opinion Pool

The linear opinion pool is a commonly used data fusion technique that is convenient due to its simplicity. The linear opinion pool is evaluated as a weighted sum of the scores for each model:

$$P_{linear}(x) = \sum_{i=1}^{n} \alpha_i p_i(x) \tag{10.11}$$

where $P_{linear}(x)$ is the probability of the combined system, α_i are weights, $p_i(x)$ is the probability output by the ith model, and n is the number of models. The parameters α_i are generally chosen such that α_i is between zero and one and the sum of the α_i's is equal to one.

The linear opinion pool is appealing in that the output is a probability distribution and the weights α_i provide a rough measure for the contribution of the ith model. However, it is noted that the probability distribution of the combiner output, namely $P_{linear}(x)$, may be multimodal. This may impose a more complicated decision strategy.

10.5.1.2 Log Opinion Pool

An alternative to the linear opinion pool is the log opinion pool. If the α_i weights are constrained to lie between zero and one and sum up to one, then the log opinion pool also outputs a probability distribution. However, as opposed to the linear opinion pool, the output distribution of the log opinion pool is unimodal [44].

Table 10.3
Error correlation between models

Model	HMM	GMM	NTN	DTW
HMM	0.085	0.056	0.035	0.029
GMM	0.056	0.084	0.049	0.036
NTN	0.035	0.049	0.075	0.038
DTW	0.029	0.036	0.038	0.055

The log opinion pool consists of a weighted product of the model scores:

$$P_{log}(x) = \prod_{i=1}^{n} p_i^{\alpha_i}(x) \tag{10.12}$$

Note that with this formulation, if any model assigns a probability of zero, then the combined probability will also be zero. Hence, an individual model has the capability of a "veto," whereas in the linear opinion pool the zero probability would be averaged in with the other probabilities.

10.5.1.3 Voting Methods

Another simple method for combining the results of multiple models is to use a voting procedure. In contrast to linear and log opinion pools, which combine model scores, the voting method combines model decisions. For the case of speaker verification, these decisions will be either yes or no and the voting method will decide the overall decision. Typically, an odd number of models is used, to avoid ties, and the final decision is based on a majority rule. The voting method has been applied to handwriting recognition [45].

10.5.2 Error Correlation Analysis

The error correlation of multiple models has been analyzed in [47] for speaker verification. Here, a binary data stream was output for each model where a "1" would correspond to an error and a "0" would correspond to a correct output. The error correlation was then computed as the cross correlation of these two binary sequences.

Table 10.4
Best equal error rates between models

Model	HMM	GMM	NTN	DTW
HMM	5.6	4.7	2.7	1.0
GMM	4.7	5.0	3.2	1.4
NTN	2.7	3.2	4.3	2.0
DTW	1.0	1.4	2.0	2.3

The error correlations were estimated between HMM, GMM, NTN, and DTW models. These errors were computed on the composite scores from these models, so there is only one score for each model for each verification utterance. The errors were measured for the point where the false-reject error was zero, so that errors could be measured solely in terms of false-accepts. The analysis yielded the following error correlation matrix shown in Table 10.3. The diagonal entries in Table 10.3 correspond to the false-accept error of the model for the case of zero false-rejects.

As one would expect, the GMM and HMM have the highest correlation of errors, since they are very similar with respect to the parameters that are estimated. The HMM here uses identical mixture probabilities as the GMM, but in addition uses the transition probabilities. By inspecting this matrix of error correlation, one can presume that the combination of the GMM and HMM will yield the least benefit and the combination of the HMM and DTW will yield the most benefit. This indeed was observed, as is illustrated in Table 10.4. The HMM combined with the GMM gave an equal error rate of 4.7% whereas the combination with DTW gave an equal error rate of 1.0%.

10.5.3 Two-Model Combination

Several studies have been performed for combining discriminative information with distortion-based or statistical models in speaker recognition applications. One such evaluation was based on a text-dependent speaker verification task that used whole-word models [36]. The two modeling approaches were dynamic time warping and the neural tree network. Each modeling approach used a leave-one-out

Table 10.5
Two-model performance summary

Models	Combining Method	Application	Performance
NTN + VQ [36]	linear	identification	17.3/10.2/7.1%
NTN + DTW [36]	voting	verification	5.2/9.8/4.0%
NTN + DTW [36]	log	verification	5.2/9.8/3.3%
NTN + DTW [36]	voting	verification	5.2/9.8/4.0%
NTN + GMM [38]	linear	verification	2.4/3.2/0.9%
DTW + VQ [41]	linear	verification	4.8/3.7/2.9%

training strategy to yield four models from four training repetitions. Hence, there were a total of eight models. The two model approaches were evaluated individually, and then combined using the linear and log opinion pools along with a voting strategy. The voting technique simply recorded the number of models that exceeded their individual thresholds. For this experiment, the linear opinion pool performed the best, with an equal error rate (EER) of 2.0% as compared with the EERs for the log opinion pool and voting methods, which were 3.3% and 4.0%, respectively.

Another evaluation was performed for text-dependent speaker verification using subword models [38]. In this approach, a blind segmentation algorithm was first used to partition the utterance into its segments. Each segment has an NTN and GMM trained for it. The linear opinion pool is then used to combine the scores at the phrase level. A substantial performance gain is also observed for this application. Another study was performed to evaluate the combination of DTW and VQ with cohort normalization for speaker verification [41]. In this research, the fused model was also found to perform better than the individual models. A summary of the two-model combination approaches is given in Table 10.5. The performance column in this table contains the performance of the first model, second model, and combined model, respectively.

10.5.4   Three-Model Combination

Studies were also performed for combining more than two modeling methods [48]. Here, the weight selection criterion becomes more complicated, since the dimension of scores increases. One popular method for computing an optimal weight vector for this task is the Fisher linear discriminant.

The Fisher linear discriminant is a method for reducing the dimensionality of a pattern recognition problem [49]. By projecting all data points onto a line, the dimension of a problem can be reduced to one. The Fisher linear discriminant is defined as the linear function that maximizes the criterion function:

$$J(w) = \frac{|\hat{m}_1 - \hat{m}_2|^2}{\hat{s}_1^2 - \hat{s}_2^2} \qquad (10.13)$$

which represents the ratio of the mean separation to the sum of the variances. This can be accomplished by projecting the observation vectors onto the line defined by:

$$w = S_w^{-1}(\hat{m}_1 - \hat{m}_2) \qquad (10.14)$$

where $\hat{m}_1$ and $\hat{m}_2$ are the means of the data for class 1 and class 2 and S_w is the scatter matrix defined as:

$$S_w = \hat{\Sigma}_1 + \hat{\Sigma}_2 \qquad (10.15)$$

Here, $\hat{\Sigma}_1$ and $\hat{\Sigma}_2$ are the sample covariance matrices of the data from class 1 and class 2, respectively. The solution for w in (10.14) will maximize the criterion function of (10.13).

A more generalized method for determining the scatter matrix S_w has also been evaluated as [50]

$$S_w = s\hat{\Sigma}_1(1 - s)\hat{\Sigma}_2 \qquad (10.16)$$

which allows one to weight the sample covariance matrices for each class through the weight parameter s where $0 \leqslant s \leqslant 1$.

The Fisher linear discriminant was analyzed as a method to compute the optimal weight vector for a three-model fusion problem within speaker verification [48]. The three modeling approaches consisted of the HMM, NTN, and DTW. The performance was evaluated for three separate environments representing data within wireless telephony, landline telephony, and multimedia (PC micro-

Table 10.6
Three-model performance summary

Models	Environment	Individual Performance	3-Model Performance (Exhaustive/Fisher)
HMM/NTN/DTW	wireless	9.83/8.58/5.99%	3.71/3.75%
HMM/NTN/DTW	landline	2.29/1.44/2.34%	0.84/0.95%
HMM/NTN/DTW	multimedia	2.48/3.32/2.27%	0.03/0.85%

phone). The analysis compared the performance corresponding to the weights for the Fisher linear discriminant in addition to the optimal weights found by an exhaustive search over the weight space. For the Fisher linear discriminant, the scatter matrix was computed by using equal weight of the covariance matrix estimates for model scores of the speaker and imposter as represented in (10.15). The results of these experiments are shown in Table 10.6, which includes the individual model performance as well. In two out of three cases, the Fisher weights performed comparably to the weights found by the exhaustive search. In the case in which it did not, i.e., the multimedia environment, it was conjectured that the smaller data size used for this experiment as compared with the larger data sets used for the wireless and landline scenarios was to blame.

10.6 Summary

In this chapter, we have discussed the application of neural networks to speaker recognition. Recent research has shown neural networks to be a viable alternative to the traditional statistical models and distortion-based techniques. The research performed thus far has shown a promising future for neural networks in speaker recognition applications, particularly due to their exploitation of discriminative information.

One of the drawbacks with many of the neural-network-based approaches is that the training time can be excessive for large data sets. Additionally, convergence issues arise and scaling problems for large populations have also been encountered. Currently, there are some solutions to these problems. Hierarchical classifiers, such as decision trees and neural tree networks, provide fast training algorithms and are not subject to the problems of local minima that can inhibit convergence in other modeling approaches. Whereas a multilayer perceptron can get stuck in a local minima, i.e., leaving training patterns misclassified, a decision tree or neural tree network can always be trained to classify all training patterns correctly. Regarding the problem of scaling to large populations, research has been performed for modular approaches to large-population speaker identification applications. Here, a number of connectionist models can be trained for subpopulations of a larger population and then combined with another neural network. This has been evaluated for both a TDNN-type network [51] and a hierarchical approach [52].

Also, the concept of data fusion playing a more prominent role in speaker recognition applications has gained attention. It is clear that the criteria used to train neural-network-based models is quite different from that used to train statistical or distortion-based models. Hence, one can expect their outputs to be uncorrelated to some degree. By combining the outputs of several diverse models, one can expect a substantial improvement in performance. Many experiments in this area [36, 40] have shown a reduction in the error rate by at least a factor of two.

It is the opinion of the author that neural networks will play a more prominent role in speaker verification as opposed to speaker identification. Speaker verification is not subject to the scaling problems encountered in identification, as the performance is generally independent of population size. Also, in terms of performance, neural networks have been shown in numerous cases to outperform the distortion-based or statistical models for speaker verification. This is primarily due to the power of supervised training algorithms for incorporating discriminative information, as compared with the cohort normalization techniques that are used for nondiscriminative modeling approaches.

Acknowledgments

I would like to thank my colleagues at T-NETIX/SpeakEZ and the CAIP Center at Rutgers University for numerous fruitful discussions contributing to the material in this chapter.

References

[1] Furui, S., "Cepstral Analysis Technique for Automatic Speaker Verification," *IEEE Trans. on Acoustics, Speech, and Signal Processing*, Vol. ASSP-29, April 1981, pp. 254–272.

[2] Rosenberg, A. E., C. H. Lee, and S. Gokeen, "Connected Word Talker Recognition Using Whole Word Hidden Markov Models," *Proc. ICASSP*, 1991, pp. 381–384.

[3] Naik, J. M., L. P. Netsch, and G. R. Doddington, "Speaker Verification over Long Distance Telephone Lines," *Proc. ICASSP*, 1989, pp. 524–527.

[4] Rosenberg, A. E., C. H. Lee, and F. K. Soong, "Sub-Word Unit Talker Verification Using Hidden Markov Models," *Proc. ICASSP*, 1990, pp. 269–272.

[5] Matsui, T., and S. Furui, "Speaker Adaptation of Tied-Mixture-Based Phoneme Models for Text-Prompted Speaker Recognition," *Proc. ICASSP*, 1994, pp. 1125–1128.

[6] Soong, F. K., A. E. Rosenberg, L. R. Rabiner, and B. H. Juang, "A Vector Quantization Approach to Speaker Recognition," *Proc. ICASSP*, 1985, pp. 387–390.

[7] Reynolds, D., "Speaker Identification and Verification Using Gaussian Mixture Models," *Speech Communications*, Vol. 17, August 1995, pp. 91–108.

[8] Castellano, P., and S. Sridharan, "A Two Stage Fuzzy Decision Classifier for Speaker Identification," *Speech Communications*, Vol. 18, February 1996, pp. 139–149.

[9] Atal, B. S., "Effectiveness of Linear Prediction Characteristics of the Speech Wave for Automatic Speaker Identification and Verification," *J. Acoust. Soc. of Am.*, Vol. 55, June 1974, pp. 1304–1312.

[10] Katagiri, S., B. H. Juang, and A. Biem, "Discriminative Feature Extraction," in R. J. Mammone, ed., *Neural Networks for Speech and Vision Processing*, Chapman and Hall, 1993.

[11] Juang, B. H., and S. Katagiri, "Discriminative Learning for Minimum Error Classification," *IEEE Trans. on Signal Processing*, Vol. 40, No. 12, December 1992, pp. 3043–3054.

[12] Liu, C. S., C. H. Lee, B. H. Juang, and A. E. Rosenberg, "Speaker Recognition Based on Minimum Error Discriminative Training," *Proc. ICASSP*, 1994, pp. 325–328.

[13] Higgins, A., and L. Bahler, "Text-Independent Speaker Verification by Discriminator Counting," *Proc. ICASSP*, 1991, pp. 405–408.

[14] Rosenberg, A. E., J. Delong, C. H. Lee, B. H. Juang, and F. K. Soong, "The Use of Cohort Normalized Scores for Speaker Recognition," *Proc. ICSLP-92*, October 1992, pp. 599–602.

[15] Farrell, K. R., R. J. Mammone, and K. T. Assaleh, "Speaker Recognition Using Neural Networks and Conventional Classifiers," *IEEE Trans. Speech and Audio Processing*, Vol. 2, No. 1, Part II, 1994.

[16] Naik, J. M., and D. M. Lubensky, "A Hybrid HMM-MLP Speaker Verification Algorithm for Telephone Speech," *Proc. ICASSP*, 1994, pp. 153–156.

[17] Delacretaz, D. P., and J. Hennebert, "Text-Prompted Speaker Verification Experiments with Phoneme Specific MLPs," *Proc. ICASSP*, 1998, pp. 777–780.

[18] Oglesby, J., and J. S. Mason, "Radial Basis Function Networks for Speaker Recognition," *Proc. ICASSP*, 1991, pp. 393–396.

[19] Hattori, H., "Text-Independent Speaker Verification Using Neural Networks," *Proc. ESCA Workshop Automatic Speaker Recognition, Identification and Verification*, Martigny, Switzerland, 1994.

[20] Tsoi, A. C., D. Shrimpton, B. Watson, and A. Black, "Application of Artificial Neural Network Techniques to Speaker Verification," *Proc. ESCA Workshop Automatic Speaker Recognition, Identification and Verification*, Martigny, Switzerland, 1994.

[21] Oglesby, J., and J. S. Mason, "Optimization of Neural Models for Speaker Identification," *Proc. ICASSP*, 1990, pp. 261–264.

[22] Bennani, Y., and P. Gallinari, "A Connectionist Approach for Speaker Identification," *Proc. ICASSP*, 1990, pp. 265–268.

[23] Bennani, Y., and P. Gallinari, "On the Use of TDNN-Extracted Features Information in Talker Identification," *Proc. ICASSP*, 1991, pp. 385–388.

[24] Farrell, K. R., and R. J. Mammone, "Speaker Identification Using Neural Tree Networks," *Proc. ICASSP*, 1994, pp. 165–168.

[25] Morgan, D. P., and C. L. Scofield, *Neural Networks and Speech Processing*. Norwell, Mass.: Kluwer Academic, 1991.

[26] Rumelhart, D. E., and J. L. McClelland, *Parallel Distributed Processing*. Cambridge, Mass.: MIT Cambridge Press, 1986.

[27] Castellano, P., and S. Sridharan, "Text-Independent Speaker Identification with Functional-Link Neural Networks," *Proc. ESCA Workshop Automatic Speaker Recognition, Identification and Verification*, Martigny, Switzerland, 1994.

[28] Lastrucci, L., M. Gori, and G. Soda, "Neural Autoassociators for Phoneme-Based Speaker Verification," *Proc. ESCA Workshop on Automatic Speaker Recognition, Identification and Verification*, Martigny, Switzerland, 1994.

[29] Broomhead, D., and D. Lowe, "Multivariable Functional Interpolation and Adaptive Networks," *Complex Systems*, Vol. 3, 1988, pp. 269–303.

[30] Waibel, A., T. Hanazawa, and G. Hinton, "Phoneme Recognition with Time Delay Neural Networks," *IEEE Trans. on Acoustics, Speech, and Signal Processing*, Vol. ASSP-37, March 1989, pp. 328–339.

[31] Kohonen, T., *Self Organization and Associative Memory*. New York: Springer-Verlag, 1989.

[32] Quinlan, J., "Induction of Decision Trees," *Machine Learning*, Vol. 1, 1986, pp. 81–106.

[33] Breiman, L., J. H. Friedman, R. A. Olshen, and C. J. Stone, *Classification and Regression Trees*, Belmont, Calif.: Wadsworth, 1984.

[34] Quinlan, J., "Simplifying Decision Trees," in G. Gaines and J. Boose, eds., *Knowledge Acquisition for Knowledge-Based Systems*, London: Academic Press, 1988.

[35] Sankar, A., and R. J. Mammone, "Growing and Pruning Neural Tree Networks," *IEEE Trans. on Computers*, Vol. C-42, March 1993, pp. 221–229.

[36] Farrell, K. R., "Text-Dependent Speaker Verification Using Data Fusion," *Proc, ICASSP*, 1995, pp. 349–352.

[37] Liou, H., and R. J. Mammone, "Text-Dependent Speaker Verification Using Sub-Word Neural Tree Networks," *Proc. Conf. Automatic Inspection and Identification of Humans*, SPIE, July 1994.

[38] Sharma, M., and R. J. Mammone, "Subword-Based Text-Dependent Speaker Verification System with User Selectable Passwords," *Proc. ICASSP*, 1996, pp. 93–96.

[39] Soong, F. K., and A. E. Rosenberg, "On the Use of Instantaneous and Transitional Spectral Information in Speaker Recognition," *IEEE Trans. on Acoustics, Speech, and Signal Processing*, Vol. ASSP-36, June 1988, pp. 871–879.

[40] Farrell, K. R., S. Kosonocky, and R. J. Mammone, "Neural Tree Network/Vector Quantization Probability Estimators for Speaker Recognition," *Proceedings of Neural Networks for Signal Processing*, September, 1994.

[41] Schalkwyk, J., N. Jain, and E. Barnard, "Speaker Verification with Low Storage Requirements," *Proc. ICASSP*, 1996, pp. 693–696.

[42] Lachenbruch, P. A., and M. R. Mickey, "Estimation of Error Rates in Discriminant Analysis," *Technometrics*, Vol. 10, February 1968, pp. 1–11.

[43] Perrone, M. P., and L. N. Cooper, "When Networks Disagree: Ensemble Methods for Hybrid Neural Networks," in R. J. Mammone, ed., *Neural Networks for Speech and Vision Processing*, Chapman and Hall, 1993.

[44] Benediktsson, J. A., and P. H. Swain, "Consensus Theoretic Classification Methods," *IEEE Trans. on Systems, Man and Cybernetics*, Vol. 22, No. 4, 1992, pp. 688–704.

[45] Xu, L., A. Krzyzak, and C. Y. Suen, "Methods of Combining Multiple Classifiers and Their Applications to Hand-written Character Recognition," *IEEE Trans. on Systems, Man and Cybernetics*, Vol. 23, No. 3, 1992, pp. 418–435.

[46] Ho, T. K., J. J. Hull, and S. N. Srihari, "Decision Combination in Multiple Classifier Systems," *IEEE Trans. on Pattern Analysis and Machine Intelligence*, Vol. 16, No. 1, 1994, pp. 66–75.

[47] Farrell, K. R., R. P. Ramachandran, and R. J. Mammone, "An Analysis of Data Fusion Methods for Speaker Verification," *Proc. ICASSP*, 1998, pp. 1129–1132.

[48] Farrell, K. R., "Model Combination and Weight Selection Criteria for Speaker Verification," *Neural Networks for Signal Processing, IEEE*, August, 1999.

[49] Duda, R. O., and P. E. Hart, *Pattern Classification and Scene Analysis*. New York: Wiley, 1973.

[50] Fukunaga, K., *An Introduction to Statistical Pattern Recognition*. San Diego, Calif.: Academic Press, 1990.

[51] Bennani, Y., "Speaker Identification Through a Modular Connectionist Architecture: Evaluation on the Timit Database," *Proc. ICSLP-92*, October 1992, pp. 599–602.

[52] Rudasi, L., and S. A. Zahorian, "Text-Independent Talker Identification with Neural Networks," *Proc. ICASSP*, 1991, pp. 389–392.

11

Neural Networks for Voice Conversion

Bayya Yegnanarayana, Malayath Narendranath

11.1 Introduction: Speech and Speaker Characteristics

The speech signal contains information about both the intended message and the identity of the speaker. Extracting the message from the signal is the main research activity in the area of speech recognition [1, 2]. The area of speaker recognition and verification deals with techniques to extract speaker-specific information from the speech signal [3–5]. Human beings are able to effortlessly recognize a familiar speaker from his or her speech. It is necessary to identify and extract this speaker-specific information contained in the speech signal in order to perform voice conversion. Voice conversion involves transformation of the speaker characteristics in a speech uttered by the source speaker so as to generate speech with the characteristics of the target speaker.

In the development of a voice conversion system, speaker-specific knowledge is identified and extracted by analyzing speech data from the source and the target speakers. Using this knowledge, the mapping involved in transforming the characteristics of the source speaker into those of the target speaker is captured. This may be viewed as the learning phase. In the transformation phase, the parameters extracted from the speech of the source speaker are transformed using the mapping function captured in the learning phase. Finally, the speech with the voice characteristics of the target speaker is synthesized using the transformed parameters.

Voice conversion can be speaker dependent or speaker independent. In both the cases the identity of the target speaker is fixed. In speaker-dependent voice conversion, the source speaker is also fixed. The task is to transform the voice characteristics of the speech of the source speaker to those of the target speaker. In speaker-independent voice conversion, the task is to transform the voice characteristics of *any* speaker, so that the transformed speech sounds like that of the target speaker. Incorporation of speaker-specific characteristics into a text-to-speech system to produce speech with the desired (target) voice can also be considered as a speaker-independent voice conversion.

In developing a voice conversion system, we must first identify the factors in the speech signal that are responsible for giving individuality

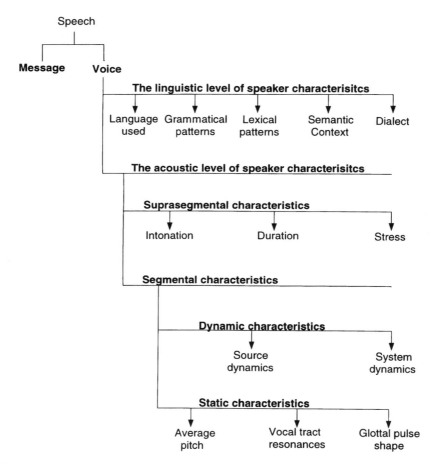

Figure 11.1 Line diagram showing various speaker-dependent knowledge sources.

to the speech of a speaker. Speaker characteristics exist at various levels. Figure 11.1 shows the knowledge sources used at various levels for producing and perceiving voice characteristics. At the highest level, namely the linguistic level, we use factors such as language, dialect, syntactic structures, and semantic context for the identification of a speaker from his or her speech. The characteristics of a speaker at this level are difficult to analyze and model, although these characteristics are used by human beings for recognizing speakers from spontaneous speech.

The factors at the acoustic level can be extracted directly from the speech waveform. The acoustic-level characterization can be divided further into suprasegmental and segmental levels. At the suprasegmental level, prosodic features such as intonation, duration, and stress carry significant speaker-specific information. After the linguistic factors, the prosodic features are the most important speaker-specific characteristics that human beings use in recognizing speakers. At the segmental level, the source and system characteristics of the speech production mechanism reflect the speaker characteristics. The source characteristics refer to the physiology of the vocal folds. The system characteristics refer to the shape and size (mainly the effective length) of the vocal tract. The segmental speaker characteristics have both dynamic and static parts. The dynamic part of the speech production includes both the vocal tract system dynamics and the glottal source dynamics. These dynamic features are determined by the sound units and are thus dictated by the text to a large extent. Static speaker characteristics refer to the shape of the glottal pulse, average pitch, average length of the vocal tract system, and the physical characteristics of the nasal tract.

Speaker characteristics at the linguistic level are significant, especially in spontaneous speech. In a reading style the speaker characteristics at the linguistic level are significantly reduced in the speech data. The interspeaker variations at the suprasegmental level can be attributed to several complex mental phenomena. These variations have no relation to any physical system. The prosodic characteristics of a speaker at the suprasegmental level are normally derived by analyzing large amounts of speech data. Thus, acquisition of prosodic knowledge involves significant manual effort [6, 7]. In general, analysis and modeling of speaker characteristics at the linguistic and suprasegmental levels are difficult tasks.

Segmental characteristics are directly related to the physiology of the speech production system. Therefore, at the segmental level it is possible to transform the features of the source speaker into features corresponding to the target speaker. The problem of voice conversion using the information at the segmental level can be understood from the nature of the speech production mechanism and its manifestation in the speech signal. During normal speech production, the time-varying excitation source produces time-varying fundamental frequency (F0). This is called pitch contour or intonation contour. The average F0 of a male speaker is significantly lower than the average F0 of a female speaker. Thus, in order to perform a voice transformation across two voices, the average F0 must be appropriately modified. There will be significant differences in the shapes of the glottal pulses for different speakers [8]. Figure 11.2 shows the approximate glottal pulses derived for segments of male and female voices. A glottal pulse is characterized by the closed and open glottis intervals, besides the rate of closure near the instant of significant excitation [9, 10]. Typically, female voices will have much shorter intervals of the closed glottis region compared with male voices. Also, the rate of closure is normally much higher for male than for female voices.

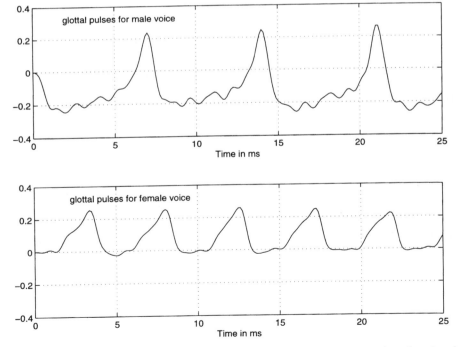

Figure 11.2 Glottal pulses for a sequence of male and female voices indicating closed and open phases in each cycle.

Apart from the interspeaker variations in the source characteristics, the vocal tract system also contributes to speaker variability. The interspeaker variations in the vocal tract system are manifested as variations in the formant frequencies (vocal tract resonances). The dynamics of the vocal tract system is reflected in the form of formant transitions. The formant frequencies are significantly higher for female speech in comparison with those for male speech. Figure 11.3 shows the linear prediction spectra for a segment of voiced speech for male and female voices. The locations of the spectral peaks indicate the formant frequencies.

For realizing voice conversion, the formants extracted from the speech of the source speaker have to be appropriately transformed using a mapping function. This mapping function has to be captured from a limited amount of formant data. The function should be able to modify the formants extracted from any utterance of the source speaker. In addition, the modification should not introduce distortion in the form of discontinuities in the formant transitions in continuous speech.

To capture the mapping of the vocal tract system characteristics between two speakers, the vocal tract system must be represented in a

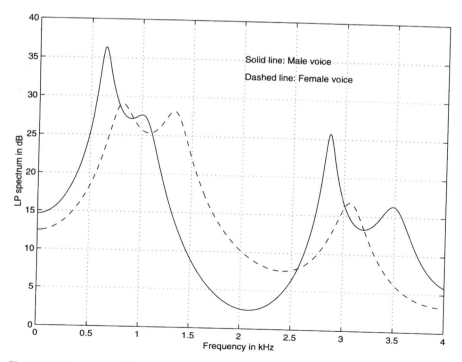

Figure 11.3 Linear prediction spectra for a voiced segment of male and female voices.

suitable manner. One can use the envelope of the short-time spectrum of a speech segment to represent the vocal tract system. But this representation is not motivated by the mechanism of speech production. If the vocal tract system is modeled by a linear time-varying system represented by a set of time-varying parameters, the corresponding representation takes into account the speech production mechanism to some extent. But the effectiveness of these representations depends on how well the source and system characteristics can be separated. From a transformation point of view, it is desirable to represent the vocal tract system using articulatory parameters. But generally it is difficult to extract the articulatory parameters from the speech signal. Hence, as a compromise between spectral and articulatory parameters, formants may be used for representing the vocal tract system. Formants are the resonances of the vocal tract system, and thus they are closely related to the physiology of speech production. At the same time, in comparison with the articulatory parameters, formants are easier to extract from the speech signal.

From a speech perception point of view, voiced segments, especially vowels, carry significant speaker-specific information compared with unvoiced consonants [11, 12]. There are two main reasons for this: (1) vowel sounds are spectrally well defined and thus carry significant information of the vocal tract shape [1] and (2) consonant sounds are dynamic in nature, and their durations are usually less than the durations of vowels. While perceiving consonants, a listener pays more attention to the message part than to the speaker characteristics. This argument need not be valid for consonants such as laterals and nasals, which carry significant speaker-specific information.

Even though in the above discussion speaker characteristics are attributed to the segmental factors, the real voice characteristics of a speaker are present mainly in the manner of production acquired by the speaker over years. These learned factors may be present either throughout an utterance (gross prosodic features) or only in specific segments (segment-specific prosodic features) of an utterance. For voice transformation, both the gross and the segment-specific prosodic features of the target speaker have to be incorporated into the synthesis, in addition to the segmental characteristics of the speaker.

As mentioned earlier, speaker characteristics largely manifest at the suprasegmental level of speech. Important suprasegmental features are intonation, stress, and duration. Pauses and speaking rate also reflect speaker characteristics [7, 13]. The range of variability in the suprasegmental features is so large that it is difficult to derive the transformation characteristics of these features. Of all the supra-segmental features, intonation provides maximum speaker-specific information. Syllable duration and word duration also vary from speaker to speaker. Also, each speaker may have unique characteristics at the suprasegmental level in producing certain syllable sequences or words or phrases. Identifying such unique utterances for each speaker and representing speaker-dependent suprasegmental knowledge is a challenging task.

Transformation of speaker-specific characteristics at the supra-segmental level can be carried out only through a set of rules derived by analyzing a large number of speech utterances [6, 7, 14]. This is a difficult and time-consuming task. On the other hand, it is possible to capture the transformations of parameters at the segmental level, such as fundamental frequency (F0), glottal pulse shape, spectral parameters, formants, etc. It is generally difficult to extract the source parameters, which describe the glottal pulse shape, from speech data. What can effectively be transformed is the vocal tract shape, since it is possible to capture the mapping function of the vocal tract parameters from the source speaker to the target speaker. But this requires suitable choice of parameters to represent the vocal tract system. It is interesting to note that articulatory parameters give a direct representation of the vocal tract shape and are suitable for capturing the mapping function. But they are more difficult to extract from the speech signal. On the other hand, parameters such as linear prediction coefficients (LPC) and spectral band energies are easy to extract but are difficult to use for deriving the mapping function. Formants are a good choice, since they provide a good acoustic description of the vocal tract shape, and they can also be extracted from the speech signal. It has been observed that the generalization of the captured mapping is more effective if the parameters are closer to the description of the vocal tract shape [15].

We first attempt to obtain a transformation between the vocal tract systems of the source and target speakers using a single linear trans-formation. The linear transformation is derived using the formants extracted from isolated utterances of vowels. Error between the trans-formed formants and the desired formants of the target speaker can be reduced significantly by using piecewise linear transformations. But piecewise linear transformations are capable of transforming formants extracted from steady vowels only. This transformation will introduce discontinuities in the formant transition regions. Thus the studies on linear formant transformation and piecewise linear transformation establish that the formant transformation between two speakers is highly nonlinear.

A feedforward neural network with nonlinear processing elements is capable of capturing arbitrary functional relationships [16]. Therefore, such a network may be useful in capturing the inherently nonlinear transformation of the formants. The network can be trained using the formant data extracted from isolated utterances of vowels for two speakers. Even though such a network can be shown to transform formant transitions without introducing discontinuities, the transformed formant transitions are not as smooth as the desired transitions of the target speaker. The failure of the network in transforming formant transitions is due to lack of generalization capability of the network for this pattern mapping task. The problem can be circumvented by using representative data to train the neural network. The generalization capability of the network can be verified using synthetic test data.

A trained neural network can be used for voice conversion. The following are the tasks involved in developing a voice conversion system:

1. Data collection for study of speaker-specific characteristics;
2. Acoustic-phonetic analysis of speech data;
3. Identification of speaker-dependent features;
4. Signal processing algorithms for features extraction;
5. Development of techniques for capturing the mapping function between speaker-dependent features of two speakers;
6. Synthesis of speech incorporating the desired voice characteristics;
7. Evaluation of quality of converted speech.

In Section 11.2 we will briefly review some of the attempts at voice conversion. The transformation of the characteristics of the vocal tract system characteristics is developed in Section 11.3, which also discusses the generalization capability of the transformation captured by the neural network. In Section 11.4 we will present the details of implementation of a voice conversion system using the transformation captured by a neural network. Issues related to source parameters and evaluation of the voice conversion system are also discussed.

11.2 Studies In Voice Conversion

The first attempt at voice conversion was reported in [17], in which Atal and Hanauer described the application of linear prediction vocoder for modifying the characteristics of a male voice into those of a female voice. The parameters used for modification of the speaker characteristics are pitch and formant frequencies and their bandwidths. Seneff [18] suggested a method to modify the speaking rate and pitch of a speech signal without extracting the pitch information separately. In these studies voice conversion was demonstrated as a possible application of the methods of speech processing proposed by the authors.

The work of Childers et al. [19] can be considered as one of the first attempts to focus on the problem of voice conversion in its own right. In this method, utterances of the same sentence by a male and a female speaker were analyzed to extract the speaker-dependent information [8, 19]. Using electroglottograph (EGG) measurements, the time intervals of the closed- and open-phase parts of the glottal waveform [20] were measured for different segments of utterances of the source and the target speakers. A scale factor was derived using the average values of the closed- and open-phase durations of the two speakers. The first three formants for the corresponding segments in the source and target speaker's utterances were extracted. From the formant data, the scale factors for the three formants were derived. The scale factor for pitch was also derived from the average pitch values of the source and target speakers. For voice conversion, the roots of the LPC polynomial for each frame of the utterance of the source speaker were modified. The roots corresponding to the first three formants were shifted in the z-plane using the scale factors derived during the analysis phase. The LPCs were recomputed from the modified roots. Speech was synthesized using the modified LPCs and average pitch. The open- and closed-phase intervals modified by the scale factors were used in the Fant's model of glottal excitation for voiced segments. Random noise with appropriate gain was used for unvoiced segments. The synthesized speech was of good quality, and possessed the characteristics of the target speaker's voice. What was strikingly lacking in the approach was the transformation of the speaker characteristics at the suprasegmental level.

In the work of Slifka and Anderson [21], the scale factors for modifying the LPC roots were computed statistically. The method was, however, not suitable for transforming the dynamic characteristics of the vocal tract system.

The method for voice conversion proposed by Abe et al. [22] considers pitch, energy, and spectral parameters as speaker-dependent features. Spectral parameters were extracted from the utterances of source and target speakers, and vector quantized. Similarly, the pitch values were scalar quantized. The correspondence between frames of the source and target speakers was established using a dynamic time-warping algorithm. Using the result of the warping algorithm the correspondence between a code vector of the source speaker and the code vectors of the target speaker was established. For each code vector of the source speaker, the distribution of the corresponding code vectors of the target speaker was obtained. The distribution was used to represent each vector in the source speaker's codebook as a linear weighted sum of the vectors in the target speaker's codebook. This correspondence is termed as mapping codebook. In the case of pitch frequency and gain, scalar quantization was used and the mapping codebooks for these parameters were defined based on the maximum occurrence in the histogram. In the transformation phase, the parameters extracted from the speech of the source speaker were quantized using the source speaker's codebook. Using the mapping codebook, the corresponding vectors in the target speaker's codebook were determined. Speech was synthesized using the transformed parameter vectors.

In a later work reported by Mizuno and Abe, formant frequencies were modified using piecewise linear transformation rules to realize voice transformation [23, 24]. The basic methodology is the same as that suggested by Abe et al. in [22], except for the following points: (1) instead of using spectral parameters, formants were used, (2) spectral tilt was also considered for conversion, and (3), instead of a mapping codebook, piecewise linear formant transformation rules were used to transform the formant frequencies and the spectral tilt.

In a cross-language voice conversion, the objective is to preserve the voice characteristics of the source speaker, when speech is translated from one language to another language and synthesized in the target language. In a translation from Japanese to English, Abe et al. used the technique of cross-language voice conversion [25]. The translated text was synthesized using the MITalk system so that the speech sounded like that of the Japanese speaker. In this case the target speaker was a Japanese and the MITalk system represented the source speaker. The voice transformation was realized using the mapping codebook technique [22].

Abe described a voice conversion system using segments of speech units [26]. Speech of the source speaker was given to a speech recognition system for segmentation and labeling. Speech segments identified by the speech recognition system were replaced by the speech segments uttered by the target speaker. This system has the drawback that it depends on the performance of the speech recognition system.

In the pitch-synchronous overlap/add (PSOLA) method [27], a source-system decomposition was used to perform prosodic and spectral transformations [27, 28]. Prosodic modifications were applied on the excitation signal using the TD-PSOLA [29] technique. For spectral transformation, sentences uttered by the source and target speakers were time aligned by Dynamic Time Warping (DTW). This provides a mapping between the acoustic spaces of the two speakers. From this mapping, the required spectral transformation was learned. First the acoustic space of the reference speaker was partitioned by means of vector quantization (VQ). The transformations associated with different classes were derived in the training phase. Two methods were investigated for learning such a transformation, namely linear multivariate regression (LMR) and dynamic frequency warping (DFW). Cepstral coefficients were used to represent each of the analysis frames of the input speech. The class to which the cepstral vector belongs was then identified by finding the nearest code-vector. Then the transform related to this class was applied to the cepstral vector. This can be either the linear transformation (captured by the LMR technique) or the warping function (captured by the DFW technique). An LPC parameter vector was derived from the transformed cepstral or spectral vector, which was used in the synthesis of speech to reflect the voice characteristics of the target speaker. This method worked well for short words. But for sentences, due to poor time alignments, the quality of the spectral transformation was poor.

An effort to make detailed prosodic modifications was reported by Arslan and Talkin [30]. In their work, the labeling provided by an HMM was used to carry out segment-dependent durational transformation. They also modified the pitch range of the source speaker to match that of the target speaker.

Stylianou et al. proposed a probabilistic framework for transforming the spectral parameters [31]. They used the HNM (harmonic + noise model) decomposition for parameter estimation. The source speaker's parameter distribution is modeled by a Gaussian mixture model (GMM). The transformation is represented by a continuous parametric function that takes into account the probabilistic classification provided by the

Table 11.1
Categories of approaches for voice conversion

Basis	Parameters/Features	Where These Approaches Are Adopted
Speech production	Articulators, Glottal vibration	. . .
Acoustic features	Formants Glottal Wave, Pitch	[8, 9, 23, 24]
Parametric representation	Spectral Cepstral, LPCs	[17, 18, 21]
Speech recognition	VQ, GMM, HMM	[22, 25, 26, 30, 31, 32]
Perception	Prosodic & Syllabic features	[27, 28]

GMM. This method was extended by Kain and Macon [32] by modeling the transformation as a joint probability density function of the source and the target spectral vectors.

From the review of work in the area of voice conversion, one may organize the various attempts into one or more of the broad categories listed in Table 11.1. In the speech-production-based approaches, the variations in the vocal tract system and excitation source are represented in the movements of the articulators and glottal vibration. Voice transformation determined at this level is bound to be accurate, since the production parameters are used to synthesize the speech. However, it is difficult to extract these articulatory parameters from the speech signal. Approaches based on acoustic features of speech production use parameters such as formants, pitch, and glottal waveform. It is possible to extract these features from the speech signal. These features are closer to speech production mechanisms than to the spectral parameters extracted directly from the signal using signal processing methods. Parametric representations of speech, such as spectrum, cepstrum, and LPCs, are motivated by the applications such as speech coding. Since the parametric representation is available, attempts have been made to use these representations for voice conversion as well [17–18]. Most of the attempts of voice conversion at the parametric level involve determining mapping of parameters from source speaker to target speaker. The parametric representation of speech is compressed into models based on vector quantization, Gaussian mixture models, and Hidden Markov models, mainly for applications in speech recognition. Transformation of the compressed information has also been explored for voice conversion [22, 25–26]. Finally, methods of voice conversion based on perception require analysis of data to determine speaker-dependent features at the segmental and suprasegmental levels. For synthesis, emphasis is given mainly to those features that are characteristic of a given speaker. It is important to note that in the synthesis-based methods it is not necessary to transform all segments of speech. The difficulty lies in determining the segments containing significant speaker-specific information. These methods are useful to produce speech with specific speaker characteristic for a text-to-speech synthesis.

The work reported by Childers et al. [8] provides a good model for synthesizing speech from a set of acoustic parameters. In this method, the acoustic parameters were extracted from the speech of the source speaker and were modified using scale factors. The problem of learning transformations of the speaker-dependent parameters of the source speaker to those of the target speaker was not addressed. In the following sections the issue of learning transformations is addressed. First some linear and piecewise linear methods of transformation of formant parameters are discussed. Then it is shown that the inherently nonlinear transformation can be captured effectively using a neural network.

11.3 Neural Networks for Transformation of Vocal Tract Shapes

In this section we consider methods to capture the transformation of the vocal tract shapes from the source speaker to the target speaker [33]. As discussed earlier, formants are used for representing the vocal tract shape, as they can provide better generalization of the mapping function derived from the training data [15], and at the same time, they can be extracted from the speech signal. A simple linear transformation of the formants gives large error between the desired (target) formant data and the transformed data. This is mainly due to nonlinear mapping implicit in the transformations of the vocal tract shapes. Thus there is a need for capturing the mapping function. Neural network models can be exploited for this purpose. We now discuss the significance of representative training data for improving the generalization of the mapping function captured by the network.

11.3.1   Linear Approximation of Formant Transformation

We first discuss methods to capture the relation between the formants derived from the speech data of two speakers using linear transformation. Data for this study consists of isolated utterances of vowels /ī/, /ē/, /ā/, /ō/, and /ū/ from five male and five female speakers. Each of these vowels was repeated by every speaker twenty times. Fifteen sets of the data were used as the training set and the remaining as the test set. The first three formants were extracted using a method based on the properties of the minimum phase group delay functions [34]. This method is not based on the model of the vocal tract system as in the case of linear prediction analysis.

A single linear network was trained using the LMS algorithm [35] to transform all the formants. A male speaker was considered as the source speaker and a female speaker as the target speaker. Percentage errors between the source and the target formants before and after the application of the linear transformation are shown as Case A and Case B, respectively, in Table 11.2. From the table we note that, even though the linear transformation brings the source formants closer to the target formants, the error is still large, especially in F_1 and F_2. Also in some cases (for example, F_2 of vowel /ō/ and F_1 of vowel /ē/), the error increases after transformation.

To reduce the error in the transformation, separate linear networks were trained for each of the three formants. Thus the formant transformation now consists of three scale factors. The scaling factors for the

Table 11.2

Percentage error between target (transformed) and source formants. Case A: Raw formant data. Case B: Single linear formant transformation. Case C: Formant-dependent linear transformation. Case D: Piecewise linear transformation. Case E: Neural network transformation.

Vowels	Case A			Case B			Case C			Case D			Case E		
	F_1	F_2	F_3	F_1	F_2	F_3	F_1	F_2	F_3	F_1	F_2	F_3	F_1	F_2	F_3
/ī/	27.1	14.3	4.8	9.0	3.6	6.8	12.0	4.8	5.1	3.8	1.7	1.7	4.0	2.7	2.1
/ē/	7.0	32.9	18.8	19.8	16.2	5.4	16.2	6.8	4.6	4.2	2.3	3.2	1.0	2.4	2.3
/ā/	42.0	21.0	10.1	25.3	18.0	3.8	21.0	4.5	4.7	9.2	1.2	3.4	4.1	1.4	2.4
/ō/	7.3	4.9	5.4	9.9	12.0	5.6	7.0	3.2	4.2	3.3	1.9	2.3	3.0	2.1	1.3
/ū/	21.5	4.2	9.3	14.4	11.2	3.7	11.0	2.6	3.9	4.8	2.5	1.9	3.2	2.4	2.0

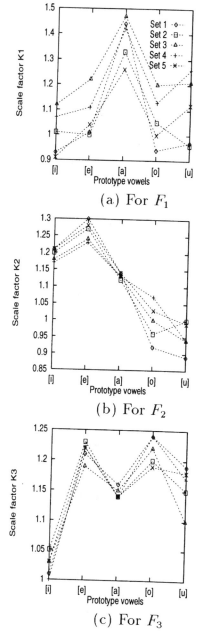

(a) For F_1

(b) For F_2

(c) For F_3

Figure 11.4 Scale factors for piecewise linear formant transformation function for five vowels. The figure shows the scale factors for five sets of speakers, each set corresponding to one male-female speaker pair.

three formants are different for different vowels. The Case C in Table 11.2 shows the error between the transformed and the target formants. Comparing Case B with Case C, we note that there is an overall reduction in error in the formant-dependent transformation over the single linear transformation.

The transformation of vocal tract system can be improved significantly using a separate linear transformation for each vowel. Thus, considering five vowels and three formants, 15 linear functions were captured. Figure 11.4 shows the scale factors for the resulting piecewise linear transformation function. The figure shows the scale factors for five different pairs of source (male) and target (female) speakers. The scale factors depend both on the formant and on the type of the vowel. But variations of the scale factors show similar trends for all of the male-female speaker pairs.

A notable deviation from the uniform scaling of the formants is the large-scale factors for the first formant of the open vowel /ā/ compared with the scale factors for the closed vowels /ī/ and /ū/. Likewise, the scale factor for the second formant is high for the front vowels /ī/ and /ē/. For the back vowels /ō/ and /ū/ the values of the scale factors for the second formant are less than 1 in some cases. This means that the second formant frequency for the back vowels /ō/ and /ū/ is higher for male speakers than for female speakers. These observations are consistent with the results reported in [36].

In using the piecewise linear transformation, it is necessary to first identify the vowel. A simple classification scheme was used for recognizing the vowel by comparing the formant vector (consisting of the three formant values) of the test speech segment with the mean formant vector of each vowel. Results of the error analysis for the piecewise linear transformation is shown as Case D in Table 11.2. We note that the error is significantly lower for the piecewise linear transformation compared with Case C.

Figure 11.5 shows the effect of transforming some formant transitions using the piecewise linear formant transformation derived from steady vowels.

The first column shows the formant contours of the source speaker for the vowel sequences /āī/, /āū/ and /ōī/. The third column shows the corresponding formant transitions of the target speaker. The source formant transitions were transformed using the piecewise linear transformation and the results are shown in the second column of the figure. The transformed formant transitions show discontinuities. These discontinuities are significant in the third formant for the vowel

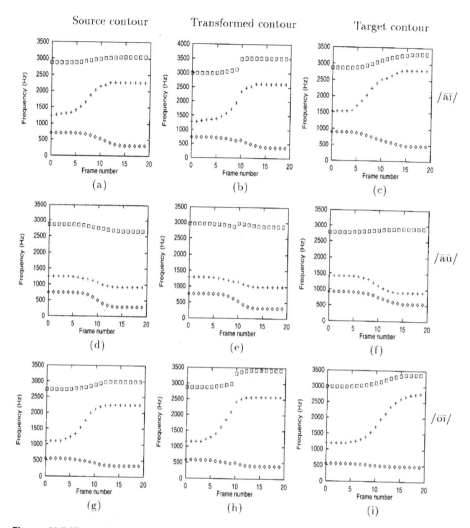

Figure 11.5 Illustration of the problem of formant discontinuity: (a), (d), and (g) show the source (male) formant contours corresponding to the vowel sequences /āī/, /āū/, and /ōī/; (b), (e), and (h) show the corresponding transformed formant contours using the piecewise linear transformation; (c), (f), and (i) show the corresponding target formant contours (female).

sequences /āī/ and /ōī/. These discontinuities occur due to the use of two separate linear functions to transform the two different parts of the formant transition. Such a discontinuous formant contour gives poor quality speech from synthesis.

There are two main difficulties in using the piecewise linear formant transformations. First, we need to identify the class of the sound to apply the appropriate transformation to the formants. Second, the piecewise linear transformation results in a discontinuous mapping function for the vocal tract transformation as shown in Figure 11.4. In the following section the capability of a feedforward neural network in circumventing these problems is demonstrated.

11.3.2 Neural Network Models

In the previous section we noted that the transformation of formants from male to female voices or vice versa is highly nonlinear. A multilayer feedforward neural network with nonlinear processing elements can be used to capture the nonlinear mapping function for the formants of the source and target speakers. Formants extracted from isolated utterances of vowels can be used to train the network. The network transforms a formant vector without knowing the class of the input vector. Such a network is useful for transforming formant transitions also. A smooth transition will be obtained as in natural speech, although there may be significant differences from the target formant transition.

A neural network is trained with the formant data of the source speaker as input and that of the target speaker as the desired output. The training can be done using the standard back-propagation algorithm [16]. The training data consists of formants derived from isolated utterances of vowels by the speakers, as described in the previous section. A multilayer feedforward neural network with two hidden layers is used, with three units in each of the input, hidden, and output layers.

Figure 11.6 shows the transformation captured by the network. The transformation is shown in the form of scale factors by which the formants of the source male speaker are scaled by the network to derive the formants of the target female speaker. It is interesting to note that the scale factors captured by the network are similar to those derived using the piecewise linear approximation shown in Figure 11.4. Therefore, formants extracted from steady vowels are transformed in a similar fashion by both the piecewise linear transformation and the neural network. But the way in which the transformation is carried out by the linear transformation and the neural network is significantly different. In the case of piecewise linear transformation, the transformation is described by a set of fifteen scale factors. In order to transform a formant vector, we need to know the class of the speech segment

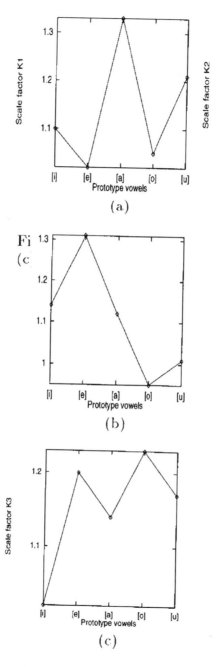

Figure 11.6 The scale factors learned by the network: (a), (b), and (c) show the scale factors for first, second, and third formants.

from which the formant vector is extracted. Moreover, in the piecewise linear transformation, there may be discontinuities in the transformed formant contours, although the source formant transitions are smooth. In the case of the neural network, since the formant transformation is captured as a single continuous nonlinear function, it is expected that the transformed formant transition will be smooth. To evaluate the performance of the network, formants extracted from the test utterances of the male source speaker are transformed using the neural network to get the formants of the target female speaker. The column for Case E in Table 11.2 shows the percentage error between the formants of the target speaker and the transformed formant data. It is clear that the neural network transformation reduces the error significantly compared with the linear and piecewise transformations discussed earlier.

Let us examine the capability of the network to transform the formant transitions. For this purpose, formants from the vowel sequences /āī/, /āū/, and /ōī/ are extracted from the utterances of the source and the target speakers. Figure 11.7 illustrates the transformed formant contours. The first column of the figure shows the formant transitions for the source speaker. The second column shows the formant transitions obtained from the neural network. The corresponding target formant transitions are shown in the third column of the figure. From the figure, it is clear that the network is capable of transforming formant transitions without introducing discontinuities. But the transitions in the formant contours do not match the target contours. The deviations are more pronounced in the second formant for the sounds /āī/ and /ōī/.

11.3.3 Generalization

The objective in using a neural network is to be able to capture the implicit nonlinear mapping between the shapes of the vocal tracts of the source and target speakers. The mapping is to be captured using a set of training data collected from the speech segments of the speakers. The training set consists of pairs of formant data; each pair corresponds to the same sound produced by both speakers. For this purpose the training data set is to be prepared by manually selecting the matched pair of speech segments from the source and target speakers. From the set of training samples, if the network is able to capture the nonlinear mapping function between the vocal tract shapes, then it is said to have generalized from the training data. It is easier to collect the matching pairs of segments from steady vowel sounds than from the transitions.

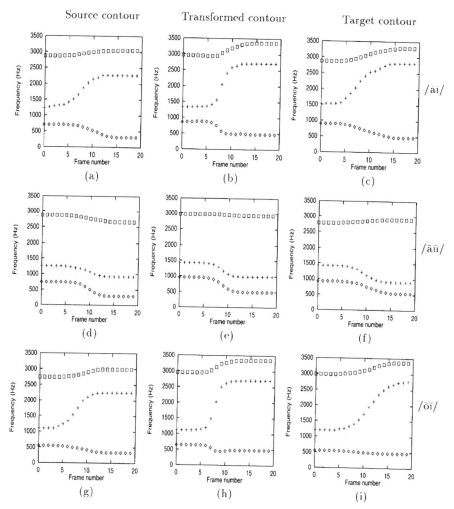

Figure 11.7 Illustration of the problems in transforming formant contours: (a), (d), and (g) show the source (male) formant contours corresponding to the vowel sequences /āī/, /āū/, and /ōī/; (b), (e), and (h) show the corresponding transformed formant contours using the trained network; (c), (f), and (i) show the corresponding target formant contours (female).

The network can be said to have generalized better if it is able to transform the transition regions also. We examine that issue in this section. Note that generalization here refers to the mapping task.

The failure to faithfully transform the formant transitions by a neural network trained with formant data extracted from the steady vowels may be attributed to poor generalization by the network. The generalization capability of the network is influenced by the following four factors [16]:

1. *Training data:* This refers to how well the training data set represents the input-output mapping function.
2. *Architecture of the network:* The architecture refers mainly to the size and structure of the network. If one uses a network size that is too large, it may lead to memorization of the training examples, thus resulting in poor generalization.
3. *Training methodology:* The training algorithm, especially the stopping criterion, affects the performance of the network. For example, in the back-propagation algorithm, overtraining of the network may lead to poor generalization.
4. *The inherent complexity of the problem:* The complexity of the mapping function also affects the generalization performance.

If the features involved for the smooth behavior of the mapping function are deep in the input data, then it may be difficult for the neural network to capture the mapping function effectively [37, 38]. The poor generalization of the network trained using the formant data of only steady vowels may be attributed to the nonrepresentative nature of the training data. Table 11.3 shows the mean (M) and variance (V) of the

Table 11.3

Mean (M) and variance (V) of the formant frequencies extracted from steady vowels uttered in isolation by the source and the target speakers

Vowels	Source speaker						Target speaker					
	F_1		F_2		F_3		F_1		F_2		F_3	
	M	V	M	V	M	V	M	V	M	V	M	V
/ī/	728	16	1244	48	2773	17	922	32	1419	25	2895	48
/ē/	554	24	1919	24	2587	17	525	21	2540	61	3055	99
/ā/	333	15	2245	37	2985	40	470	19	2722	37	3298	110
/ō/	549	9	1098	18	2738	21	592	22	1099	29	2893	63
/ū/	385	8	1059	51	2595	18	542	27	1079	28	2833	65

Table 11.4

Mean and variance of the formant frequencies extracted from steady vowels occurring in continuous speech

Vowels	Source speaker						Target speaker					
	F_1		F_2		F_3		F_1		F_2		F_3	
	M	V	M	V	M	V	M	V	M	V	M	V
/ī/	687	86	1334	92	2379	145	873	92	1556	115	2773	73
/ē/	475	54	1980	89	2674	72	590	114	2218	200	2890	122
/ā/	324	26	2194	59	2792	59	403	28	2586	94	3173	132
/ō/	534	81	1226	104	2459	96	628	126	1232	135	2735	158
/ū/	372	31	1191	116	2436	85	462	46	1176	139	2894	75

formant data extracted from isolated utterances of vowels; Table 11.4 shows the same for the data extracted from continuous speech. Comparing Table 11.3 and Table 11.4, we note that the variance of the training data set significantly increases when the formants are extracted from continuous speech. Therefore, the data derived from continuous speech is more representative (covers more areas of the feature space) and hence is more appropriate for training a network for better generalization.

For extracting formants from continuous speech, speech data from the source and target speakers was segmented manually to mark the vowel regions. The first three formants were extracted from frames having maximum energy in each of the steady vowel regions. The formant data was collected from 50 sentences. Five hundred pairs of formant vectors were collected for training the neural network. The data set captures the natural variability of the formants in continuous speech.

A multilayer feedforward neural network with two hidden layers was used to capture the mapping function from the formant data derived from continuous speech. The network consists of three units each in the input and output layers, and eight units in each of the two hidden layers. In the training phase, the formant values extracted from the steady vowel regions were used. Figure 11.8 illustrates the steps in training the network. After training, it is expected that the neural network will capture the mapping function, which is capable of transforming not only the steady vowels but also the transitions.

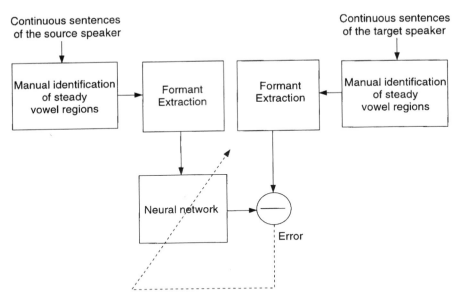

Figure 11.8 Training procedure of a feedforward network using the formants extracted from steady vowels occurring in continuous speech.

In order to achieve good generalization, a cross-validation check was used while training the network [16]. The cross-validation check prevents overfitting of the training data by the network. In this method the training set is partitioned into two sets, a set for estimation and another set for validation. The network is trained with the data in the estimation set. After each iteration, the network is tested on the data in the validation set. The error in the output of the network for the estimation set is called the training error; the error for the validation set is called the generalization error. Figure 11.9 shows the plots of the training and the generalization errors as a function of the iteration number. The figure shows that even though the training error reduces, the generalization error shows a rising trend after a certain number of iterations.

This indicates that after a certain number of iterations, the network parameters are adjusted to overfit the data in the training set. Hence the training is stopped when the generalization error starts increasing, even though further training will result in reduced training error.

Generalization capability of the network refers to the ability to transform formant transitions without introducing discontinuities, even

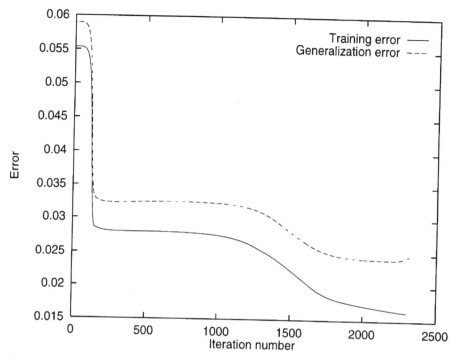

Figure 11.9 Variation of the training and generalization errors as training progresses.

though the network was trained using formants extracted from mostly steady sounds (vowels). One can extract the formants from utterances of the source speaker for vowel sequences and transform them using the trained network. The error between the transformed and the target formant contours gives a measure of the generalization capability of the network. Since the source and the target speech are normally warped in time, the target formants cannot be directly compared with the transformed formants. Therefore, synthetic formant transition contours may be generated and used for testing the generalization capability of the trained network.

Let the source and target formant data in the training set be represented as $F_{vs}(n)$ and $F_{vt}(n)$, respectively, for $n = 1, 2, \ldots N$, where v can be one of the five vowels /ā/, /ē/, /ī/, /ō/, or /ū/, and N represents the number of formant pairs used for training. The subcripts s and t refer to the source and target speakers, respectively. The mean formant vector for each of the vowels is computed for both the source and the target speakers in the following way.

$$\overline{F}_{vs} = \frac{1}{N}\sum_{n=1}^{N} F_{vs}(n)$$

$$\overline{F}_{vt} = \frac{1}{N}\sum_{n=1}^{N} F_{vt}(n)$$

Following this notation, the mean vectors for the vowels /ā/ and /ī/ for the source speaker are representated as $\overline{F}_{as}$ and $\overline{F}_{is}$, respectively, and for the vowel sequence as $\overline{F}_{ais}$. A source formant transition, corresponding to the vowel sequence /āī/, for example, is derived by interpolating $\overline{F}_{as}(n)$ and $\overline{F}_{is}(n)$ by a monotonically increasing/decreasing function. We represent this synthetic formant contour by $\overline{F}_{ais}(n)$, for $n = 1, 2 \ldots l$, where l is the number of points used to interpolate $\overline{F}_{as}(n)$ and $\overline{F}_{is}(n)$. The corresponding target formant contour is derived by interpolating $\overline{F}_{at}(n)$ and $\overline{F}_{it}(n)$ by the same function which was used to interpolate the source mean formant vectors, and is represented by $\overline{F}_{ait}(n)$, for $n = 1, 2 \ldots l$. Now the synthetic source formant transition

Table 11.5

Table showing the result of using representative training data in improving the generalization capability of a network

Vowel sequences	Generalization error given by a network trained using formants extracted from isolated utterances of vowels			Generalization error given by the network trained using formants extracted from vowels occurring in continuous speech		
	F_1	F_2	F_3	F_1	F_2	F_3
/āē/	8.8	11.8	5.0	1.7	6.6	2.6
/āī/	13.9	11.8	4.4	1.8	5.8	3.1
/āō/	3.3	11.7	7.2	2.1	4.1	1.4
/āū/	6.7	15.5	3.6	1.9	6.1	3.0
/ēī/	8.7	7.3	5.8	1.3	5.0	1.0
/ēō/	6.9	13.3	6.5	1.8	3.1	1.4
/ēū/	11.0	11.8	4.0	0.6	2.4	1.4
/īō/	7.6	14.0	5.6	1.9	3.8	1.9
/īū/	11.2	11.4	3.1	1.0	4.3	1.3
/ōū/	26.8	23.1	7.9	1.2	3.0	1.5

Source contour   Transformed contour   Target contour

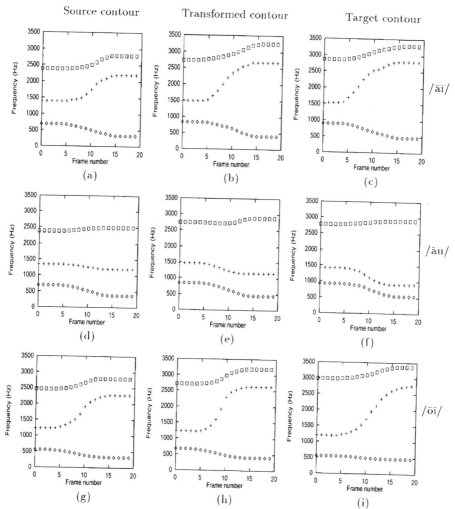

Figure 11.10 Illustration of the capability of a neural network to faithfully transform formant transitions: (a), (d), and (g) show the formant contour corresponding to the vowel sequences /āī/, /āū/, and /ōī/ extracted from the speech of the source speaker (male); (b), (e), and (h) show the corresponding transformed formant contour using the trained network; (c), (f), and (i) show the corresponding target formant contours (female).

$\overline{F}_{ais}(n), n = 1, 2 \ldots l$, corresponding to the vowel sequence /āī/ is transformed using the trained network. The transformed formant transition is represented by $\overline{F}_{ai}(n)$, for $n = 1, 2 \ldots l$. The generalization error for the formant transition corresponding to the vowel sequence

/āī/ is given by

$$E_{ai} = \frac{1}{l} \sum_{n=1}^{l} \frac{|\overline{F}_{ai}(n) - \overline{F}_{ait}(n)|}{max(\overline{F}_{ai}(n), \overline{F}_{ait}(n))}. \tag{11.1}$$

The total generalization error is obtained by adding the generalization errors for all possible vowel-to-vowel formant transitions. Table 11.5 shows the generalization errors for the two networks, one trained using isolated utterances of steady voiced sounds and the other using the steady vowels extracted from continuous speech. The results show a significant reduction in the generalization error for the second case. Figure 11.10 shows the result of transforming formants of the source speaker for the vowel sequences /āī/, /āū/, and /ōī/. The transformed formant transitions are closer to those of the target speaker in Figure 11.10 compared with the formant transitions obtained in Figure 11.7.

11.4 Implementation of Voice Conversion

11.4.1 Voice Transformation System

There are two phases in the development of a voice conversion system, a learning phase and a transformation phase. In the learning phase, factors that are responsible for voice personality are identified, and the speaker-specific knowledge is acquired and represented in a suitable form. In the transformation phase, the given speech signal is modified using the knowledge acquired during the learning phase. In this section we describe the issues involved in the development of the transformation phase. We focus particularly on incorporation and testing of the transformation of the vocal tract system. Linear prediction analysis is used to represent the vocal tract system, since it is convenient to synthesize speech using LPCs. The LPCs extracted from the speech of the source speaker are modified to reflect the characteristics of the target speaker. The modified LPCs are then used to synthesize speech. The quality of speech after voice conversion depends on the excitation source used for synthesis. Simple linear transformation of the average pitch may not be adequate, as interspeaker variations are significant at the suprasegmental level. Therefore, for assessing the quality of the vocal tract transformation performed at the segmental level, the source speaker's voice characteristics at the suprasegmental level must be masked. This is accomplished by normalizing the intonation features between the two speakers. Finally, the performance of the conversion system needs to be evaluated for several pairs of speakers to study the significance of the vocal tract transformation.

In the transformation phase of voice conversion, the voiced/ unvoiced class of each of the analysis frames of the speech of the source speaker is determined. The first three formants are extracted using an algorithm based on the properties of minimum phase group delay functions [34]. The formants are modified using the trained neural network. It is reasonable to assume that speaker-specific information is mainly in the voiced segments of speech [11]. Hence, only the parameters of the voiced segments are modified.

Direct synthesis using the transformed formants produces poor quality speech due to lack of bandwidth and spectral slope information. Therefore, an LPC vocoder is used as a synthesis model, where the vocal tract system is represented as a time-varying all-pole digital filter. The filter is represented by

$$H(z) = \frac{1}{A(z)} = \frac{G}{1 + \sum_{i=1}^{p} a_i z^{-i}} \tag{11.2}$$

where p is the order of the all-pole system, G is the gain, and $a_i z$ are the LPCs. This filter is excited with random noise during unvoiced frames and with a train of periodic glottal pulses during voiced frames to generate synthetic speech.

The complex roots of the linear prediction polynomial $A(z)$ represent the resonances (formants) of the vocal tract system. For a complex conjugate root $re^{j\theta}$, the vocal tract resonant frequency (F) and bandwidth (B) are given by

$$F = \frac{\theta}{2\pi T} \tag{11.3}$$

$$B = \frac{-\log r}{\pi T} \tag{11.4}$$

where T is the sampling period. Formant transformation can be realized merely by shifting the angle θ of the root in the z-plane. From (11.3) it is

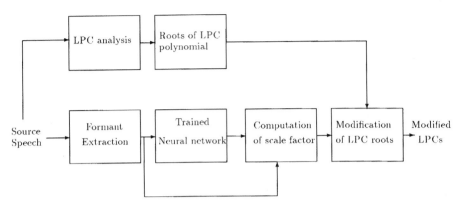

Figure 11.11 Block diagram showing the various steps involved in the modification of LPCs for formant transformation.

evident that θ is directly proportional to the formant frequency. Hence, by shifting the complex pole in the z-plane, we can implement a formant transformation. Figure 11.11 shows the computations for modification of the LPCs.

The ratios of the transformed formants and the formants of the source speaker are represented by a set of three scale factors α_i, for $i = 1, 2, 3$, corresponding to the three formants. For each frame of speech data, a set of three values is obtained for the three scale factors. These scale factors vary significantly across an utterance, as shown in Figure 11.12 for the all-voiced sentence "We were away a year ago." This shows the highly nonlinear nature of the formant transformation between speakers. These scale factors are used to modify the LPCs extracted for each frame of speech data.

To derive a representation for the transformed vocal tract system, the LPCs of the source speaker are modified as shown in Figure 11.13. The complex roots of the LP polynomial closer to the formants (derived from minimum phase group delay method) are shifted using the scale factors. The real roots of the LP polynomial are left unaltered. Other complex roots are modified using the scale factors corresponding to the nearest formant frequencies.

The LPCs for the synthesis filter are obtained by modifying the LPCs extracted from the speech of the source speaker. For the extraction of LPCs, pitch, and gain, a sliding window of size 25.6 ms and a shift of 6.4 ms was used. A 10th-order linear prediction was used for the analysis to obtain the inverse filter $A(z)$.

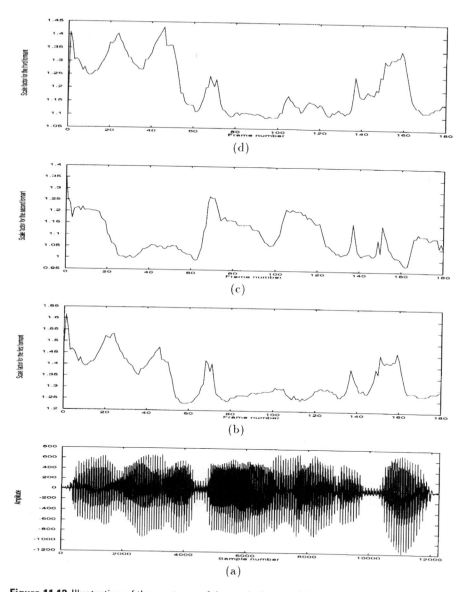

Figure 11.12 Illustration of the contours of the scale factors: (a) is a speech waveform for the utterance "We were away a year ago"; the plots in (b), (c), and (d) show the variation of the scale factors for the first, second, and third formants, respectively.

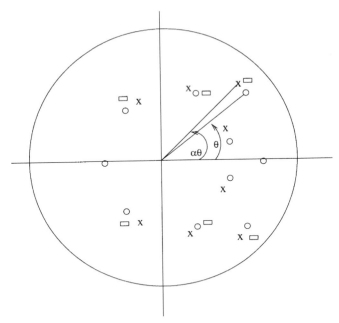

x   Modified roots obtained by multiplying with scale factors derived from NN model

o   Roots obtained from LP analysis

▭   Roots corresponding to formants derived from min phase group delay method.

Figure 11.13 Illustration of modification of roots of LP polynomial.

11.4.2 Normalization of Intonational Features

To assess the effect of the transformation, the voice quality of the transformed utterances is compared with the same sentences uttered by the target speaker. Since the converted speech is synthetic, it is difficult to compare it with the natural utterance of the target speaker. Hence, speech is synthesized using the pitch, gain, and LPCs extracted from the test utterance of the target speaker using an LPC vocoder. To evaluate the effectiveness of the transformation at the segmental level, the interspeaker variations at the suprasegmental level must be eliminated from the test utterance. At the suprasegmental level, intonation is a significant feature of an individual speaker. Therefore, the intonation pattern of the test utterance of the target speaker is modified so that it matches that of the source speaker.

A normalization procedure can be developed using the

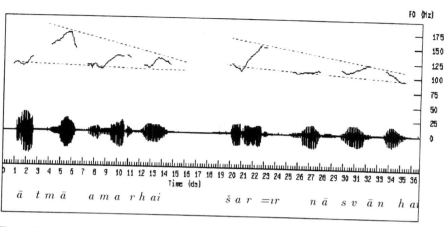

Figure 11.14 Pitch pattern of a typical declarative sentence, *ātmāamar hai šarīr nāsvān hai,* spoken by a native Hindi speaker.

characteristics of the intonation patterns of speech in reading style. Important characteristics of the intonation patterns are declination, local fall-rise, resetting, and tapering effect [7, 39]. These features represent important linguistic information as well as characteristics of the individual's voice. Figure 11.14 illustrates the intonation pattern for a typical declarative sentence.

The F0 contour of an utterance is characterized by a few target points. The target points are the local maxima and minima of F0 which cause rise and fall in the F0 movements. The local minima and maxima are called valleys and peaks, respectively. If two imaginary lines are drawn in a declarative sentence, one connecting all the peaks and the other all the valleys, it is possible to say that the F0 contour drifts down as a function of time until the occurrence of a major syntactic or semantic break (such as at the end of /ātmā amar hai/ in the example), which is also marked by a significant pause of about 300 ms. The lines show an upward trend for interrogative sentences. The difference between the F0 values measured at a valley and the following peak is called the F0 range. The F0 range is another important feature that carries significant speaker-specific information as well as speaking style [40–42].

If we assume that the rate of fall of F0 values is constant, then we can model the valleys and peaks as points on two separate lines, the base line and the top line, respectively. Thus if P_1 and P_2 are the two peak F0 values measured at times T_1 and T_2, then the equation of the top line [7] becomes:

$$P(t) = P_1 + \frac{P_2 - P_1}{T_2 - T_1}(t - T_1) \qquad (11.5)$$

Thus, once we model the pitch contour using the above equation, it is possible to predict the F0 value of any peak if we know the position (time) at which it occurs. Similarly the base line is modeled by the following equation.

$$V(t) = V_1 + \frac{V_2 - V_1}{T_2 - T_1}(t - T_1) \qquad (11.6)$$

where V_1 and V_2 are the F0 values at the two valley instants T_1 and T_2.

From the energy contours of the target and the source utterances, the vowel nuclei are identified using an algorithm described in [43]. The pitch values at these vowel nuclei for the source and the target speakers are represented by $F0_s(n)$ and $F0_t(n)$, respectively, for $n = 1, 2 \ldots N$, where N is the number of syllables in the sentence. The perceptually significant feature in an intonation pattern is the relative values of the F0 measured at the vowel nuclei and not the absolute F0 values. Hence, the F0 ranges at the vowel nuclei of successive syllables are modified. $F0_t(2)$ is modified so that the range between $F0_t(1)$ and $F0_t(2)$ measured in semitones becomes equal to the range between $F0_s(1)$ and $F0_s(2)$. This modification is continued until we modify the range between $F0_t(N - 1)$ and $F0_t(N)$. From this data, the F0 values of the other voiced frames are computed using a cubic spline interpolation [44].

11.4.3 Evaluation of Voice Transformation

In this section, studies on the voice conversion between pairs of speakers are described. Speech data for 50 sentences spoken by two male and two female speakers was collected. Formant transformations were captured using a neural network.

Linear pitch transformation can also be obtained using the procedure described below. Pitch was extracted from the training set of isolated utterances of vowels. Let the pitch frequency corresponding to the source and the target speakers be represented by $F0_s(n)$ and $F0_t(n)$, respectively, for $n = 1, 2 \ldots N$, where N is the total number of training values of F0. A linear transformation is used to determine the scale factor for the average pitch. For evaluation, the F0 contours are extracted from the test utterances of both the source and the target speakers. The source pitch contours are transformed using the linear pitch transformation. Table 11.6 shows the error in the average F0 of the source and the target pitch contours before and after the application of the transformation function. From the table it is clear that the average pitch transformation is able to modify the average pitch of the source speaker to match satisfactorily that of the target speaker.

The learned formant and pitch transformations were used to carry out the following four types of voice transformations.

Case A: Male to Female

Case B: Female to Male

Case C: Male to Male

Case D: Female to Female

The transformed speech was obtained for the following conditions.

1. *Average pitch modification:* Speech with modified average pitch and original vocal tract system characteristics of the source speaker.

2. *Formant transformation:* Speech with original pitch and the transformed formants.

3. *Average pitch and formant transformation:* Speech with modified average pitch and transformed formants.

Table 11.6
The error analysis on the linear F0 transformation.

Vowels	Error in percentage between the target and the source F0	Error in percentage between the transformed F0 and the target F0
/ī/	51.2	9.2
/ē/	45.4	7.3
/ā/	50.9	8.2
/ō/	40.4	7.1
/ū/	47.8	6.9

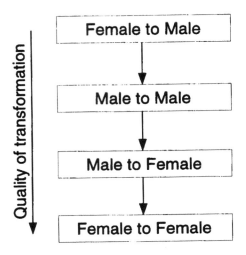

Figure 11.15 Figure showing the order in which the quality of voice conversion decreases depending on the type of conversion being attempted.

These were compared with the speech synthesized from the natural utterances of the target speaker. The pitch contour used to synthesize this was normalized to match that of the source pitch contour. Comparison by informal listening indicates that for the third condition, the characteristics of the target speaker are perceived clearly in the transformed speech.

The studies also show that whenever the target speaker is a female, the conversion quality becomes poor. This is consistent with the observations in [8, 45]. The reason for this can be attributed to the general problems in synthesizing female speech [46]. The quality of voice conversion was found to degrade in the order shown in Figure 11.15.

11.5 Summary

Voice characteristics are present at segmental, suprasegmental, and linguistic levels. The speaker characteristics at the suprasegmental and linguistic levels are learned features acquired over a period of time. Hence, these features are difficult to derive from the speech data. Speaker characteristics at the segmental level can be attributed to the speech production mechanism and are related to the characteristics of the source and system components of the speech production process. The interspeaker variations of the vocal tract system can be modeled as a transformation operation. Likewise, the inter-speaker variations in the source can also be captured in the form of simple transformations. In a voice conversion task, speech synthesized using the transformed source and system parameters of a speaker should reflect the characteristics of the target speaker. A major issue in the transformation task is to obtain a suitable representation of the vocal tract system. Formants, the resonances of the vocal tract system, seem to provide a good representation of the shape of the vocal tract; at the same time, they can be extracted from the speech signal using available signal processing methods. A linear transformation of each of the formants corresponding to the source speaker into the formants of the target speaker's voice appears to be a straightforward approach for voice conversion. But a simple linear transformation is not adequate to capture the transformations of all the vocal tract shapes. Piecewise linear transformations can take care of the variations in the transformations for each vowel category. The match between the transformed formant values and the true formant values of the target speaker is significantly better in the piecewise linear transformation over the linear transformation. But the piecewise linear transformations introduce discontinuities in the formant transitions in continuous speech. This is because the transformation of formants even for steady vowels is highly nonlinear. Hence, variations in the vowel transitions cannot be mapped easily using linear transformations. A multilayered feedforward neural network can be used to capture the complex nonlinear transformation of the formants. A neural network trained using steady vowel data may not be able to capture the transformation of the formants well, although the transitions of the transformed formants may be smooth. For better generalization of the transformation captured by a neural network even for formant transitions, suitable training data needs to be collected. The training data should represent the variations of formants in continuous speech for the source and target speakers. The generalization performance of a trained neural network can be evaluated using synthetic formant transition data for the source and target speakers.

In implementing a voice conversion system, it is necessary to have a suitable model for synthesis, which incorporates the transformed formant parameters and some source parameters into a linear time-varying digital filter. The performance of the voice conversion system can be evaluated at the component levels, such as the transformation of system characteristics (formants), source characteristics (pitch), and suprasegmental characteristics, or by the perception of the overall speech quality.

Voice conversion is a challenging task, since many factors responsible for the conversion are not known. In fact, the quality of a voice conversion system will improve significantly if one understands how the prosodic knowledge of a speaker can be modified to correspond to that of the target speaker. There are no clues at present on how to capture the prosodic knowledge and represent it. It is also not known how to obtain a transformation of the prosodic knowledge between two speakers.

References

[1] Rabiner, L. R., and B. H. Juang, *Fundamentals of Speech Recognition*, Englewood Cliffs, N.J.: Prentice Hall, 1993.

[2] O'Shaughnessy, D., *Speech Communication*, Massachusetts, Addison-Wesley, 1987.

[3] Doddington, G. R., "Speaker Recognition: Identifying People by Their Voices," *Proc. IEEE*, Vol. 73, No. 11, 1985, pp. 1651–1664.

[4] Atal, B. S., "Automatic Recognition of Speakers from Their Voices," *Proc. IEEE*, Vol. 64, No. 4, 1976, pp. 460–475.

[5] Gish, H., and M. Schmidt, "Text-Independent Speaker Identification," *IEEE Signal Processing Magazine*, Vol. 11, 1994, pp. 18–32.

[6] Rajesh Kumar, S. R., "Significance of Duration Knowledge for a Text-to-Speech System in an Indian Language," M.S. thesis, Indian Institute of Technology, Madras, India, 1990.

[7] Madhukumar, A. S., "Intonation Knowledge for Speech Systems for an Indian Language," Ph.D. thesis, Indian Institute of Technology, Madras, India, 1993.

[8] Childers, D. G., K. Wu, D. M. Hicks, and B. Yegnanarayana, "Voice conversion," *Speech Communication*, Vol. 8, 1989, pp. 147–158.

[9] Ananthapadmanabha, T. V., and B. Yegnanarayana, "Epoch Extraction from Linear Prediction Residual for Identification of Closed Glottis Interval," *IEEE Trans. Acoust. Speech Signal Processing*, Vol. ASSP-27, No. 4, August 1979, pp. 309–319.

[10] Smits, R. L. H. M., and B. Yegnanarayana, "Determination of Significant Excitation in Speech Using Group-Dealy Functions," *IEEE Trans. Speech and Audio Processing*, Vol. 3, No. 5, September 1995, pp. 325–333.

[11] Eatock, J. P., and J. S. Mason, "A Quantitative Assessment of the Relative Speaker Discriminating Properties of Phonemes," *Proc. Int. Conf. Acoust. Speech Signal Processing*, 1994, pp. 133–136.

[12] Eatock, J. P., and J. S. Mason, "Automatically Focusing on Good Discriminating Speech Segments in Speaker Recognition," *Proc. Int. Conf. Spoken Language Processing*, 1990, pp. 133–136.

[13] Yegnanarayana, B., S. P. Wagh, and S. Rajendran, "Speaker Verification System Using Prosodic Features," *Proc. Int. Conf. Spoken Language Processing*, September 1994, pp. 1867–1870.

[14] Ramachandran, V. R., "Coarticulation Knowledge for a Text-to-Speech System for an Indian language," M.S. thesis, Indian Institute of Technology, Madras, India, 1993.

[15] Adabala, N., "Generalization Capability of Feedforward Neural Networks for Pattern Recognition," M.S. thesis, Indian Institute of Technology, Madras, India, 1996.

[16] Haykin, S., *Neural Networks*, New York: Macmillan, 1994.

[17] Atal, B. S., and S. L. Hanauer, "Speech Analysis and Synthesis by Linear Prediction of the Speech Wave," *J. Acoust. Soc. Am.*, Vol. 50, 1971, pp. 637–655.

[18] Seneff, S., "System to Independently Modify Excitation and/or Spectrum of Speech Waveform with out Explicit Pitch Extraction," *IEEE Trans. Acoust. Speech Signal Processing*, Vol. 30, 1982, pp. 566–578.

[19] Childers, D. G., B. Yegnanarayana, and K. Wu, "Voice Conversion: Factors Responsible for Voice Quality," *Int. Conf. Acoust. Speech Signal Processing*, 1985, pp. 19.10.1–19.10.4.

[20] Fant, G., "Glottal Flow: Models and Interaction," *J. Phonetics*, Vol. 4, No. 3–4, 1986, pp. 393–399.

[21] Slifka, J., and T. R. Anderson, "Speaker Modification with LPC Pole Analysis," *Int. Conf. Acoust. Speech Signal Processing*, 1995, pp. 644–647.

[22] Abe, M., S. Nakamura, K. Shikano, and H. Kuwabara, "Voice Conversion Through Vector Quantization," *Int. Conf. Acoust. Speech Signal Processing*, 1988, pp. 655–658.

[23] Mizuno, H., and M. Abe, "Voice Conversion Based on Piecewise Linear Conversion Rules of Formant Frequency and Spectrum Tilt," *Proc. 1994 Int. Symp. Speech, Image Processing and Neural Networks*, Hong Kong, 1994, pp. 469–472.

[24] Mizuno, H., and M. Abe, "Voice Conversion Algorithm Based on Piecewise Linear Conversion Rules of Formant Frequency and Spectrum Tilt," *Speech Communication*, Vol. 16, 1989, pp. 153–164.

[25] Abe, M., K. Shikano, and H. Kuwabara, "Cross-Language Voice Conversion," *Proc. Int. Conf. Acoust. Speech Signal Processing*, 1990, pp. 345–348.

[26] Abe, M., "A Segment Based Approach to Voice Conversion," *Proc. Int. Conf. Acoust. Speech Signal Processing*, 1991, pp. 765–768.

[27] Valbret, H., E. Moulines, and J. P. Tubach, "Voice Transformation Using PSOLA Techniques," *Speech Communication*, Vol. 11, No. 2–3, 1992, pp. 175–187.

[28] Moulines, M., and J. Laroche, "Non-Parametric Techniques for Pitch-Scale and Time-Scale Modification of Speech," *Speech Communication*, Vol. 16, No. 2, 1995, pp. 175–205.

[29] Moulines, E., and F. Charpentier, "Pitch-Synchronous Waveform Processing Techniques for Text-to-Speech Synthesis Using Diphones," *Speech Communication*, Vol. 9, No. 5/6, 1990, pp. 453–467.

[30] Arslan, L. M., and D. Talkin, "Speaker Transformation Using Sentence HMM Based Alignments and Detailed Prosody Modification," *Proc. Int. Conf. Acoust. Speech Signal Processing*, 1998, pp. 289–292.

[31] Stylianou, M., O. Cappe, and E. Moulines, "Continuous Probabilistic Transform for Voice Conversion," *IEEE Trans. Speech and Audio Processing*, Vol. 6, No. 2, March 1998, pp. 131–142.

[32] Kain, A., and M. W. Macon, "Spectral Voice Conversion for Text-to-Speech Synthesis," *Proc. Int. Conf. Acoust. Speech Signal Processing*, 1998, pp. 285–288.

[33] Narendranath, M., H. A. Murthy, S. Rajendran, and B. Yegnanarayana, "Transformation of Formants for Voice Conversion Using Artificial Neural Networks," *Speech Communication*, Vol. 16, No. 2, 1995, pp. 207–216.

[34] Murthy, H. A., and B. Yegnanarayana, "Formant Extraction from Minimum Phase Group Delay Function," *Speech Communication*, Vol. 10, 1991, pp. 209–221.

[35] Widrow, B., and S. D. Stearns, *Adaptive Signal Processing*, Englewood Cliffs, N.J.: Prentice–Hall, 1985.

[36] Fant, G., A. Kruckenburg, and L. Nord, "Prosodic and Segmental Speaker Variations," *Speech Communication*, Vol. 10, 1991, pp. 521–531.

[37] Muller, B., and T. Reinhardt, *Neural Networks: An Introduction*, Berlin: Springer-Verlag, 1991.

[38] Yegnanarayana, B., *Artificial Neural Networks*, New Delhi: Prentice Hall, 1998.

[39] Ladd, D. R., "Intonational Phrasing: The Case for Recursive Prosodic Structure," *Phonology Yearbook*, Vol. 3, 1986, pp. 311–340.

[40] Brown, B. L., W. J. Strong, and A. C. Rencher, "Fifty-four Voices from Two: The Effect of Simultaneous Manipulations of Rate, Mean Fundamental Frequency and Variance of Fundamental Frequency on Rating of Personality from Speech," *J. Acoust. Soc. Am.*, Vol. 55, 1974, pp. 313–318.

[41] Scherer, K., "Personality Markers in Speech," in K. R. Scherer and H. Giles, eds., *Social Markers in Speech*, Cambridge: Cambridge University Press, 1979, pp. 147–209.

[42] Henton, C. G., "Fact and Fiction in the Description of Female and Male Pitch," *Language & Communication*, Vol. 9, 1989, pp. 299–311.

[43] Rajendran, S., and B. Yegnanarayana, "Word Boundary Hypothesization for Continuous Speech in Hindi Based on F0 Patterns," *Speech Communication*, Vol. 18, 1996, pp. 21–46.

[44] Rogers, D. F., and J. A. Adams, *Mathematical Elements for Computer Graphics*, New York: McGraw Hill, 1989.

[45] Pinto, N. B., D. G. Childers, and A. L. Lalwani, "Formant Speech Synthesis: Improving Production Quality," *IEEE Trans. Acoust. Speech Signal Processing*, Vol. 37, 1989, pp. 1870–1887.

[46] Karlsson, I., "Female Voices in Speech Synthesis," *Speech Communication*, Vol. 19, 1991, pp. 111–120.

12

Neural Networks for Speech Coding
Lizhong Wu

12.1 Introduction

Can neural networks be applied to speech coding? What kind of performance can we expect from neural networks with respect to speech coding? How should speech coders be implemented by neural networks? This chapter tries to answer these questions and discusses the fundamental issues arising from the application of speech coding to neural networks.

The outline of the chapter is as follows. Section 12.2 surveys various coding approaches after a brief introduction of source coding and neural networks. Section 12.3 shows the effectiveness of neural networks in source coding by demonstrating the quantization performances of several networks, including self-organizing mapping, associative memories, multilayer perceptrons, and recurrent networks. Section 12.4 presents several real-world speech coding applications with neural networks. It covers network architectures, learning algorithms, system designs, and coding performance. Section 12.5 provides a conclusion.

12.2 Source Coding and Neural Networks

12.2.1 Source Coding

Assume that x_i is a stationary discrete sequence from a source alphabet $\mathscr{S}$. For a fixed dimension p, let $X_i = (x_{pi}, x_{pi+1}, \ldots, x_{p(i+1)-1})^T \in \mathscr{S}^p$. An

r bits/vector coder (or quantizer) is defined by:

1. A codebook $\mathscr{W} = (W_1, \ldots, W_L)$, where $L = 2^r$, each W is a p-dimensional codevector and the index of codevector is codeword;
2. A mapping from the source space to the code space $q : \mathscr{S}^p \rightarrow \mathscr{W}$;
3. A partition $\mathscr{P} = (P_1, \ldots, P_L)$ over the source space $\mathscr{S}^p$, such that $q(X_i) = W_k$ if $X_i \in P_k$;
4. A distortion measure $d : \mathscr{S}^p \times \mathscr{W} \rightarrow (0, \infty)$, such that $d(X_i, W_k)$ measures the reproduction error of X_i with W_k. The performance of the quantizer is then estimated by the expected distortion:

$$D = \sum_i d(X_i, q(X_i)) = \sum_{k=1}^{L} \sum_{X_i \in P_k} d(X_i, W_k). \qquad (12.1)$$

Figure 12.1 shows the quantization process of a one bit/sample two-dimensional vector quantizer.

The basic design scheme for source coding is the generalized Lloyd algorithm or the LBG algorithm [1], which begins with an initial guess of the codebook $\mathscr{W}^{(0)}$, finds the optimal partition $\mathscr{P}^{(0)}$ for $\mathscr{W}^{(0)}$, finds the optimal codebook $\mathscr{W}^{(1)}$ for the partition $\mathscr{P}^{(0)}$, and continues until some

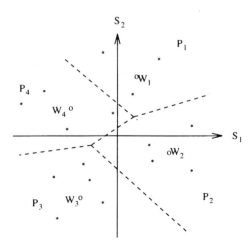

Figure 12.1 One bit/sample two-dimensional vector quantizer.

convergence criterion is met. The iterative procedure produces a codebook giving a minimal average distortion with a finite training set. To avoid the local minima in the above searching procedure, more than one initial guess of codebooks $\mathscr{W}^{(0)}$ is usually tried.

The theoretical basis for source coding is Shannon's rate-distortion theory [2]. Although it does not provide a constructive design technique, the rate-distortion function offers a theoretical performance bound for source coders.

References on source coding includes [1, 3–7].

12.2.2 Neural Networks

A neural network consists of many simple, neuronlike processing units that interact through weight connections. The state of each unit is determined by the input received from other units in the network. The architectures of the networks used in signal processing can roughly be divided into three categories: feedforward networks, feedback networks, and self-organizing networks. Feedforward networks [8] transform sets of input signals into sets of output signals. The desired input-output transformation is usually determined by external, supervised adjustment of the network parameters. In feedback networks [9], the input information defines the initial activity state of a feedback system, and after state transitions the asymptotic final state is identified as the outcome of the computation. In self-organizing networks [10], neighboring cells compete in their activities by means of mutual lateral interactions, and develop adaptively into specific detectors of different signal patterns.

Many different network structures exist together with related learning schemes, but all have in common that the properties of neural networks are determined by where the connections are and by their weights, and the learning procedure is conducted by changing the weights or adding or removing connections. Therefore, any network can be expressed in terms of a transformation function that maps inputs into outputs, as shown in Figure 12.2. Assuming that the input vector $X = (x_1, \ldots, x_p) \in \mathscr{R}^p$, the desired output vector $Y = (y_1, \ldots, y_q) \in \mathscr{R}^q$, and $\mathscr{W}$ is a weight space appropriate to the network architecture, the relationship among X, Y, and $W \in \mathscr{W}$ can be described by a function $F: \mathscr{R}^p \times \mathscr{W} \to \mathscr{R}^q$. Given weight W and input X, the output is given as $F(X, W)$, which is compared with the desired output Y. A performance measure of neural networks is the

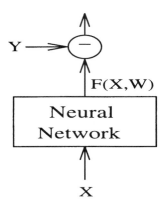

Figure 12.2 Neural network with an input X, a target Y, and a transform function $F(X, W)$.

expected cost function:

$$D(W) = E[d(Y, F(X, W))], \qquad W \in \mathscr{W} \tag{12.2}$$

The goal of network learning is to find a solution to the problem

$$W^* = \arg \min_{W \in \mathscr{W}} D(W) \tag{12.3}$$

Since the joint probability of X and Y is not known, W^* cannot be solved directly. The learning is usually conducted by optimizing performance over sets of samples. Learning schemes such as simulated annealing and genetic algorithms seek global minima over the samples. For practical applications, however, the gradient descent technique is more widely used because of its simplicity, even though it cannot guarantee a globally minimal solution.

12.2.3 Source Coding with Neural Networks

Relationships between source coding and neural network learning are obvious. The optimal performance of source coders can be described by their rate-distortion functions. Neural networks can be expressed by their mapping transform functions. While training a neural network as a source coder with a given data set and a predefined distortion measure, the learning process of the neural network can be viewed as changing its weights to minimize the Kullback-Leibler information between the transition probability of the neural network and that of the optimal source coder, so that the neural network satisfies the necessary condition for approaching the rate-distortion bound. Therefore, in theory, it is possible to train a neural network as an optimal source coder. Detailed theoretical descriptions can be found in [11].

Research on source coding with neural networks has concentrated on the three approaches described below. In Section 12.3, we will see how they perform when compared with the optimal source codes.

12.2.3.1   Vector Quantization with Kohonen Self-Organizing Feature Maps

In this type of vector quantizer, the network is formed by a set of single-layer linear neural units. Each unit corresponds to one codeword, and its connection weight directly represents each component of the code-vector. The source vector is applied to all unit inputs and the code of the current input data is then specified by the index of the unit with the least output cost function. This network is usually trained using Kohonen's relaxation algorithm [12]. We will discuss this model more thoroughly in Section 12.3.1.

12.2.3.2   Multilayer Neural Network as Front-End of a Coder

Early studies include [13, 14]. In these models, the number of hidden units should be smaller than the number of units in the input or output layer, and source vectors are applied not only to the input layer but also the output layer as training patterns. The input stage of the network learns a transformation rule for suppressing signal redundancy and the output stage learns an inverse transformation rule to reproduce the quantized signal. During coding, a quantizer is inserted at the output of the hidden layer to generate discrete finite-valued code symbols. However, this quantizer is designed independently of the neural network since the joint optimization for the quantizer violates the requirement of the activation function to be continuous, which is necessary to train the network. So the neural network in this model actually acts as a front-end for dimensionality reduction. We will discuss this model more thoroughly in Section 12.3.2.

12.2.3.3   Codebook-Excited Neural Networks

The codebook-excited neural network (CENN) [11] consists of a multilayer neural network and a trained excitation codebook. The neural network is driven by a vector selected from the excitation codebook. Both the neural network and the codebook can be trained simultaneously. We will discuss this model more thoroughly in Section 12.3.3.

12.3 Quantization Performance of Neural Networks

This section analyzes and compares the quantization performances of the three types of neural network source coding models listed above: Kohonen self-organizing feature maps, multilayer neural network front-ends, and codebook-excited neural networks.

A Gauss-Markov or first-order Gauss auto-regressive data set will be used as a benchmark for various empirical evaluations. This data set has been commonly used in the simulation of speech coding systems, since its theoretically attainable performance bound is known. The data is defined by: $X_{n+1} = \alpha X_n + G_n$, where $\{G_n\}$ is zero mean, unit variance, i.i.d Gaussian samples, and α is a correlation coefficient. For $\alpha = 0.9$, the theoretically achievable SNR with respect to the MSE measure and one BPS transmission rate is 13.2 dB [2]. In our benchmark, 120,000 samples are generated and divided into two equally sized groups, a training set and a test set.

12.3.1 Kohonen Self-Organizing Feature Maps

Kohonen self-organizing feature maps [12] are neural networks formed by a set of single-layer linear units. Physically this network has only one layer of linear neurons, due to the lateral interactions in the network; however, its topology is in effect even more complicated than that of multilayer back-propagation networks. Any neuron in the Kohonen network is also able to create an internal representation of input information in the same way as the "hidden units" in back-propagation networks eventually do [15]. It is commonly used in vector quantizer (VQ) design because of its ability to adapt and continue learning. The Kohonen algorithm can be described in the following three steps:

1. Initialize all weight vectors W_k, $k = 1, \ldots, L$, to small random values, where L is the number of neurons (the size of the codebook) in the neural network VQ;

2. For each input vector X_n, $n = 1, \ldots, N$, where N is the number of vectors to be quantized, find W^* with a minimal distance to the X_n, and update the W^* and its neighborhood denoted by N_c, with

$$\Delta W_k^{(m)} = -\eta^{(m)} \nabla_{W_k} D(X_n, W_k^{(m-1)}), \qquad k \in N_c;$$

3. Repeat step 2 until the decrease of the quantizing distortion is less than some given threshold.

The optimal design condition can be derived based on the selection of the two training parameters, the gain sequence $\eta^{(m)}$, and the size of the neighborhood set N_c. An optimal form that gives fastest learning in minimizing the distortion of neural network VQs has been derived by Wu and Fallside [16] as:

$$\eta_k = \frac{1}{2\|P_k\|} \tag{12.4}$$

$$N_c = 1. \tag{12.5}$$

The optimal gain sequence is associated with the neuron to be updated. P_k is the source space partition due to the neuron k. The size of the partition or the number of input vectors that are encoded by the neuron k is $\|P_k\|$.

Figure 12.3 compares the learning curves for the above optimal design condition with those for the conventional setup in the original Kohonen algorithm. It consists of both batch learning and continuous learning. In batch learning, the weights are updated only after each complete presentation of all input vectors. In continuous learning, the weights are updated after each input vector.

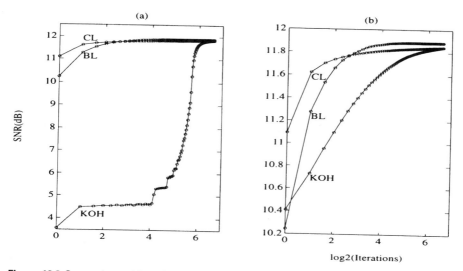

Figure 12.3 Comparison of learning curves of neural network VQs.

In Figure 12.3(a), the learning process of the original Kohonen algorithm is divided into two phases. The setup of the training parameters in each phase is [17]:

$$\text{phase 1:} \quad \begin{cases} \eta(t) = C_1\left(1 - \dfrac{t}{T_1}\right) \\[2mm] N_c(t) = (N_0 - 1)\left(1 - \dfrac{t}{T_1}\right) + 1 \end{cases} \tag{12.6}$$

$$\text{phase 2:} \quad \begin{cases} \eta(t) = C_2\left(1 - \dfrac{t - T_1}{T_2}\right) \\[2mm] N_c(t) = 1 \end{cases} \tag{12.7}$$

where T_1, T_2 are the number of iterations in phase 1 and phase 2, and C_1 and C_2 are constants. Here $T_1 = 50$, $T_2 = 49$, $C_1 = 0.1$, $C_2 = 0.01$, and N_0 is the initial neighborhood, which covers half of the neurons at the beginning. In Figure 12.3(b), the weight updating was restricted to a winning neuron. The learning process consisted of only the second phase, i.e., $T_1 = 0$, $T_2 = 99$, $C_1 = 0$, and $C_2 = 0.01$.

We see from either case, Figure 12.3(a) or (b), that the optimal design condition indeed provides the fastest learning. The results in [16] also show that the VQ using Kohonen algorithm does not perform better than the VQ using the LBG algorithm. Only when the optimal design condition is used will the rate-distortion performance of the VQ using Kohonen algorithm approach that of the VQ using the LBG algorithm, and the number of iterations will not be greater than that for the VQ with the LBG algorithm. Batch learning sweeps through all the inputs and accumulates $\nabla_W D$ before changing weights, so it is guaranteed to move in the direction of the steepest descent. The VQ with batch learning is identical to the VQ with the LBG design algorithm and slightly better than the VQ with continuous learning, but the latter converges more quickly.

Although the VQ using Kohonen's self-organizing feature maps will not outperform the VQ using the conventional LBG algorithm, the self-organizing feature maps can be extended to finite state vector quantization. A finite state vector quantizer (FSVQ) can further exploit the redundancy between the vectors (frames) of a memory source. An FSVQ is a finite collection of ordinary VQs with different codebooks. For each current input, one of the VQs is switched on. The state of the switch is determined by past input vectors. An FSVQ can be described by the following three mappings.

An encoding mapping α: $\quad \mathscr{X} \times \mathscr{S} \to \mathscr{C}$

A state mapping f: $\quad\quad \mathscr{C} \times \mathscr{S} \to \mathscr{S}$

A decoding mapping β: $\quad \mathscr{C} \times \mathscr{S} \to \hat{\mathscr{X}}$

where $\mathscr{X}$ and $\hat{\mathscr{X}}$ are an input space and a reproduction space, $\mathscr{C}$ is a code space, and $\mathscr{S}$ is a state space.

Given distortion criterion of $d(X, \hat{X})$, $X \in \mathscr{X}$, and $\hat{X} \in \hat{\mathscr{X}}$, an FSVQ is designed by finding the mappings α, β, and f, so that

$$E(d(X, \beta(\alpha(X, s), s))) = \min E(d(X, \hat{X})) \qquad (12.8)$$

where $E(\cdot)$ stands for the expectation over the $\mathscr{X}$.

A family of algorithms for the design of an FSVQ has been given by Dunham and Gray [18] and Foster, Gray, and Dunham [19]. These algorithms first design an initial set of vector quantizers together with a next-state function followed by an iterative improvement algorithm. Due to the complexity of the FSVQ, neither theoretical nor practical results have been obtained so far that demonstrate even local optimality.

In Sections 12.3.1.1 through 12.3.1.4, we introduce two neural network models for FSVQ based on Kohonen self-organizing feature maps. In the first model, besides the connections of an input layer which learn the distribution of a source, the interconnections among output neurons are simultaneously trained to capture the conditional distribution of the inputs. In the second model, although no interconnection exists among output neurons, the neurons are arranged in accordance with the connectivity of their input layer, and the neurons depend on each other.

12.3.1.1 Architecture and Training Process

A neural network FSVQ consists of K neurons. A case with $K = 5$ is shown in Figure 12.4. Each neuron has a firing or quiescent state. $\mathscr{W} = \{W_k, k = 1, \ldots, K\}$ denote codevectors and $\mathscr{C} = \{c_k, k = 1, \ldots, K\}$ are codewords. For each input, only L $(L < K)$ neurons are firing and the others are quiescent. The firing pattern of the L neurons is determined by the state of the quantizer. Associated with a state s, $s \in \mathscr{S}$, the collection of firing neurons is denoted by $\mathscr{U}_s$, and its corresponding collection of codevectors is defined as a state codebook

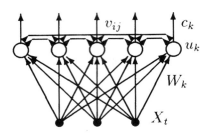

Figure 12.4 Architecture of a neural network model for an FSVQ.

$\mathcal{W}_s$. Output neurons in the first neural network FSVQ model are completely connected. The connectivity from neuron i to neuron j is v_{ij}. In general, $v_{ij} \neq v_{ji}$ for $i \neq j$ and $v_{ii} \neq 0$. For the second neural network FSVQ model, all $v_{ij} = 0$, but neurons $u_k, (k \in K)$, are arranged in order in accordance with the following equation

$$d(W_k, W_i) > d(W_k, W_j), \quad \text{if } |k - i| > |k - j|, \text{for } i, j, k \in K \quad (12.9)$$

With a current state s_t, an input X_t is encoded by

$$W_t = \arg \min_{W_k \in \mathcal{W}_{s_t}} d(X_t, W_k) \quad (12.10)$$

and c_t ($c_t \in \mathscr{C}_{s_t}$), the index of the W_t, is transmitted.

Since $\|\mathcal{W}_s\| < \|\mathcal{W}\|$, neural network FSVQs are expected to provide better performances than neural network VQs with the same number of neurons. First, an input vector can be encoded with fewer bits. Because the state of a model relies on previous inputs, which can be approximated by past codewords, a decoder can automatically track the state of an encoder and we do not need to transmit state information with additional bits. This leads to reduction of information rate by $(\log_2 K - \log_2 L)$ bits, in theory. Second, searching an optimal codeword for a current input is restricted to a smaller range, so the computational complexity in neural network FSVQs is also reduced.

The above analysis of the advantages of neural network FSVQs is based on the assumption that distortion will not increase. Since the conditional entropy $H(X_t|X_{t-1}, X_{t-2}, \ldots, X_{t-l_0})$ is less than $H(X_t)$, these advantages are true if the neural network FSVQ has learned the conditional distribution $p(X_t|X_{t-1}, X_{t-2}, \ldots, X_{t-l_0})$ during training, and employs this conditional distribution in coding.

In the first neural network FSVQ model, the connectivity $V = \{v_{ij} : i, j = 1, 2, \ldots, K\}$ is trained to reflect the conditional distribution of an input, and in the second model, the neurons with similar weights in their input layer tend to cluster, so the first model is referred to as a conditional histogram neural network FSVQ and the second one as the nearest-neighbor neural network FSVQ.

12.3.1.2 Conditional Histogram Neural Network FSVQ

For simplicity, here we discuss the first-order predictive case, i.e., $p(X_t|X_{t-1}, X_{t-2}, \ldots, X_{t-t_0}) = p(X_t|X_{t-1})$. After X being quantized, $p(X_t|X_{t-1})$ can be approximately represented by $p(c_i|c_j)$, where $c_i = \alpha(X_t, s_t)$, $c_i \in \mathscr{C}_{s_t}$, $c_j = \alpha(X_{t-1}, s_{t-1})$, $c_j \in \mathscr{C}_{s_{t-1}}$, and $\mathscr{C}_{s_t}$ and $\mathscr{C}_{s_{t-1}}$ respectively correspond to $\mathscr{W}_{s_t}$ and $\mathscr{W}_{s_{t-1}}$. If we let v_{ij} stand for the grade membership of c_i and c_j, then the definition for firing (or quiescent) neurons at time t is that they should have strong (or weak) connectivity with the neuron j. Therefore, the $\mathscr{C}_{s_t}$ can be designed by selecting L neurons with larger $v_{ij}, i \in 1, 2, \ldots, K$.

The learning rule for W is like that in neural network VQs. We concentrate on the learning for V. V can be represented by an associative memory, without hidden units, as shown in Figure 12.5. We aim to store a set of associations between inputs $\{c_{t-1}\}$ and outputs $\{c_t\}$ by modifying the weights on the connections. For the first-order prediction, each time only one of the input neurons is activated, so all the input vectors are orthogonal. A linear associative memory trained with a simple Hebbian procedure will exhibit a perfect recall.

According to the Hebbian learning rule [12], if a neuron u_i receives an input from another neuron u_j, then, if both are highly active, the weight v_{ij} from u_j to u_i should be strengthened. So we define the learning rule for the connectivity v_{ij} as

$$\Delta v_{ij} = \begin{cases} g_1 & \text{if } i = c_t, j = c_{t-1} \\ g_2 & \text{otherwise} \end{cases}$$

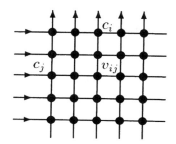

Figure 12.5 Associative memory for storing and recalling a conditional distribution $p(c_t|c_{t-1})$.

where $g_1 > g_2$. To prevent the same neurons from firing in all cases, it is necessary to impose a constraint on g_1 and g_2 to keep the sum of the weights (or the sum of their squares) constant. So when a neuron becomes more sensitive to one input, it becomes less sensitive to other inputs.

This learning process results in the following correlation matrix memory [12] between c_t and c_{t-1}:

$$V = \hat{E}(c_t c_{t-1}^T)[\hat{E}(c_{t-1} c_{t-1}^T)]^{-1}$$

For the pth-order $(p > 1)$ prediction, p input neurons are activated each time, and the input vectors to the associative memory are now no longer orthogonal. Willshaw [20] showed that if the number of active input neurons was less than the logarithm of the total number of input neurons, the probability of incorrectly activating output neurons could be made very small. In this case, an associative memory with a learning procedure like gradient descent, or an associative net with hidden units, can perform better [21].

The final purpose of learning is that both the W and V distributions can respectively approach $p(X_t)$ and $p(X_t|X_{t-1})$. Since the learning process for both W and V is carried out at the same time, their approaches $p(X_t)$ and $p(X_t|X_{t-1})$ are dependent on each other. Before W (or V) converges to a stable estimation, its restriction on the V's (or the W's) learning algorithm for the next input cannot be very accurate. This dependence between both learning processes is directly related to the difference between the number of neurons K and the size of the state codebook L. So we make the size of the state codebook variable during learning. It begins from K, gradually reduces, and finally reaches L.

12.3.1.3 Nearest-Neighbor Neural Network FSVQ

Let X_{t-1} and X_t denote two successive inputs and

$$W_{t-1} = \arg \min_{W_k \in \mathcal{W}_{s_{t-1}}} d(X_{t-1}, W_k),$$

$$W_t = \arg \min_{W_k \in \mathcal{W}_{s_t}} d(X_t, W_k).$$

Since

$$d(W_{t-1}, W_t) \leq d(W_{t-1}, X_{t-1}) + d(X_{t-1}, X_t) + d(X_t, W_t),$$

$d(W_{t-1}, W_t)$ approximately expresses $d(X_{t-1}, X_t)$. For a highly correlated source sample, $d(X_{t-1}, X_t)$ is small and we can expect that $d(W_{t-1}, W_t)$ is also small. If we define the state at time t as the last chosen neuron indexed by c_{t-1}, then the elements of a current state codebook are in the nearest neighbor of the c_{t-1}. The range of the neighborhood is equal to the size of the codebook L. So,

$$\mathscr{C}_{S_t} = \{c_{t-1} - L/2, \ldots, c_{t-1}, \ldots, c_{t-1} + L/2\}$$
$$\mathscr{W}_{S_t} = \{W_{c_{t-1}-L/2}, \ldots, W_{c_{t-1}}, \ldots, W_{c_{t-1}+L/2}\}$$

$$(12.11)$$

Kohonen's topology-preserving mapping is modified to train the nearest-neighbor neural network FSVQ. The training process consists of two phases. In the first phase, $\mathscr{W}$ is trained to learn the $p(X)$ and is kept in the order defined by (12.9). The updating rule for $\mathscr{W}$ is the same as that of the Kohonen algorithm. In the second phase, $\mathscr{W}$ is fine tuned with

$$\Delta W_t = \eta^{(m)} \nabla_W d(X_t, W_t) \qquad (12.12)$$

where W_t is defined by (12.10) and $\eta^{(m)}$ is the gain sequence.

12.3.1.4 Simulations

Table 12.1 compares the design SNR performances of neural network VQs, conditional histogram neural network FSVQs, and the nearest-

Table 12.1
Comparison of the rate-distortion performances

Vector dimension	CVQ SNR (dB)	CFSVQ-CH		CFSVQ-NN	
		SNR (dB)	K	SNR (dB)	K
2	7.967	10.597	12	10.129	10
3	9.396	11.288	20	10.582	14
4	10.261	11.467	44	10.777	22
5	10.663	12.570	256	11.092	40
6	11.118	13.297	256	11.373	88

CVQ stands for neural network VQ, CFSVQ-CH stands for conitimal histogram neural network FSVQ, and CFSVQ-NN stands for nearest-neighbor neural network FSVQ. K is the number of neurons in the networks.

Table 12.2
Generalization performance of neural network
FSVQs

Vector dimension	SNR (dB) for test set	
	CFSVQ-CH	CFSVQ-NN
2	10.437	9.987
3	11.064	10.484
4	11.175	10.602

neighbor neural network FSVQs. The number of neurons correspond-ing to the maximum SNR is also listed in the table. The maximum number of neurons was restricted to 256. Here, we assumed an initial state $c_0 = 0$. We found that the results were not sensitive to the initial state if the training set was large enough. Because of the definition of (12.12), we only considered the case of $L > 2$, (i.e., the vector dimension of the $W > 1$). As shown in the table, all neural network FSVQs yielded better SNR performances than neural network VQs. The nearest-neighbor neural network FSVQs achieved great improvement at lower dimensions, but the gap became small as the dimension increased. Conditional histogram neural network FSVQs outperformed neural network VQs by about 2 dB for all dimensions of the network. At dimension 6, the conditional histogram neural network FSVQ with 256 neurons yielded 13.297 dB SNR, which had already exceeded the Shannon bound to the rate-distortion function.[1] The performance of the conditional histogram neural network FSVQ was very near to the best result shown in [19], which was obtained by a so-called omniscient state codebook design technique. However, the number of neurons corre-sponding to an optimal performance in the conditional histogram neural network FSVQ was far less than the state size in the FSVQ with the omniscient state codebook design algorithm. This means that the conditional histogram neural network FSVQ requires less storage memory and possesses lower computational complexity.

Table 12.2 shows the generalization property of neural network FSVQs. Compared with Table 12.1, the SNR performance for a test set

---

1. The theoretical bound is given when the number of samples tends to infinity. Here, due to the finite training set, the result exceeded the theoretical bound.

was only about 0.2 dB lower than that for a training set within the given dimensions.

12.3.2 Coders with Neural Network Front-Ends

This model has been studied by many researchers. As shown in Figure 12.6, a quantizer is inserted into a hidden layer of the neural network. The quantizer encodes the hidden output vectors into discrete finite-valued code symbols. Both input layers and output layers contain at least one layer of neurons and accomplish the transformation $Y = F_{in}(S)$ and $Z = F_{out}(\hat{Y})$. $F_{in}(\cdot)$ and $F_{out}(\cdot)$ stand for the transformation functions of the input layers and the output layers respectively, S is a source vector, $\hat{Y}$ is a reproduction vector of the Y and is selected from the reproduction codebook of the quantizer, and the output vector Z is as the reconstructed vector of the S.

The network is trained using the error back-propagation algorithm. This algorithm requires all activation functions to be continuous. Since a quantizer contains winner-takes-all operations, it is impossible to train the quantizer and the neural network jointly. In [14, 22], the quantizer and the neural network were independently designed. When training the network with the error back-propagation algorithm, the quantizer was removed from the network and $\hat{Y} = Y$ was assumed. After the network $F(\cdot) = F_{in}(\cdot)F_{out}(\cdot)$ was established, a data set $\{Y\}$ was generated with $Y = F_{in}(S)$, which was used to design the quantizer $q(\cdot)$. The $q(\cdot)$ was then inserted back between $F_{in}(\cdot)$ and $F_{out}(\cdot)$.

Wu and Fallside [11] proposed a method to improve the design

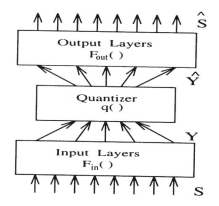

Figure 12.6 Quantizer with a neural-network-based front-end.

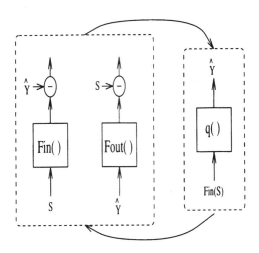

Figure 12.7 Sequentially reoptimizing the quantizer with a neural network front-end.

of this coder. As shown in Figure 12.7, after $F(\cdot)$ and $q(\cdot)$ are separately obtained, the input layer network $F_{in}(\cdot)$ and the output layer network $F_{out}(\cdot)$ are respectively trained, again using the error back-propagation algorithm. Now the effect of the quantizer is introduced into the network learning. For the input layer network, its input is S, its output is $F_{in}(S)$, and its desired output is $\hat{Y}$, where $\hat{Y} = q(F_{in}(S))$. For the output layer network, its input is $\hat{Y}$, its output is $F_{out}(\hat{Y})$, and its desired output is S. After $F_{in}(\cdot)$ and $F_{out}(\cdot)$ are updated, a new $\{Y\}$ is generated using the new $F_{in}(\cdot)$, which is used to train a new $q(\cdot)$. This procedure for sequentially reoptimizing $F_{in}(\cdot)$, $F_{out}(\cdot)$, and $q(\cdot)$ is repeated until the reduction of the distortion is smaller than a given threshold.

Table 12.3 gives a simulation result with a Gauss-Markov source. The transmission rate is one bit/sample and the size of the quantizer codebook is 2^{N_i}, where N_i is the number of input units. In this example, the SNRs for both the training and test sets are improved by about 0.55 dB after eight sequential reoptimizing iterations.

Table 12.4 evaluates the performance of the coder with a neural-network-based front-end, both with and without the reoptimizing design. The simulation is also carried out with a Gauss-Markov source. We note that even after reoptimizing its performance, it is still much worse than that of other types of coders. For example, compared with the conventional vector quantizer (whose corresponding performance is listed in the right column of Table 12.4), the SNR performance

Table 12.3
Sequentially reoptimizing the design of the coder with a neural-network-based front-end

Iter.	SNR (dB)	
	Training set	**Test set**
0	8.076	8.057
1	8.503	8.442
2	8.599	8.530
3	8.646	8.566
4	8.673	8.593
5	8.686	8.611
6	8.694	8.618
7	8.703	8.620
8	8.705	8.619
9	8.705	8.618

of the coder with a neural-network-based front-end is about 2 dB lower at all given dimensions.

The coding with a neural-network-based front-end is actually a type of transforming coder. Gray and Linde [23] have compared the performance between vector quantizers and optimized transform coders for a Gauss-Markov source and found that the full search vector quantizer generally performs better than the transform coder. So it is impossible to expect that the coder with a neural-network-based front-end can yield a good performance even if its components, the neural network and the quantizer, are optimally designed.

The neural network in the model of Figure 12.6 actually performs an auto-association [24]. An auto-associative multilayer perceptron with linear units is nothing but an indirect way of performing data compression by a Karhunen-Loeve transform. Its optimal weight values can be derived by standard linear algebra, the singular value decomposition technique. This shows that the nonlinear functions at the hidden layer are completely unnecessary.

To confirm the above results, Bourlard [25] experimentally compared the solutions both with the singular value decomposition

Table 12.4

Quantization performance of the one bit/sample coder with a neural-network-based front-end

			SNR (dB)					
			Initial design		After seq. reopt.		VQ	
N_i	N_o	N_i	Training	Test	Training	Test	Training	Test
4	2		8.547	8.527	8.828	8.793	10.261	10.122
5	2		8.702	8.682	8.742	8.716	10.663	10.440
6	2		7.847	7.810	8.835	8.694	11.118	10.830
7	2		8.665	8.581	9.024	8.916	11.437	10.979

technique and the error back-propagation algorithm. He determined the optimal weight matrices with the singular value decomposition technique and used these values as initialization of the error back-propagation algorithm. It was found that the error back-propagation algorithm was unable to improve the weights by reducing the error. Moreover, when starting the error back-propagation training algorithm several times with random weights, it always got stuck in local minima, giving higher error values. This illustrates the fact that, in the model of Figure 12.6, the linear approach is definitely preferable.

Therefore, we can conclude that when using multilayer neural networks for vector quantization or source coding, the model of Figure 12.6 is not a good choice.

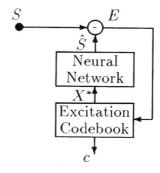

Figure 12.8 Codebook-excited neural network.

12.3.3 Codebook-Excited Neural Networks

An alternative model named the codebook-excited neural network (CENN) for source coding has been developed by Wu and Fallside [11]. The CENN consists of a multilayer neural network and a trained excitation codebook. A schematic diagram of this model is shown in Figure 12.8. The neural network is driven by a vector selected from the excitation codebook which is formed by training with entries initialized as Gaussian. For each source vector S, a vector X^* from the excitation codebook $\mathscr{X}$ which leads to

$$d(S, F(X^*)) = \min_{X \in \mathscr{X}} d(S, F(X))$$

is searched. Here, $F(\cdot)$ stands for the network transformation function and $d(\cdot)$ is a given distortion measure. The size of $\mathscr{X}$ is determined by the transmission rate r, and is equal to 2^r. After X^* is found, its index c is transmitted. At a decoding terminal, S is reconstructed by $F(X^*)$. Thus the total coding distortion over an encoded data set $\mathscr{S}$ is:

$$D = \sum_{S \in \mathscr{S}} d(S, \hat{S}) \tag{12.13}$$

$$= \sum_{S \in \mathscr{S}} d(S, F(X^*)) \tag{12.14}$$

$$= \sum_{S \in \mathscr{S}} \min_{X \in \mathscr{X}} d(S, F(X)) \tag{12.15}$$

For a given $\mathscr{S}$ and an error measure $d(\cdot)$, the quantization performance of a CENN is determined by $F(\cdot)$ and $\mathscr{X}$. The detailed learning algorithm is given in [11].

The performances of feedforward and recurrent CENNs for the Gauss-Markov source are plotted in Figure 12.9. For comparison, the theoretical performance bound, 13.2 dB, given by the Shannon lower bound to rate-distortion function [2] and the asymptotically optimal block quantizer bound [4, 26]

$$D_p(r) = \frac{e\Gamma(1+p/2)^{2/p}}{1+p/2} D_\infty(r) \tag{12.16}$$

are also plotted in the figure. The most encouraging result is the

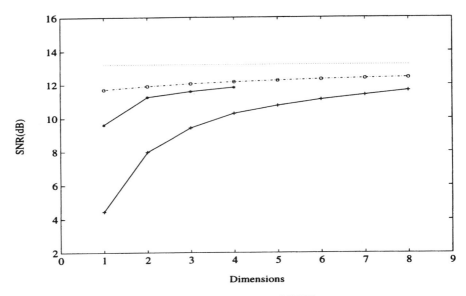

Figure 12.9 Comparison of quantization performance of CENNs.

quantization performance of recurrent CENNs. A two-dimensional quantizer has achieved about the same generalization performance as that of an eight-dimensional feedforward CENN. This shows that the recurrent connections accumulate the dependent characteristics over more than one previous input vector and exploit their redundancy.

12.3.4   Concluding Remarks

With the optimal design condition, the quantization performance of Kohonen self-organizing feature maps will be the same as that of the VQ designed using the LBG algorithm. The "topology-preserving" property of Kohonen self-organizing feature maps can be used to design predictive VQs that will achieve much better performance than ordinary VQs for highly correlated sources. This can be implemented by adding an associative memory to the output of Kohonen self-organizing feature maps.

In using multilayer neural networks for vector quantization or source coding, a multilayer neural network as the front-end of a quantizer projects the source vectors into a lower-dimensional space by removing the correlation between components. It actually performs a Karhunen-Loeve transform. Since the quantizer and the neural network cannot be trained jointly, the quantization performance of this type of neural network coder is much worse than the optimal performance, although the sequentially reoptimizing scheme can offer some improvement.

The codebook-excited neural network alleviates the above limitation. Both the neural network and the codebook can be trained with the error back-propagation algorithm simultaneously. Furthermore, the temporal information in source data can easily be taken into account by time-delayed feedback connections.

12.4 Speech Coding with Neural Networks

This section presents three real-work applications of neural networks for speech coding. It includes training algorithms, simulation results, and performance comparisons.

12.4.1 Coding Speech Spectrum with Neural Networks

The model used here is a conditional histogram FSVQ using Kohonen self-organizing feature maps as described in Section 12.3.1 where a set of single-layer networks are trained to span an appropriate region of the speech linear predictive coefficient space. The code of input speech frame is then specified by the index of the network with the least output cost function.

One of the empirical evaluations of this model is the procedure proposed by Gray and Markel [27]. Let $1/A(Z)$ be the unquantized, normalized, linear predictive spectrum of a speech frame S and let $1/A'(Z)$ be its corresponding quantized spectrum, i.e.,

$$A'(Z) = 1 + \sum_{j=1}^{p} w_{kj} Z^{-j} \qquad (12.17)$$

where

$$W_k = \arg \min_{W} d(S, W) \qquad (12.18)$$

The mean absolute log spectral measure in dB is defined as

$$d = \int_{-\pi}^{\pi} |V(\theta)| \frac{d\theta}{2\pi} \qquad (12.19)$$

where

$$V(\theta) = 10 \log \frac{|A'(e^{j\theta})|^2}{|A(e^{j\theta})|^2} \qquad (12.20)$$

A comparison of the quantized and unquantized normalized linear predictive spectra of speech is shown in Figure 12.10. From this figure, we can see that both the original and quantized spectra, which are dependent on the size of codebooks, are close, and the error with the FSVQ is smaller than that with the VQ. More discussions on this model can be found in [10].

12.4.2 Nonlinear Prediction Speech Coding

Speech prediction can be defined as follows: given p previous observations of the signal $s(t)$, $X = (s(t-1), \ldots, s(t-p))^{T}$, find a

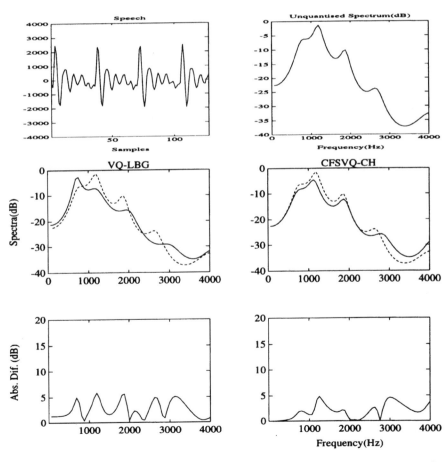

Figure 12.10 Comparison of quantized and unquantized linear predictive spectra of a speech frame.

function $g(\cdot)$ that minimizes the predictive residual

$$D = \int \int \|s - g(X)\|^2 P(X,s)dXds \qquad (12.21)$$

where $P(X,s)$ is the density function of the joint probability of X and s. The theoretical solution of (12.21) is a posterior mean estimator:

$$g(X) = \int sP_{s|X}(X,s)ds \qquad (12.22)$$

where $P_{s|X}(X,s)$ is the density function of the conditional probability of s given X.

With a multilayer neural network, if the number of input units is p and there is only one output unit, the network can be trained as a pth-order predictor. Assume that $F(\Phi,X)$ is the transfer function of the network. The aim of training is to determine the architecture of the hidden layers of the network and to adjust its weights, Φ, so that $F(\Phi,X)$ approaches the posterior mean estimator $g(X)$.

Neural-network-based predictors can be used for modeling data without any specific prior assumption about the form of nonlinearity. Their advantages have been reported by a number of researchers, e.g., Lapedes and Farber [28]. Here, we discuss a more general predictive model and show how to apply it to nonlinear predictive speech vector quantization.

12.4.2.1 A Neural Model of Nonlinear Prediction

A general structure of a neural-network-based nonlinear predictor is shown in Figure 12.11. It consists of three layers, which contain N_i, N_h, and 1 units in the input, hidden, and output layers, respectively, with N_i set to the given predictive order. To predict observations with any scale of amplitude, no nonlinear activation function is imposed on the output unit. The output of the hidden units is delayed by τ and fed back to the inputs of the hidden units via a weighting matrix W. This predictor can be described by the equations:

$$Y(t) = f(WY(t-\tau) + VX(t))$$
$$z(t) = UY(t) \qquad (12.23)$$

where U is the weight vector between the output unit and the hidden layer and V is the weight matrix between the hidden layer and the input layer. $X(t) = (s(t-1), \ldots, s(t-N_i))^T$ is the input vector, $Y(t)$ the hidden vector, and $z(t)$ the output variable; $f(\cdot)$ is a differentiable nonlinear function and τ is a time delay factor.

In (12.23), if $W = 0$, the network is of the feedforward type. Moreover, if $N_h = N_i$, $V = I$, where I is the identity matrix, $f(X) = X$, $W = 0$, and $U = \alpha$, then $z(t) = \alpha X(t)$, and the predictor becomes the linear one.

12.4.2.2 Nonlinear Predictive Vector Quantization

A nonlinear predictive VQ is formed by a set of predictors $\{F(\Phi_k, X), \ k = 1, \ldots, K\}$. The number of predictors K is equal to 2^r, r (in bits) being the size of the quantizer. The performance of predictive quantizers is evaluated by their distortion-rate functions. In the nonlinear predictive case, each predictor is trained to cover certain regions of the nonlinear predictive parameter space. The nonlinear predictive quantizer is therefore expected to cope with the non-stationary nature of the predicted signal and to improve the predictive performance over that of an individual predictor. During quantization, each frame of speech is successively applied to all the predictors in the quantizer. The predictor with the least predictive error, averaged over the whole frame, is then selected to quantize the current frame. The process can be described by the equation

$$c = \arg \min_{1 \leq k \leq K} \sum_{t}^{N} \|s(t) - F(\Phi_k, X(t))\|^2 \qquad (12.24)$$

where N is the frame length. The nonlinear prediction of the speech frame is then represented by a code symbol of r-bit length instead of the weight parameters of the selected predictor.

The training process of the nonlinear predictive quantizer is the same as that for a single predictor, except that the cost function becomes

$$D = \sum_{k}^{K} \sum_{S(t) \in k} \sum_{t}^{N} \|s(t) - F(\Phi_k, X(t))\|^2 \qquad (12.25)$$

and the updating equation becomes

$$\Delta\Phi_k = \eta \sum_{S(t) \in k} \sum_{t}^{N} \|s(t) - F(\Phi_k, X(t))\| \nabla_{\Phi_k} z_k(t) \qquad (12.26)$$

where $S(t) = \{s(t), t = 1, 2, \ldots, N.\}$.

Unlike linear prediction, however, the output of a nonlinear predictor cannot be simply expressed as the sum of both the zero-state response and the zero-input response. The memory effect of the previous selected predictor cannot be directly removed by subtracting the current frame speech with the zero-input response. The memory of the previous frame should be taken into account during the predictor selection for the current frame. The initial value $Y(t)$ for the current frame is therefore set to the last value $Y(t)$ of the previous frame.

Long-Term (Pitch) Nonlinear Speech Prediction

The speech signal consists of two types of redundancy or correlation: a short-term correlation between successive speech samples and a long-term correlation between adjacent pitch periods. Consequently, the linear prediction speech production model contains two time-varying linear predictive filters, a short-term one and a long-term one. The former supplies the spectral envelope of the speech signal and the latter reproduces the spectral fine structure. The long-term linear prediction of speech has been studied, for example, by Ramachandran and Kabal [29].

The long-term nonlinear prediction can also be implemented by neural networks. The structure chosen for the long-term nonlinear predictive model is the same as that for the short-term one shown in Figure 12.11, but the input is delayed by one pitch period before being applied to the input layer. The current sample is thus predicted from the samples around its last pitch period. The long-term nonlinear predictor is expressed by the following equation:[2]

$$Y_l(t) = f(W_l Y_l(t - \tau) + V_l X_l(t - T))$$
$$z_l(t) = U_l Y_l(t), \tag{12.27}$$

where T is the pitch period, which is found from the short-time autocorrelation function of the short-term residual [30]:

$$r_s(t) = s(t) - z_s(t) \tag{12.28}$$

Long-term prediction is always accompanied by short-term prediction in speech coding. The long-term predictor can be connected to

---

2. We use the subscript l to identify the long-term predictor and s the short-term predictor.

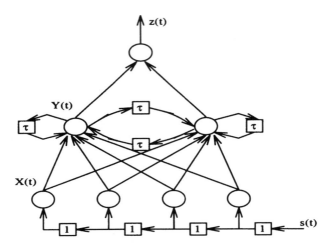

Figure 12.11 Neural-network-based nonlinear predictor.

the short-term one in cascade or parallel. In the cascade form, the short-term residual is delayed by one pitch period and fed to the long-term predictor. In the parallel form, the original speech is delayed by one pitch period and applied to the long-term predictor. So, in (12.27),

$$X_l(t) = (r_s(t - N_{li}/2), \ldots, r_s(t + N_{li}/2))^T \qquad (12.29)$$

for the cascade form and

$$X_l(t) = (s(t - N_{li}/2), \ldots, s(t + N_{li}/2))^T \qquad (12.30)$$

for the parallel form. The final residual of the combined short-term and long-term prediction for both the cascade and parallel forms is,

$$r(t) = s(t) - z_s(t) - z_l(t) \qquad (12.31)$$

The short-term and long-term predictors are trained in sequence. First, the short-term predictor is trained to minimize the short-term residual, $r_s(t)$. With the short-time autocorrelation function of $r_s(t)$, the pitch period is then estimated. Finally, the long-term predictor is trained to reduce the combined short-term and long-term residual $r(t)$.

12.4.2.3 Nonlinear Predictive Quantization Performance

Besides network architecture, two parameters, the predictive order, and the total number of weights, also affect the property of predictors. We compared the predictive quantization performance between different architectures under either of the following two conditions: (1) same predictive order and (2) same number of weights.

We have used the following notations: $LP(p)$ stands for a pth-order linear predictor, and $NLP(N_i - N_h)$ stands for a nonlinear predictor with N_i input units and N_h hidden units with recurrent neural network architecture. In all of our studies, N_h is set to two.

The nonlinear predictive quantizer is evaluated with continuous speech data. The speech samples were from the TIMIT database [31] and were prefiltered to 8 kHz. The training set consists of ten different spoken sentences by ten different speakers (five females). The test set consists of four different spoken sentences by four speakers (two females) not included in the training set.

Figure 12.12 compares the predictive gain-rate functions of the linear and nonlinear predictive VQs. The prediction gain is measured by the total speech energy over the total predictive error (in dB). The sizes of the VQs vary from one bit to six bits. The nonlinear predictive VQ is formed by 2^{r_s} $NLP(10 - 2)$ recurrent neural networks, where r_s (in bits) is the size of the short-term predictive VQ. The frame length was set to 256 samples without overlapping between frames. Therefore, the transmitted rate is $r_s/256$ bits/sample for the nonlinear predictive information.

The linear predictive VQ was designed using the approach developed in [32]. Each frame consists of 256 or 384 Hamming-windowed samples and is overlapped by 128 samples. Therefore, the transmitted rate is $r_s/128$ bits/sample for the frame size of 256 and $r_s/256$ bits/sample for the frame size of 384.

Figure 12.12 also shows the predictive gain-rate functions of the combined short-term and long-term VQs. The sizes of long-term predictive VQs are also from one bit to six bits. The long-term predictive VQ is cascaded with a six bit, short-term predictive VQ. Each predictor in the long-term nonlinear predictive VQ is an $NLP(3 - 2)$ recurrent neural network, and that in the linear predictive VQ is an $LP(3)$ linear predictor.

The pitch information is estimated using the short-time auto-correlation function of the short-term predictive residuals. It is updated every 64 samples. The long-term predictor is switched on only if the pitch period was not equal to zero.

The design of the long-term linear predictive VQ was based on [29, 33]. The predictor parameters were vector-quantized using a one-step identification/compression technique. A codebook of long-term

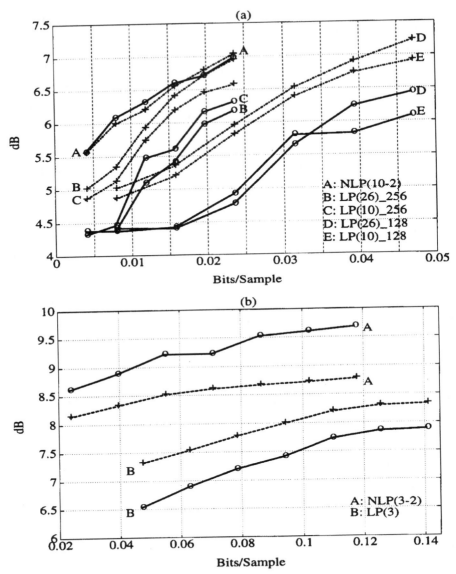

Figure 12.12 Comparison of the predictive gain-rate functions of linear and nonlinear predictive VQs.

predictor parameters is exhaustively searched to identify which predictor minimizes the short-term residual.

The frame length of long-term nonlinear prediction is 64 samples, as is the processing step-size of long-term linear prediction. Therefore, their transmitted rates are the same and equal to $r_l/64$ bits/sample, where r_l is the size of the long-term predictive VQ in bits.

From inspection of Figure 12.12, we can conclude that all nonlinear predictive quantizers outperform linear ones either with the same predictive order or with equal numbers of weights. The test performance of linear predictive VQs is 0.5–1 dB worse than the training performance. In contrast, with the same training and test data sets, Figure 12.12 shows that the test performance of nonlinear predictive VQ is close to, or even better than, the training performance.[3]

12.4.3 Code-Excited Nonlinear Predictive Speech Coding

In a code-excited linear predictive speech coding (CELP) system [34], speech is synthesized by passing an excitation vector through a cascade of short-term and long-term linear predictive filters. The excitation codebook usually performs the codevector search in an analysis-by-synthesis manner [35]. The excitation codebook may either be stochastic, i.e., pseudo-randomly populated, or predesigned over some training data.

In a code-excited nonlinear predictive speech coding system, the predictive filters are nonlinear. Here, we investigate the problems caused by such a replacement and show how to achieve better coding performance based on the capability of nonlinear predictive vector quantization.

12.4.3.1 Nonlinear Predictive Filter Tolerance for an Excitation Disturbance

In the prediction mode, the inputs are the original speech samples. In the synthesis mode, however, the predictive filter receives only the estimations of speech. The difference between the estimation and the original speech signal is referred to as an excitation or input disturbance. We will analyze the tolerance of a nonlinear predictive filter with respect to excitation disturbance.

Assume that the input of a given predictive filter is changed from X_0 to X with $D_x = \|X - X_0\|^2$, and their corresponding outputs are respectively z and z_0 with $D_z = \|z - z_0\|^2$.

In the linear prediction case, $z = \alpha^T X$, where $\alpha = (\alpha_1, \alpha_2, \ldots, \alpha_p)^T$

---

3. Both the training and test data sets are randomly chosen from the TIMIT database. Compared to the training data set, the test data set contains more voiced sounds, which usually achieves better predictive gain than unvoiced speech. Therefore, the test performance is better than the training performance.

is the pth-order linear predictive coefficient vector. We easily find

$$D_z = \|z - z_0\|^2 \leq \|\alpha\|^2 D_x \qquad (12.32)$$

In the nonlinear prediction case using the model shown in Figure 12.11, we have, [36, 37],

$$D_z \leq \rho^2 D_x \qquad (12.33)$$

where

$$\rho = \frac{\gamma\|U\|\|V\|}{\gamma\|W\| - 1} \qquad (12.34)$$

$\|U\|$, $\|V\|$, and $\|W\|$ are the Euclidean norms of U, V, and W respectively, $\gamma = \max|f'(O^*(t))|$, $O^*(t) \in (O(t), O_0(t))$, and $O(t) = WY(t - r) + VX(t)$.

Assume that two predictive filters have sensitivity parameters of ρ_1 and ρ_2. If they are connected in cascade, the sensitivity parameter of the combined filter is $\rho = \rho_1\rho_2$, and if they are connected in parallel, the sensitivity parameter of the combined filter is $\rho = \rho_1 + \rho_2$.

The upper bounds of the output variations of linear and nonlinear predictive filters, respectively, due to an input disturbance are given in (12.32) and (12.33).

As shown in [36, 37],

$$0 \leq \gamma\|W\| < 1$$

or

$$(\gamma\|W\| - 1)^2 < 1$$

If $(\gamma\|W\| - 1)^2$ is close to zero, then ρ becomes very large. Therefore, the output will greatly deviate from its original value, even for a small input disturbance.

To deal with the poor tolerance of the nonlinear predictive filter for an input disturbance in a nonlinear predictive coding system, two approaches can be applied:

1. As an accompanying training criterion, minimize the sensitive parameters at the same time as the predictive error is minimized during the design of nonlinear VQs, (see [36]);

2. Instead of using the stochastic codebook, use the trained excitation codevectors in the code-excited nonlinear predictive coder as discussed in [37].

12.4.3.2 Gain-Adaptive Nonlinear Predictive Coding

Nonlinear predictive neural networks can be trained to fit the signal with any scale of amplitudes at their outputs since no nonlinear activation function is imposed on their output units; nonetheless, a gain normalization step may still be desirable to smooth the outputs and reduce the output variances so as to improve the adaptive ability of coding. Unlike linear predictive speech coding, the gain term is placed at the output of the nonlinear predictive filter as shown in Figure 12.13 instead of at the output of the excitation codebook (both places are equivalent to the linear case). Such an architecture leads to $g(t)$ being independent of the $Z_p(t)$ and its optimal solution can be found as

$$g(t) = \frac{[S(t)]^T[Z_e(t) + Z_l(t) + Z_s(t)]}{\|Z_e(t) + Z_l(t) + Z_s(t)\|^2} \tag{12.35}$$

This results in the coding error

$$D = \|S(t) - g(t)[Z_e(t) + Z_l(t) + Z_s(t)]\|^2$$

$$= \|S(t)\|^2 - \frac{\{[S(t)]^T[Z_e(t) + Z_l(t) + Z_s(t)]\}^2}{\|Z_e(t) + Z_l(t) + Z_s(t)\|^2} \tag{12.36}$$

Therefore, the gain-adaptive nonlinear predictive speech coder can be trained by minimizing the above D or maximizing the second term of (12.37).

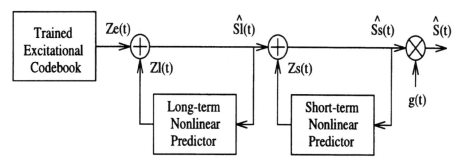

Figure 12.13 Trained code-excited nonlinear predictive (TCENLP) speech coder.

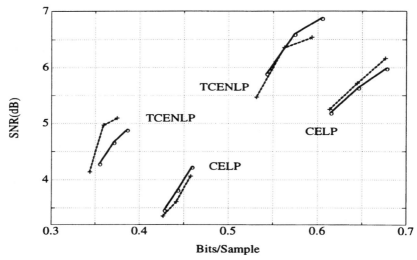

Figure 12.14 Comparison of the coding performance of the TCENLP and CELP.

Table 12.5
Bit allocation in speech coders

Parameters	Bits/sample
Code of short-term predictive VQ	$6/N_{sp}$
Code of long-term predictive VQ	$6/64$
Code of excitation codebook	B_e/N_e
Pitch	$B_p/64$
Gain	$5/N_e$

N_{sp}, the frame length of short-term prediction, is equal to 128 samples for nonlinear predictive coders and 256 for linear ones. B_e is the size of the excitation codebook and equals 7, 8, or 9 bits. B_p is the code-length of pitch-information, which is 8 bits if the pitch-period is not equal to zero and one bit for zero pitch-period. N_e is the length of excitation vectors which is set to 32 or 64.

12.4.3.3   Coding Performance

Figure 12.14 compares the coding performances of the trained code-excited nonlinear predictive speech coder (TCENLP) to that of the code-excited linear predictive speech coder (CELP), using the same training and test speech data sets as described in Section 12.4.2.3.

Table 12.5 lists the bit allocation in the speech coders. In all the simulations, the transmitted rate varies from 0.34 bits/sample to 0.68 bits/sample, i.e., from 2,720 BPS to 5,440 BPS.

The excitation codebook was initialized by Gaussian samples and trained over the training set. To observe the sensitivity of the excitation codebook, its size was varied from seven to nine bits, and the dimension of codevectors was set to 32 or 64. Because the pitch period was limited to between 20 and 148 samples (from 2.5 to 18.5 ms), it could be coded with seven bits without loss of information. The gain value was quantized with a Max-Lloyd quantizer [3, 38]. Its design was based on the same training data set. We found that the coding performance can be improved by sequentially reoptimizing the excitation codebook and the gain codebook [1]. This was accomplished by recursive adjustment of the entries of each codebook. One codebook was fixed when the other was updated.

The CELP speech coder design was based on [33, 35, 39]. The short-term and long-term linear predictive quantizers were described in Section 12.4.2.3. The excitation codebook was formed by Gaussian samples and fully searched during coding.

Subjective evaluation was carried out to compare speech quality of the nonlinear and linear predictive coders. A mean opinion score with five-point categories was used to evaluate the speech quality. The rating of speech quality, as described by its corresponding listening effort, is shown in Table 12.6. Ten subjects participated in the experiment and were asked to judge three sections of speech. These three sections were, respectively, the 8 kHz-downsampled original speech from the TIMIT database, the reconstructed speech from a 4,840 BPS TCENLP speech coder, and the reconstructed speech from a 4,920 BPS CELP speech

Table 12.6
Rating of opinion scores of speech quality and listening effort [40]

Rating	Quality	Listening effort
5	Excellent	Complete relaxation possible, no effort required
4	Good	Attention necessary, no appreciable effort required
3	Fair	Moderate effort required
2	Poor	Considerable effort required
1	Bad	No meaning understood with any feasible effort

coder. The speech was from the test data set. It consisted of four different sentences, each spoken by a different speaker. On average, the nonlinear predictive coder scored 3.00 points and the linear one 2.65 points. As a reference, the original speech scored 4.00 points. Among the ten subjects, eight judged that the nonlinear predictive coder produced higher-quality reconstructed speech. Only one subject gave an inverse judgment. One subject said that the two kinds of reconstructed speech were not significantly different.

12.4.4 Concluding Remarks

For a non-Gaussian signal, such as speech, nonlinear prediction will always achieve better predictive gain than linear prediction.

Nonlinear predictive vector quantization implemented by a set of neural networks outperforms the linear counterpart by 2−2.5 dB in the predictive gain when the transmitted rate is between 360 BPS and 940 BPS. The trained excitation codebook nonlinear predictive speech coders (TCENLP) achieve better coding performance than conventional CELP speech coders and have performance advantages of about 1.5 dB in SNR and 0.35 point in a five-category mean opinion scale for a coder with about 4,800 BPS.

The complexity of the excitation codebook search procedure in the TCENLP coder is the same as that of the CELP. More computational effort is required in neural computation of nonlinear predictive vector quantization. Compared with the single mapping of linear prediction, the nonlinear predictor network needs twice the mappings, first from the inputs to the hidden units and then from the hidden units to the output. Because we use the network that contains only two hidden units, increase in computation is not very high.

12.5 Conclusions

This chapter has demonstrated the speech coding applications of neural networks. In neural networks, knowledge or constraints are not encoded in individual units, rules, or procedures, but distributed across many simple computing units. Uncertainty is modeled not as likelihoods or probability density functions of a simple unit, but by the pattern of activity in many units. The computing units are simple in nature, and knowledge is not programmed into any individual unit's function; rather, it lies in the connections and interactions between linked processing elements.

The chapter also showed the advantages and disadvantages of the neural network approach over traditional stochastic models. Unlike stochastic models, neural network models rely on the availability of good training or learning strategies. Neural network learning seeks to optimize or organize a network of processing elements. However, neural network models need not make assumptions about the underlying probability distributions. They can be trained based on any optimization criteria such as maximum mutual information, minimum discriminant information, etc. Multilayer neural networks can be trained to generate rather complex nonlinear classifiers or mapping functions. Temporal features can easily be taken into account with time-delayed and recurrent connections. The simplicity and uniformity of the underlying processing element makes neural network models attractive for hardware implementations, which enables the operation of a network to be simulated efficiently.

References

[1] Linde, Y., A. Buzo, and R. Gray, "An Algorithm for Vector Quantizer Design," *IEEE Trans. on Communications*, Vol. COM-28, January 1980, pp. 84–95.

[2] Berger, T., *Rate Distortion Theory: A Mathematical Basis for Data Compression.* Englewood Cliffs, N.J.: Prentice-Hall, 1971.

[3] Lloyd, S., "Least Squares Quantization in PCM," Bell Laboratory Technical Note, *IEEE Trans. on Information Theory*, Vol. IT-28, No. 2, March 1982, pp. 129-137.

[4] Gersho, A., "Asymptotically Optimal Block Quantization," *IEEE Trans. on Information Theory*, Vol. IT-25, July 1979, pp. 373–380.

[5] Gray, R., "Vector Quantizers," *IEEE ASSP Magazine*, April 1984, pp. 4–29.

[6] Makhoul, J., S. Rouces, and H. Gish, "Vector Quantization in Speech Coding," *IEEE Proceedings*, Vol. 73, November 1985, pp. 1551–1588.

[7] Gersho, A., and R. Gray, *Vector Quantization and Signal Compression.* Kluwer Academic, 1995.

[8] Rumelhart, D., G. Hinton, and R. Williams, "Learning Internal Representations by Error Propagation," in D. Rumelhart and J. McClelland, eds., *Parallel Distributed Processing: Exploration in the Microstructure of Cognition*, Cambridge, Mass.: MIT Press, 1986, pp. 319–362.

[9] Tank, D., and J. Hopfield, "Neural Computation by Concentrating Information in Time," *Proc. Natl. Acad. Sci. USA*, Vol. 84, April 1987, pp. 1896–1900.

[10] T. Kohonen, "The Self-Organizing Map," *Proc. of the IEEE*, Vol. 78, September 1990, pp. 1464–1480.

[11] Wu, L., and F. Fallside, "Source Coding and Vector Quantization with Codebook-Excited Neural Networks," *Computer Speech and Language*, Vol. 6, July 1992, pp. 243–276.

[12] Kohonen, T., *Self-Organisation and Associative Memory*, 3d ed., New York: Springer Verlag, 1988.

[13] Cottrell, G., P. Munro, and D. Zipser, "Image Data Compression by Back-propagation: An Example of Extensional Programming," Tech. Rep. ICS 8702, Institute for Cognitive Science, University of California, San Diego, February 1987.

[14] Robinson, A., and F. Fallside, "Static and Dynamic Error Propagation Networks with Application to Speech Coding," in D. Anderson, ed., *Proceedings of Neural Information Processing Systems*, Denver: American Institute of Physics, 1987.

[15] Kohonen, T., "The 'Neural' Phonetic Typewriter," *IEEE Computer*, Vol. 21, No. 3, 1988, pp. 413–424.

[16] Wu, L., and F. Fallside, "On the Design of Connectionist Vector Quantizers," *Computer Speech and Language*, Vol. 5, July 1991, pp. 207–230.

[17] Brauer, P., and P. Knagenhjelm, "Infrastructure in Kohonen maps," *Proc. Int. Conf. Acoustics, Speech and Signal Processing*, 1989, pp. 647–650.

[18] Dunham, M., and R. Gray, "An Algorithm for the Design of Labelled-Transition Finite-State Vector Quantizers," *IEEE Trans. on Communications*, Vol. COM-33, January 1985, pp. 83–89.

[19] Foster, J., R. Gray, and M. Dunham, "Finite-State Vector Quantization for Waveform Coding," *IEEE Trans. on Information Theory*, Vol. IT-31, May 1985, pp. 348–359.

[20] D. Willshaw, "Holography, Associative Memory, and Inductive Generalisation," in G. E. Hinton and J. A. Anderson, eds., *Parallel Models of Associative Memory*, Hillsdale, N.J.: Erlbaum, 1981.

[21] Hopfield, J., "Learning Algorithms and Probability Distributions in Feed-forward and Feed-back Networks," *Proc. Natl. Acad. Sci. USA*, Vol. 84, December 1987, pp. 8429–8433.

[22] Sonehara, N., M. Kawato, S. Miyake, and K. Nakane, "Image Data Compression Using a Neural Network Model," in *Proc. Int. Joint Conference on Neural Networks*, 1989, pp. II35–II40.

[23] Gray, R., and Y. Linde, "Vector Quantizers and Predictive Quantizers for Gauss-Markov Sources," *IEEE Trans. on Communications*, Vol. COM-30, February 1982, pp. 381–389.

[24] Kosko, B., "Bidirectional Associative Memories," *IEEE Trans. on System, Man and Cybernetics*, Vol. 18, January/February 1988, pp. 49–60.

[25] Bourlard, H., "Neural Networks: Theory and Parallels with Conventional Algorithm," in *Lecture Notes on Neural Networks Summer School*, Cambridge: Cambridge University Engineering Department, 1991.

[26] Yamada, Y., S. Tazaki, and R. Gray, "Asymptotic Performance of Block Quantizers with Difference Distortion Measures," *IEEE Trans. on Information Theory*, Vol. IT-26, 1980, pp. 6–14.

[27] Gray, A. Jr., and J. Markel, "Distance Measures for Speech Processing," *IEEE Trans. on Acoustics, Speech and Signal Processing*, Vol. ASSP-24, October 1976, pp. 380–391.

[28] Lapedes, A., and R. Farber, "Nonlinear Signal Processing Using Neural Networks: Prediction and System Modelling," Tech. Rep. LA-UR-87-2662, Los Alamos National Laboratory, Los Alamos, New Mexico 87545, June 1987.

[29] Ramachandran, R. P., and P. Kabal, "Pitch Prediction Filters in Speech Coding," *IEEE Trans. on Acoustics, Speech and Signal Processing*, Vol. ASSP-37, April 1989, pp. 467–478.

[30] Rabiner, L., and R. Schafer, eds., *Digital Processing of Speech Signals.* Englewood Cliffs, N.J.: Prentice-Hall, 1978.

[31] Seneff, S., and V. W. Zue, "Transcription and Alignment of the TIMIT Database," in J. S. Garofolo, ed., *Getting Started with the DARPA TIMIT CD-ROM: An Acoustic Phonetic Continuous Speech Database*, Gaithersburgh, Md.: National Institute of Standards and Technology, 1988.

[32] Buzo, A., A. Gray, Jr., R. Gray, and J. Markel, "Speech Coding Based upon Vector Quantization," *IEEE Trans. on Acoustics, Speech and Signal Processing*, Vol. ASSP-28, October 1980, pp. 562–574.

[33] Davidson, G., and A. Gersho, "Real-Time Vector Excitation Coding of Speech at 4,800 bps," *Proc. Int. Conf. on Acoustics, Speech and Signal Processing*, 1987, pp. 2189–2192.

[34] Schroeder, M., and B. Atal, "Code-Excited Linear Prediction (CELP): High-quality Speech at Very Low Bit Rates," *Proc. Int. Conf. on Acoustics, Speech and Signal Processing*, 1985, pp. 937–940.

[35] Kroon, P., and E. F. Deprettere, "A Class of Analysis-by-Synthesis Predictive Coders for High Quality Speech Coding at Rates Between 4.8 and 16 kbits/s," *IEEE Journal on Selected Areas in Communications*, Vol. 6, February 1988, pp. 353–363.

[36] Wu, L., and J. Moody, "A Smoothing Regularizer for Feedforward and Recurrent Neural Networks," *Neural Computation*, Vol. 8, No. 3, 1996, pp. 463–491.

[37] Wu, L., M. Niranjan, and F. Fallside, "Fully Vector-Quantised Neural Network-Based Code-Excited, Nonlinear Predictive Speech Coding," *IEEE Trans. on Speech and Audio Processing*, Vol. 2, October 1994.

[38] Max, J., "Quantizing for Minimum Distortion," *IRE Transactions on Information Theory*, Vol. IT-6, March 1960, pp. 7–12.

[39] Trancoso, I., and B. Atal, "Efficient Procedures for Finding the Optimum Innovation in Stochastic Coders," *Proc. Int. Conf. Acoustics, Speech and Signal Processing*, 1986, pp. 2375–2378.

[40] Natvig, J., "Evaluation of Six Medium Bit-Rate Coders for the Pan-European Digital Mobile Radio System," *IEEE Journal on Selected Areas in Communications*, Vol. 6, February 1988, pp. 324–331.

13

Networks for Speech Enhancement

Eric A. Wan, Alex T. Nelson

13.1 Introduction

This chapter presents an overview of several approaches to neural speech enhancement. Speech enhancement is concerned with the processing of noisy and corrupted speech to improve the quality or intelligibility of the signal. While the general field is well established, in recent years several distinct neural-network-based frameworks have emerged in the literature. These include time-domain filtering, transform-domain mapping, state-dependent model switching, and online iterative approaches. The goal of this chapter is to explore the differences in the capabilities and assumptions of the various methods. Where appropriate, a review of traditional (nonneural) techniques is provided for context. Rather than giving a detailed description of each algorithm with explicit experimental comparisons, the chapter high-lights qualitatively those features that distinguish each approach. Potential areas of future research are also suggested, along with more general recommendations for the field.

13.1.1 Background

Speech enhancement is motivated by the need to improve the performance of voice communications systems in noisy condi-tions. Applications range from front-ends for speech recognition systems to enhancement of telecommunications in aviation, military, teleconferencing, and cellular environments. The goal is either to improve the perceived quality of the speech or to increase its intelligibility.

Improving quality can be important for reducing listener fatigue in high-stress and high-noise environments. In the recording industry, improving the quality of recorded speech may be desirable even if the noise level is low to begin with. It is also a key way for telecommunications companies to increase customer satisfaction.

Intelligibility can be measured in terms of speech recognition performance. Enhancement preprocessing techniques, however, have not proven successful at improving human recognition rates.[1] On the other hand, improving intelligibility has been clearly demonstrated in the domain of automatic speech recognition (ASR) systems, which are severely affected by even low levels of noise. In this case, speech enhancement preprocessing can greatly improve the performance of ASR systems in noisy environments.

13.1.2 Model Structure

A fairly general characterization of noisy speech is given by the following model structure:

$$y_k = h(x_k, n_k), \tag{13.1}$$

where x_k is the clean speech signal, $h(\cdot)$ is the communications channel, and n_k is a noise process. The degraded speech signal is represented by y_k. For generality, we assume that the communications channel can have a nonlinear effect on the speech. However, with the exception of a few cases (e.g., nonlinear distortion due to switching on telephone networks, and noise amplification due to automatic gain control), the channel can usually be replaced by a linear convolution with impulse response h_k. If the noise is additive, this yields

$$y_k = h_k * x_k + n_k, \tag{13.2}$$

where n_k now includes channel effects. In most applications, n_k is statistically independent of the speech signal x_k.

We should note here that some of the methods covered in this

---

1. One study *did* show an improvement in human intelligibility, but only by severely reducing the quality of the speech [57].

chapter do not compensate for channel distortion but are designed purely for removing the noise signal n_k. This is appropriate for improving quality when it is assumed that the channel was designed properly for clean speech; note that our ears are insensitive to small phase distortions or global spectral shifts. On the other hand, compensation of the channel may be critical to improve robustness in ASR systems. For example, the simple act of changing the recording microphone can drastically affect recognition accuracy.

Note that the model excludes multiple-microphone systems (y_k is a scalar), which employ beam-forming [1–2] and noise cancellation [3] techniques. Beam formers adapt the gains of a microphone array to place nulls at the noise source, while noise cancellers assume the availability of a separate reference to the noise signal. Although these methods can be extremely effective, they are restrictive in their requirements on the availability and placement of multiple microphones. This chapter considers only single-microphone methods.

13.1.3   Chapter Overview

Speech enhancement using nonneural techniques has a long and diverse history, and continues to be an area of active research. While the field is too extensive to provide a full review here, many of the neural approaches we will discuss are direct extensions of more traditional methods. We will therefore provide some review of nonneural techniques throughout the chapter in order to put the various neural methods into proper context. For a more comprehensive review, see [4–5].

Our goal is to characterize the basic speech enhancement frameworks in which neural networks have been utilized. We will focus on the assumptions and capabilities of each framework rather than on a rigorous treatment of specific algorithms. Direct comparisons between algorithms and between types of algorithms are avoided for two reasons. First, the existing literature has not made use of standard benchmark data to allow for experimental comparisons across methods. Second, such comparisons would not be fruitful in the context of this chapter because of the wide variation in operating assumptions, processing times, and capabilities between the different methods. We therefore will discuss the individual merits of the different approaches, explicate the assumptions made, and provide insight into the associated strengths and weaknesses. In addition, we will describe possible refinements to existing methods, and critical areas for future research.

Sections 13.2 and 13.3 provide a discussion of two different paradigms for speech processing: (1) time-domain filtering, which includes direct nonlinear filtering as well as extended Kalman filtering based on a predictive model, and (2) transform-domain mappings, which are nonlinear generalizations of spectral approaches. Section 13.4 discusses refinements for explicitly addressing the nonstationarity of speech. These refinements provide a mechanism for using different models for different parts of the speech signal. In these first sections, the neural network plays the role of a nonlinear mapping which is trained using a data set of clean speech and artificially corrupted noisy speech. In Section 13.5 we discuss methods for adapting speech enhancers online to the specific noisy speech of interest. These methods do not rely on a training set of clean data, and therefore avoid the generalization problems often associated with the choice of training data. We conclude the chapter by summarizing the assumptions, advantages, and disadvantages of each of the frameworks presented, and by suggesting some directions for the field.

13.2 Neural Time-Domain Filtering Methods

We begin with two basic approaches for processing the noisy speech signals in the time domain. The first method forms a direct mapping from noisy speech to clean speech, while the second approach uses a predictive model in conjunction with an extended Kalman filtering approach.

13.2.1 Direct Time-Domain Mapping

The earliest and most straightforward use of neural networks for speech enhancement is as a direct nonlinear time-domain filter. This is illustrated in Figure 13.1, in which a multilayer network is used to map a windowed segment of the noisy speech to an estimate of the clean speech. The number of inputs depends on the sampling rate of the speech signal and typically is set to cover 5 to 10 ms of data. The number of outputs is usually equal to the number of inputs. To train the network, clean speech is artificially corrupted to create noisy input data. The clean speech signal is used as a target that is time-aligned with the inputs. Standard back-propagation or any variety of training methods can be

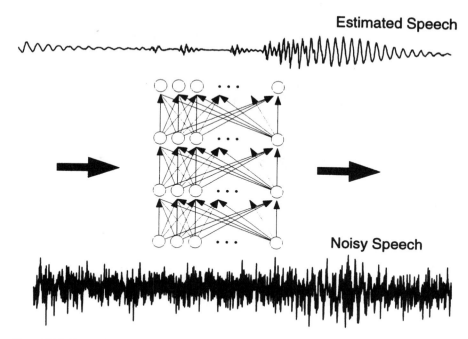

Figure 13.1 Illustration of a neural network filter which maps an input data vector to an output vector.

employed to minimize the mean squared error (MSE) between the target and the output of the network.

Data is presented by sliding the input window across the noisy speech. At each step, the window is shifted by an increment, L, between 1 and the window length M. When $L = M$, the estimation window moves along without overlap. For increments $L < M$, the resultant overlapping windows provide redundant estimates.

In the case of a single time-step increment, $L = 1$ (which most closely corresponds to traditional filter implementations), the network topology could be simplified to have only a single output. However, it has generally been recognized that using multiple outputs aids in the training process. The extra outputs balance the forward and backward flow of signals during training and allow for a greater number of shared hidden units to be used with improved generalization. After training, the estimate can be taken from one of the centermost outputs, discarding the rest.

A sample experimental result using the direct time-domain approach is shown in Figure 13.2. A neural network with 41 inputs, two hidden layers of 41 units each, and 41 outputs was used

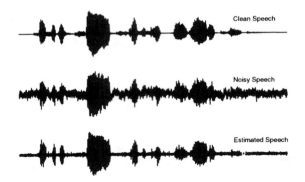

Figure 13.2 Experimental test set results using a direct time-domain neural filter. Original SNR was 3.0 dB with pink noise. Improvement was 7.04 dB.

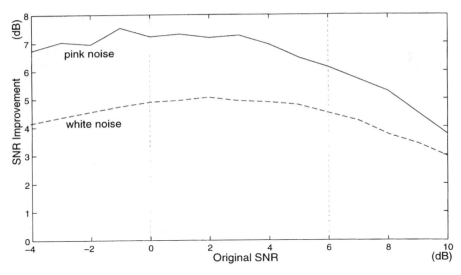

Figure 13.3 SNR improvement of a direct time-domain neural filter trained on 35 different sentences mixed with pink noise. During training, SNRs were varied between 0 and 6 dB.

$(41:41:41:41)$. The network was trained on 35 different TIMIT sentences from different speakers. To produce the noisy inputs, pink noise was added to the speech at randomly selected SNRs, chosen uniformly between 0 and 6 dB.[2] The network was tested using a TIMIT

---

2. The network was trained using standard back-propagation for 35 epochs with a step size of 0.01, followed by 5 epochs with a step size of 0.001. The presentation order of the training patterns was randomized.

speaker not in the training set. The shifting increment was $L = 1$, and the center output was used to generate the estimates. Figure 13.3 shows the network's performance on this sentence over a range of initial SNRs, both on pink noise (characteristic of the training data) and white noise (not characteristic of the training data). While impressive performance is achieved for SNRs within the training set range, note the falloff in improvement for other SNRs (as well as for white noise).

A number of researchers have reported superior results over linear filtering by using methods similar to the one described above [6–9]. Tamura [9] gives a detailed analysis of the role of the different layers in the networks, suggesting that the hidden layers provide a transformed representation of the signal and noise which facilitates their separation. Use of a neural network also allows for compensation of nonlinear channel effects, and some relaxation of the requirement that the additive noise be independent of the signal. In one application, Le and Mason report results on the method for noise introduced by a low bit-rate CELP encoder [7].

An additional variation on the filtering method results from restricting the number of units at a hidden layer to be less than the number of input or output units. This can provide noise suppression through dimensionality reduction similar to Ephraim's method based on signal subspace embedding [10]. Ephraim's idea is that the clean speech resides in a low-dimensional subspace of the noisy speech space; after first removing the dimensions that contain only noise, enhancement is performed in the subspace. Discussion of four-layer networks used for dimensionality reduction can be found in [11–12] (although results for speech enhancement have not been reported).

In still another variation, researchers at Defense Group Incorporated have implemented a recurrent structure by feeding back the network filter outputs.[3] Their system is a hybrid that uses traditional enhancement methods to preprocess the data before feeding it into the neural network. Reported results appear favorable in comparison with to a number of other traditional methods.

The advantage of time-domain filtering is the ease and efficiency of implementation. Effectively, the neural network approximates the conditional expectation $E[x_k|\mathbf{y}_k]$, where x_k is the clean speech and $\mathbf{y}_k = [y_{k-M/2}, y_{k-M/2+1}, \cdots, y_{k+M/2}]$ is the windowed noisy input. Note that the conditional expectation corresponds to a linear estimator only when all signal statistics are Gaussian (clearly unrealistic for speech

---

3. See http://www.ca.defgrp.com.

signals and real-world noise sources). This is a strong motivation for the use of nonlinear neural networks.

Once trained, a single fixed neural network is used to provide speech estimates. However, this also underscores the disadvantages of the approach. Using a fixed network implies a single, fixed expectation inferred from the entire training set. The corresponding conditional probability density can be written as

$$\rho(x_k|y_k) = \frac{\rho(y_k|x_k)\rho(x_k)}{\rho(y_k)} \tag{13.3}$$

Thus, assuming that the expectation is constant is equivalent to assuming that both x_k and y_k have constant density functions. In other words, both the noise and the speech signals would have to be stationary processes. This is clearly not the case. While training on a variety of different SNRs and speakers can greatly improve generalization (as seen in the experiment in this section), this does not explicitly account for the nonstationarity. Some researchers have incorporated pitch information as additional inputs to attempt to account somewhat for the nonstationarity of the speech. Moakes reported on this in the context of radial basis function networks [8]. Variations in the speech and noise statistics can also be addressed by using an estimate of the time-specific SNR as an additional input to the network. Related "switching"-based methods are addressed in Section 13.4.

In general, the direct time-domain filtering approach is most applicable for reducing fixed noise types, or for compensating a distortion that is associated with a specific recording or communication channel. For example, the latter case was considered by Dahl and Claesson [6], who trained a neural network on a specific speaker and noise environment for a carphone application.

13.2.2 Extended Kalman Filtering with Predictive Models

An alternative approach to time-domain filtering combines predictive modeling with a state-space estimation procedure. The underlying assumption is that the noisy speech y_k can be accurately modeled as a nonlinear autoregression (AR) with both process and additive observation noise:

$$x_k = f(x_{k-1}, ... x_{k-M}, \mathbf{w}) + v_k \tag{13.4}$$

$$y_k = x_k + n_k, \tag{13.5}$$

where x_k corresponds to the true underlying speech signal driven by process noise v_k, and $f(\cdot)$ is a nonlinear (neural network) function of past values of x_k and parameters $\mathbf{w}$. As before, y_k is the corrupted speech signal, which contains additive noise n_k. Both v_k and n_k are initially assumed to be white (though not necessarily Gaussian). In principle, a channel could be incorporated by replacing (13.5) with (13.1); this would require an additional function approximator to represent $h(\cdot)$. For simplicity, we will ignore channel effects in this section.

Note that if $f(\cdot)$ is linear, the AR formulation corresponds to the classic linear predictive coding (LPC) model of speech. It has been demonstrated (see Tishby [13]) that neural models are better at capturing the dynamics of speech than simple linear models.

The model parameters $\mathbf{w}$ can be found by training the neural network in a predictive mode on clean speech. As in the previous section, we consider using only a single network for the entire speech training set. This is primarily for the sake of explaining the extended Kalman filtering (EKF) method. The use of different predictive models on short-term windows to account for the nonstationarity of speech will be discussed in Section 13.4.

Given a linear model $f(\cdot)$, the well-known Kalman filter algorithm [14] optimally combines noisy observations y_k at each time step with predictions $\hat{x}_k^-$ (based on previous observations) to produce the linear least squares estimate of the speech $\hat{x}_k$. In the linear case with Gaussian statistics, the estimates are the minimum mean squared estimates. With no prior information on x, they reduce to the maximum-likelihood estimates. To apply the Kalman filter, we must first put (13.4) and (13.5) in state-space form:

$$\mathbf{x}_k = F[\mathbf{x}_{k-1}] + Bv_k, \qquad (13.6)$$

$$y_k = C\mathbf{x}_k + n_k, \qquad (13.7)$$

where

$$\mathbf{x_k} = \begin{bmatrix} x_k \\ x_{k-1} \\ \vdots \\ x_{k-M+1} \end{bmatrix}, \qquad F[\mathbf{x}_k] = \begin{bmatrix} f(x_k,\ldots,x_{k-M+1};\mathbf{w}) \\ x_k \\ \vdots \\ x_{k-M+2} \end{bmatrix},$$

$$C = [1 \quad 0 \quad \cdots \quad 0], \qquad B = C^T \qquad (13.8)$$

Because the neural network model is nonlinear, the Kalman filter cannot be applied directly, but requires a linearization of the nonlinear model at each time step. The resulting algorithm is known as the extended Kalman filter (EKF) [14], and effectively approximates the nonlinear function with a time-varying linear one. Letting $\sigma_{v,k}^2$ and $\sigma_{n,k}^2$ represent the variances of the process noise v_k and observation noise n_k, respectively, the EKF algorithm is as follows:

$$\hat{\mathbf{x}}_k^- = F[\hat{\mathbf{x}}_{k-1}, \hat{\mathbf{w}}_{k-1}] \tag{13.9}$$

$$P_{\hat{\mathbf{x}},k}^- = AP_{\hat{\mathbf{x}},k-1}A^T + B\sigma_{v,k}^2 B^T \quad, \text{ where } \quad A = \frac{\partial F[\hat{\mathbf{x}}_{k-1}, \hat{\mathbf{w}}]}{\partial \hat{\mathbf{x}}_{k-1}} \tag{13.10}$$

$$K_k = P_{\hat{\mathbf{x}},k}^- C^T (CP_{\hat{\mathbf{x}},k}^- C^T + \sigma_{n,k}^2)^{-1} \tag{13.11}$$

$$P_{\hat{\mathbf{x}},k} = (I - K_k C)P_{\hat{\mathbf{x}},k}^- \tag{13.12}$$

$$\hat{\mathbf{x}}_k = \hat{\mathbf{x}}_k^- + K_k(y_k - C\hat{\mathbf{x}}_k^-). \tag{13.13}$$

Often referred to as the "Time-update," (13.9) and (13.10) produce the a priori prediction of the next value of the state $\mathbf{x}_k$, along with the error covariance of this prediction, $P_{\hat{\mathbf{x}},k}^- = Cov(\mathbf{x}_k - \hat{\mathbf{x}}_k^-)$; (13.13) combines the prediction $\hat{\mathbf{x}}_k^-$ with the noisy observation y_k, using the Kalman gain term K_k to provide the optimal tradeoff. The Kalman gain makes use of the error covariance of the prediction $(P_{\hat{\mathbf{x}},k}^-)$ and the noise variance of the observation $(\sigma_{n,k}^2)$ to weight the influences of the prediction and the observation. The error covariance of the new a posteriori estimate $\hat{\mathbf{x}}_k$ is computed in (13.12), and is necessary to continue the recursion. These last three equations, (13.11), (13.12), and (13.13), are collectively called the "measurement update." Note that the state-space and Kalman equations can be modified to accommodate colored noise [14], or to have a fixed lag to produce a noncausal estimate [15].

The linearization in (13.10) is required for propagation of the error covariances $P_{\hat{\mathbf{x}},k}$ and $P_{\hat{\mathbf{x}},k}^-$, but results in a suboptimal filter. In fact, for a stationary signal with fixed SNR, the direct nonlinear time-domain filtering of the previous subsection may provide a better estimate. However, with the EKF, the time-specific SNR[4] enters through the parameter $\sigma_{n,k}^2$. Thus the EKF avoids the need to train over a range of possible noise levels (only clean speech is needed for training). This enables the EKF to easily handle cases in which the noise is

---

4. Note that SNR $= (\sigma_y^2 - \sigma_n^2)/\sigma_n^2$.

nonstationary, or where a broad range of noise levels might be encountered.

The downside of this is that both the variance of the speech innovations $\sigma_{v,k}^2$ and the noise variance $\sigma_{n,k}^2$ (or equivalently, the SNR) must be estimated online from the noisy data. The speech innovations variance $\sigma_{v,k}^2$ may be estimated from the expression for the inverse Fourier transform of the signal power spectrum given an LPC model [16]. Alternatively, an expression may be derived by noting the relationship between the minimum mean squared prediction error for clean speech versus speech with additive noise [17]. Estimating the innovations variance also requires knowledge of the SNR. Maximum-likelihood-based methods for estimating these quantities can be derived from the expectation maximization (EM) approach [18]. Methods for estimating the SNR (or full spectrum of the noise n_k), which are motivated from speech perception, are discussed in the Section 13.3.

An additional advantage of the predictive approach is that the state space (or AR) model provides for a more compact representation. The EKF forms a recursive structure, and thus requires fewer inputs than the direct filtering approach. This tradeoff is analogous to that seen for linear FIR (finite-impulse-response) filters versus IIR (infinite impulse response) filters.

While this EKF method provides a reasonable alternative to the direct filtering method, it should be noted that the approach as it appears in this section has not been reported on in the literature (perhaps because of its assumption of stationarity). Consequently, we do not include experimental results here. The basic technique, however, will form the basis for two important variations that also account for the nonstationarity of speech, to be discussed in Sections 13.4 and 13.5.

13.3 Neural Transform-Domain Methods

One perspective on time-domain processing discussed in the previous section is that the hidden layers of a multilayer network provide a transformed representation of the signal and noise that facilitates their separation. In transform-domain approaches, the new representation is explicitly provided by preprocessing in an attempt to reduce the complexity of the neural network's task. In addition, perceptual or ASR constraints are more easily incorporated.

The recent traditional speech enhancement literature has been largely dominated by transform-based methods, and by the DFT-based

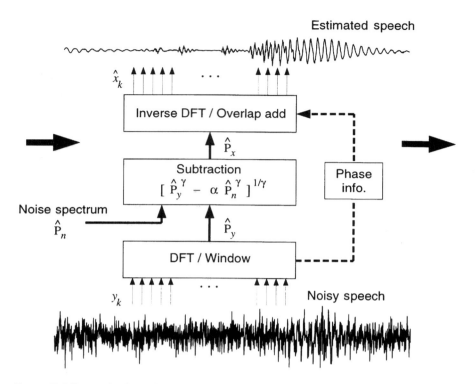

Figure 13.4 Spectral subtraction enhancement. The estimated noise spectrum is subtracted from the spectrum of a window of noisy speech. The noisy phase is combined with the result before computing the inverse DFT to produce the enhanced speech waveform.

spectral subtraction method [19] in particular. Because of its importance, we will start with a brief review of the approach. This will help to motivate the use of neural networks with transforms that are themselves nonlinear.

13.3.1 Spectral Subtraction

The popularity of spectral subtraction is largely due to its relative simplicity and ease of implementation. As shown in Figure 13.4, the short-term power spectrum $\hat{P}_y$ (magnitude squared of the short-term Fourier transform) of the noisy signal is computed, and an *estimate* of the short-term noise spectrum $\hat{P}_n$ is subtracted out to produce the estimated spectrum $\hat{P}_x$ of the clean speech. Explicitly,

$$\hat{P}_x = [\hat{P}_y^\gamma - \alpha\hat{P}_n^\gamma]^{1/\gamma} \tag{13.14}$$

where the scaling factor α allows for emphasis or deemphasis of the noise estimate, and γ allows for several variants, including power subtraction ($\gamma = 1$) and magnitude subtraction ($\gamma = 0.5$).

The estimate $\hat{\mathbf{P}}_x$ is combined with the phase from the original noisy signal to produce an estimate of the Fourier transform of x. Finally, the inverse Fourier transform is applied with the overlap-and-add method to construct a time-domain estimate of the speech waveform $\hat{x}$. The assumption is that the phase information is not important (perceptually), so only an estimate of the magnitude of the speech is required. Central to the linear spectral subtraction method is the additivity of the speech and noise spectra in the Fourier transform domain, allowing for simple linear subtraction of the noise spectrum estimate.[5]

Typically, the noise spectrum $\hat{\mathbf{P}}_n$ is approximated from a window of the signal where no speech is present. This requires that the speech be accurately segmented into speech and nonspeech parts. A related approach developed by Hirsch [20] estimates the noise level within a frequency subband by taking a histogram of spectral magnitudes over several successive time windows. The assumption is that the most frequently occurring value represents the magnitude of the noise in that band. Hence, care must be taken to compute the histogram over segments that have a sufficient number of nonspeech segments. Another approach based on assumptions of a bimodal distribution of the total histogram of the logarithmic spectral energies is presented in [21–22].

Regardless of how the noise statistics are estimated, the true short-term spectrum of the noise for the specific segment being processed will always have finite variance (this is true even for a stationary signal). Thus the noise estimate $\hat{\mathbf{P}}_n$ will always over or under estimate the true noise level. This represents a fundamental problem with spectral subtraction and other transform-based methods. The consequence is that when $\hat{\mathbf{P}}_y$ (which includes the true short-term noise signal) is near the level of the estimated noise spectrum, spectral subtraction results in some randomly located negative values of the quantity $\hat{\mathbf{P}}_y - \hat{\mathbf{P}}_n$. These negative values are clipped at zero to give a valid power spectrum, resulting in a series of annoying low-level tones ("musical noise") throughout the estimated signal $\hat{x}$.

While a great deal of work has been done to try to reduce these effects [1, 5, 23–26], they can only be eliminated completely if the

---

5. Spectral subtraction can also be interpreted as a spectral-based noise cancellation system, in which the noise reference (spectrum) is derived from an earlier segment of the noisy signal.

analysis window is increased to be of infinite length. This would allow the "short-term" noise spectrum to converge to the true spectrum. On the other hand, speech dictates that a finite short-term window be used to account for nonstationarities in the signal.

Note also that spectral subtraction only achieves a maximum-likelihood estimate of the (stationary) speech signal when the analysis window goes to infinity. This is in contrast to the Kalman approach, which can provide maximum-likelihood estimates using only finite length input windows. However, to be optimal, the KF requires prior knowledge of the true autoregressive model $f(\cdot)$. In practice, this model can only be estimated. On the other hand, spectral subtraction is model free, but sacrifices maximum-likelihood estimates in order to accommodate the nonstationarity of the signals.

13.3.2 Neural Transform-Domain Mappings

While the Fourier transform domain is computationally convenient, other domains such as the log power spectral, cepstral, and LPC domains are often more desirable to work in. For example, the log spectral domain is thought to offer a more perceptually relevant measure of distortion than the spectral domain, and also allows homomorphic filtering [27] to better compensate for channel distortion. If the goal of the enhancement is to improve perceptual quality, then there are clear advantages to minimizing distortion in a perceptually relevant domain. Furthermore, if the enhancement is for a front-end to an ASR system, the transform domain chosen for recognition usually dictates the domain in which speech enhancement occurs.

Unfortunately, removing noise in these alternative transform-domains can no longer be done with a simple linear subtraction of the noise from the speech. Because the transformed noise and speech are combined in a nonlinear way (e.g., through a log function), a nonlinear form of "subtraction" is required. The necessary nonlinear function depends on the transform domain and can be approximated by a neural network. This neural network subtraction or transform-domain mapping is illustrated in Figure 13.5. As in the time-domain approach, a representative training set of clean speech and corresponding noisy speech is used to train the network. However, one advantage of the transform-domain is that time-alignment of the inputs and targets is not as critical as with a time-domain mapping. This allows for generation of training data in more realistic settings using less controlled recording devices.

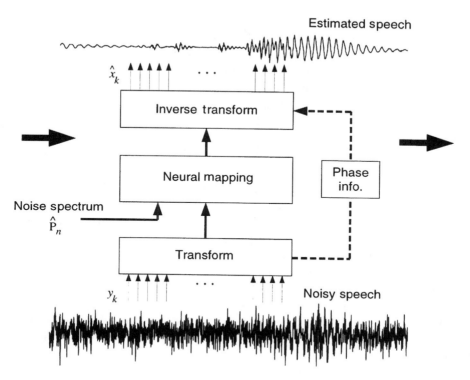

Figure 13.5 Transform-domain enhancement. A fixed transform is used to preprocess the data before it is modified. A neural network is used to perform the nonlinear enhancement mapping. Postprocessing transforms the signal back into the time-domain.

Although neural-network subtraction allows for the use of nonlinear transforms, this flexibility comes at considerable cost. In classic spectral subtraction, the "mapping" is simply a subtraction and is functionally independent of the noise level and noise spectrum. This is precisely due to the linearity of the Fourier transform. However, if another transform is used, the neural network must be able to provide different mappings for different noise types and levels. This can be attempted by incorporating additional inputs into the network which encode estimates of the SNR and/or noise distribution in some way. Clearly, this requires that the training set includes a representative sample of the noise levels and distributions that are expected to be encountered by the final system.

During operation, errors in the noise spectrum estimates will result in degradation of performance. The problems associated with estimating the noise statistics are fundamentally the same as in classic spectral subtraction. However, artifacts such as musical noise may be less severe due to properties of the transform domain chosen. For example, the variance in the short-term spectral estimation using an LPC analysis will be lower than that for a direct DFT approach.[6] Also, the nature of the nonlinear mapping is less likely to produce values that require the kind of strict truncation used in linear spectral subtraction. Finally, the use of a nonlinear mapping allows for a fair amount of freedom in the choice of transform domain, which can therefore be chosen for its inherent robustness and perceptual qualities. The associated enhancement technique might then be less affected by variations in the noise spectral estimation.

Like the noise source, the distribution of the speech signal also affects the neural-network mapping. Training with a single network averages across all speech signals and effectively assumes stationarity of speech (both within a speech signal and between different speakers). This assumption is the same as in the direct time-domain approaches. However, it is not fully understood whether the assumption is more severe in the time domain versus some transform domain. Approaches that skirt this issue through the use of multiple networks will be discussed in the next sections.

A number of researchers have performed preliminary investigations based on neural transform domain mappings. We summarize some of this work here.

The use of a feedforward neural network in the log power spectral domain has been demonstrated by Xie in [28] (an estimate of the mean of the speech signal and the noise variance provide additional input to the network). Informal listening tests showed favorable acceptance (no musical noise was reported). On a simple HMM speech recognition system (speaker-dependent 100-word digit database), using the method as a preprocessor resulted in a substantial improvement in recognition rates, as shown in Table 13.1.

In similar work, Sorensen designed a system for compensating F-16 jet noise based on a mapping in a cepstral domain [29]. Impressive results were reported for an ASR experiment (510-word test set for multiple speakers), with the recognition rates summarized in Table 13.2.

---

6. In general, parametric spectral estimation methods have lower variance than nonparametric methods.

Table 13.1

Recognition Accuracy (%) Comparison Between the Case of No Neural Preprocessing and the Case of Neural Transform Domain Mapping in the Log Power Spectral Domain

F16 Noise (SNR)	−6 dB	0 dB	6 dB	12 dB	18 dB
No Processing	16	45	81	94	99
Neural preprocessing	50	89	96	95	97
Difference	34	44	15	1	−2

Car Noise (SNR)	−6 dB	0 dB	6 dB	12 dB	18 dB
No Processing	Differ-ence	16	22	56	36
Neural preprocessing		38		92	

Table 13.2

Recognition Accuracy (%) Comparison Between the Case of No Neural Preprocessing and the Case of Neural Transform Domain Mapping in the Capstrum Domain

F16 Noise (SNR)	0 dB	3 dB	9 dB	15 dB	21 dB	∞ dB
No Processing	Differ-ence	14.1	65.3	17.8	65.0	33.1
Neural preprocessing		79.4		82.8		86.9

Use of a time-delay neural network (TDNN) for Mel-scaled spectral estimation was reported by Dawson [30] (though performance is hard to evaluate).

Moon et al. reported mapping LPC coefficients with an Elman net used in conjunction with a TDNN speech recognizer [31]. They also proposed "coordinated" training in which the recognition error was used (through backpropagation) to directly modify the preprocessor enhancement network.

Finally, Wan et al. [32] reported on experiments in which feedforward neural networks were used to modify RASTA processing [33]. In this work, as in spectral subtraction, a Fourier transform was used, but neural networks form nonlinear filters to map the trajectory of noisy magnitude spectral components, over several analysis windows, to an estimate of the clean spectrum. While noticeable improvement in noise reduction was achieved, often greater signal distortion and more musical noise was noted.

13.4 State-Dependent Model Switching Methods

As discussed in Sections 13.2 and 13.3, both the time-domain filtering and transform-domain mapping methods place unwarranted assumptions on the stationarity of the speech signal and noise source. A reasonable way to avoid stationarity assumptions is to replace the fixed neural models with time-varying models. This can be done by creating different fixed models for the different dynamical regimes of the signal, and then switching between these models during the speech enhancement process. Two approaches for this are discussed in this section.

13.4.1 Classification Switched Models

For either the direct time-domain or transform-domain mapping methods, the simplest approach is to first classify the noisy speech signal and then select the appropriate mapping out of a set of trained networks. This is illustrated in Figure 13.6, where the speech is divided into

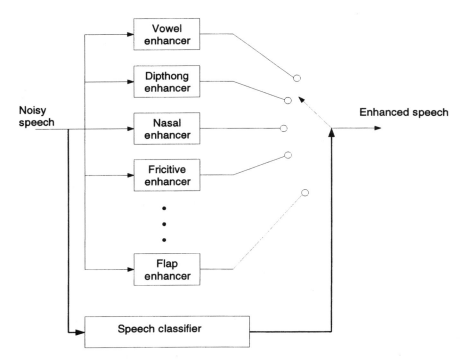

Figure 13.6 Switching model based on signal classification.

different sound classes. Different noise sources and different noise levels can similarly be used for classification. For each class, a separate neural network is trained in advance using data from that class. The added complexity of this approach is in devising the appropriate classifiers (which may also be neural networks [34–35]) as well as the larger number of networks required. In addition, issues regarding transients between the switching of states must also be addressed.

While the idea for switching methods was originally proposed to improve traditional speech enhancement systems [36], the idea can also be used for neural speech enhancement. A variation by Tamura [37] changes only the last layer of a time-domain filter network for the different classifications. In general, these approaches have not been fully explored, and to the best of our knowledge they have yet to be used within the neural transform-domain mapping framework.

13.4.2 Hybrid HMM and EKF

A principled switching method reported by Lee [38] incorporates the extended Kalman filtering approach to speech enhancement discussed in Section 13.2.2. The method is a straight forward neural extension of work by Ephraim [39], which utilizes a hidden markov model (HMM) to automate the process of dividing the received speech signal into various classes (see [40] for a review on HMM systems). In this case, each state of the HMM corresponds to a model-based filter that provides a maximum-likelihood estimate $\hat{x}_k$ given that the observations y_k belong to class i. The overall estimate is a weighted average of class-based estimates, given by

$$\hat{x}_k = \sum_i p(class_i|y_k) \cdot [\hat{x}_k|y_k, class_i], \qquad (13.15)$$

where $p(class_i|y_k)$ is the probability of being in class i given the window of noisy observations y_k, and the second term in the sum represents the maximum-likelihood estimate of the speech given class i and the data. The posterior class probability $p(class_i|y_k)$ is easily calculated using standard "forward-backward" recursive formulas for HMMs [41].

Alternatively, the estimate $\hat{x}_k$ may be simply taken as the estimate for the single filter whose posterior class probability is maximum:

$$\hat{x}_k = [\hat{x}_k|class_m] \text{ given } p(class_m|k) \geqslant p(class_i|k) \ \ \forall i. \qquad (13.16)$$

This variation is more strongly related to the classification-based switching methods described at the start of this section.

Within each class, the maximum-likelihood estimation for speech is provided by an EKF using an autoregressive model (as in Section 13.2.2) for that class. The different autoregressive neural models are trained using a set of clean speech data. In addition, the necessary speech innovations variance $\hat{\sigma}_v^2$ is estimated from the clean speech for each class. One of the more appealing aspects of this process is that both the HMM and neural AR models may be trained in coordination within a maximum-likelihood framework using an approach based on the Baum reestimation algorithm [41]. Details on training can be found in Lee [38] for the neural case, and more thoroughly in Ephraim [39] for the linear case. Lee reports results for systems using eight states across various initial noise levels (see Figure 13.7 for a sample experiment).[7] More extensive evaluations are necessary to fully judge the potential performance of this method.

The HMM approach provides a principled means for constructing state-switching systems. However, one drawback is the need to choose a sufficient number of states in order to divide the speech into relatively stationary classes. The implementation is also fairly complex in comparison with the other methods. In addition, it is assumed that the HMM will still provide correct posterior class probabilities on noisy speech (even though it is trained on noise-free speech).

While the use of a Kalman filtering approach for each state of the HMM is justified from the maximum-likelihood perspective, a hybrid system in which each state involves a spectral mapping method also deserves investigation.

13.5 Online Iterative Methods

All the methods described up to this point require a prior training set for learning the appropriate neural mapping(s). Generalization is required across speech classes, various speakers, noise types, and signal-to-noise ratios. The situation is ameliorated somewhat by the use of the state-dependent techniques. Nevertheless, these methods still make extremely strong demands on the ability of the neural models to generalize

---

7. We would like to thank Dr. Ki Yong Lee for supplying us with experimental data and details on his work.

Clean Speech

Noisy Speech

Estimated Speech

Figure 13.7 Experiments with hybrid HMM/EKF. SNR increased from 13.0 dB to 14.45 dB. HMM has 8 states, neural networks have 12 inputs, 2 hidden nodes, and 1 output. 8 minutes of clean Japanese speech is used for training.

from the training set to real-world environments. The methods described in this section avoid the generalization issue entirely by making use of only the noisy data at hand. No explicit clean training set is used; rather, the network adapts to enhance the given noisy speech signal of interest.

13.5.1 Online Predictive Enhancement

A very simple approach to using the noisy data stems from an early linear speech enhancement technique. The idea is that if the noise is additive, then an adaptive linear predictor will not be able to predict the noise for time lags beyond the correlation length of the noise [3, 42]. For white noise, a linear one-step-ahead predictor trained on the noisy data can learn to predict only the signal, not the noise.[8]

Extending these ideas to neural predictors is straightforward. The network is simply adapted online, where the input is a tapped delay-vector of noisy speech and the target for training is the same signal time-advanced. This is illustrated in Figure 13.8. Unfortunately, the use of neural predictors in this fashion is problematic because a neural network will start to model the noise process over a finite data segment. The advantage gained by potentially improving the prediction of speech is offset by the disadvantage of predicting the noise.

One possible way to reduce this problem is to limit the flexibility of the neural predictor (by choosing a smaller architecture) until it is

---

8. Conversely, adaptive line enhancement techniques work by removing harmonic signals that can be predicted at long time horizons. In this case, speech plays the role

Estimated speech

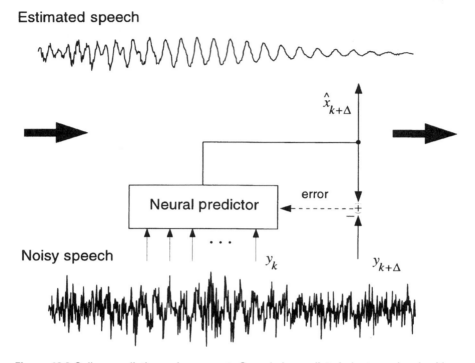

Figure 13.8 Online predictive enhancement. Speech is predicted Δ-steps-ahead with a neural network.

unable to predict the noise but can still predict the speech. This is done in [43], using minimum description length ideas to select the neural architecture. In general, online predictive approaches to speech enhancement have received less interest than more recent techniques. Unvoiced speech is problematic, and performance is also degraded for real-world noise sources (where the noise autocorrelation may be longer than the speech).

13.5.2 Maximum-Likelihood Estimation and Dual Kalman Filtering

A better-motivated approach is to consider the problem of finding the maximum-likelihood estimates of the speech and the model parameters given the noisy data. However, even for linear models, this represents a difficult nonlinear optimization problem. Lim and Oppenheim [16], proposed finding an approximate maximum a posteriori estimation solution by iteratively Wiener filtering and using a least-squares approach to fit an LPC model. Since then, a number of researchers have proposed variations on this method, which include using the Expectation-Maximization (EM) algorithm [44–47], accommodating colored noise [48], and placing perceptual constraints on the iterative search [49].

Wan and Nelson [17] have proposed a related approach, where neural autoregressive models are used. The speech model is the same nonlinear autoregression as in (13.4) and (13.5) of Section 13.2.2. However, to avoid using a single model to describe the entire non-stationary speech signal (or requiring the complexity of model-switching methods), the speech is windowed into approximately stationary segments, with a different model used for each segment. With a state-space representation of the speech model, the EKF method discussed in Section 13.4.2 gives the maximum-likelihood estimate of the speech, assuming the model is known. However, as no clean data set is used, the model parameters themselves must now be learned online from the noisy data for each window of speech. To allow the simultaneous estimation of the speech model and speech signal, a separate set of state equations for the parameters of the neural network (weight vector **w**) is formulated:

$$w_k = w_{k-1} + \alpha_k \tag{13.17}$$

$$y_k = f(x_{k-1}, w_k) + v_k + n_k, \tag{13.18}$$

where the state transition is simply an identity matrix and the covariance of α_k is selected to improve convergence. The neural network $f(x_{k-1}, w_k)$ plays the role of a time-varying nonlinear observation on **w**. An EKF can now be written to compute the maximum-likelihood estimate of the model, assuming that the state **x** is known. The use of the EKF for weight estimation can also be related to recursive least squares (RLS), where the covariance for α_k plays the role of the "forgetting factor" in RLS [50]. Hence, this method represents an efficient second-order online optimization method.

This weight EKF can be run in parallel with the EKF for state estimation, resulting in the *dual extended Kalman filter* (Dual EKF) [51], shown in Figure 13.9. At each time step, the current estimate of **x** is used by the weight filter, and the current estimate of **w** is used by the state filter.[9] This provides a very effective method for solving the

---

of the unpredictable "noise" [3].

9. This optimization approach relates to work done by Nelson [58] for system identification, and to Matthews' neural approach [59] to the recursive prediction

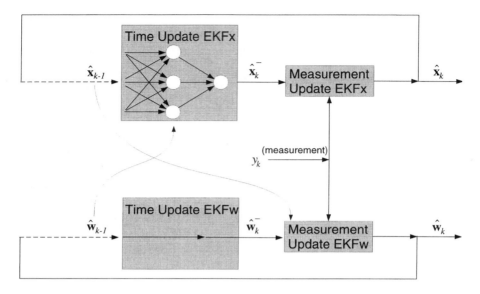

Figure 13.9 The Dual Extend Kalman Filter (Dual EKF). EKFx and EKFw represent the filters for the states and the weights, respectively.

maximum-likelihood estimates for the speech signal given only the noise source. Additional issues related to recurrent training, error coupling, the relationship of the algorithm to EM, as well as a two-observation form of the weight EKF are discussed in [51–52].

The result of applying the Dual EKF to a speech signal corrupted with simulated nonstationary bursting noise is shown in Figure 13.10. The method was applied to successive 64 ms (512 point) windows of the signal, with a new window starting every 8 ms (64 points). A normalized Hamming window was used to emphasize data in the center of the window and deemphasize data in the periphery.[10] Feedforward networks with ten inputs, four hidden units, and one output were used. Weights typically converged in less than 20 epochs. The results in the figure were computed assuming both σ_v^2 and σ_n^2 were known. The average SNR is improved by 9.94 dB, with little resultant distortion. When σ_n^2 and σ_v^2 are estimated using only the noisy signal,[11] similar

---

error algorithm [60].

10. The standard EKF equations are also modified to reflect this windowing in the weight estimation.
11. The estimation of the noise variances was performed for each window as described in [17]. This technique effectively provides an approximate "short-term" SNR as opposed to the histogram techniques (see Section 13.3.1), which average the noise

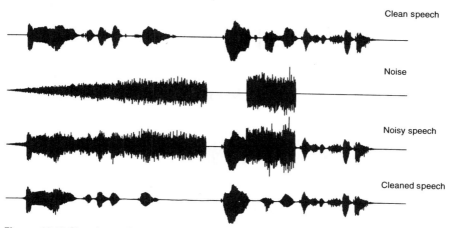

Clean speech

Noise

Noisy speech

Cleaned speech

Figure 13.10 Cleaning noisy speech with the dual EKF. Nonstationary white noise was generated artificially and added to the speech to create the noisy signal y. The SNR improvement is 9.94 dB.

results are achieved with an SNR improvement of 8.50 dB. In comparison, classical techniques of spectral subtraction [19] and adaptive RASTA processing [53] achieve SNR improvements of only .65 and 1.26 dB, respectively. Experiments in which real-world colored noise is added to the signal have also been performed. An advantage of the Kalman framework is that colored noise can be elegantly addressed by incorporating an additional state-space representation of the noise process. This modification affects both the state estimation and the weight estimation equations.

In principle, this method can accommodate any speaker, noise, or noise level encountered. In this sense, it is more in the spirit of spectral subtraction, which works independently of the type of signal it is estimating. However, like spectral subtraction, the dual EKF algorithm requires estimation of noise statistics.

While the approach does away with the need for a training set, there is considerable computational cost in training the neural networks online. Furthermore, the windowing of the data, which addresses the nonstationarity issue, also limits the size of the network structures that can be used. While for small windows of speech, compact models are sufficient (e.g., vocoder technology), this also questions whether the approach fully utilizes the flexibility of neural modeling.

statistics over long segments of speech.

A possible direction of research that addresses some of these issues is an intermediate approach that makes some use of pretrained models. This would be a state-dependent approach which selects among pretrained class-based models using an HMM (see Section 13.4.2), and then adapts the selected model online to the noisy data.[12] This could produce faster convergence, avoid the need to explicitly window the data, and allow larger networks to be used.

13.5.3 Noise-Regularized Adaptive Filtering

This approach involves a window-based and iterative process that is similar to the dual EKF method, but does not use an AR model for the speech. Rather, the approach utilizes the same architecture as in the direct time-domain mapping filters of Section 13.2.1 and Figure 13.1, while still avoiding the need for a clean data set to train the network.

Recall that the direct filtering approaches attempt to map a noisy vector of speech $\mathbf{y}_k = \left[y_{k-M/2} \cdots y_k \cdots y_{k+M/2}\right]^T$ directly to an estimate of the speech signal $\hat{x}_k = f(\mathbf{y}_k)$. The neural network, $f(\cdot)$, is trained by minimizing the mean squared error (MSE) cost,

$$\min_f E[(x_k - f(\mathbf{y}_k))^2] \tag{13.19}$$

where the corresponding optimal solution is given by the the conditional mean. (For illustrative purposes we will consider single-output filters, though the approach can also be extended to the MIMO case). In Section 13.2.1, training was performed by assuming that the clean signal x_k is known. However, consider the expansion:

$$E[(x_k - f(\mathbf{y_k}))^2] = E[(y_k - f(\mathbf{y}_k))^2] + 2E[n_k \cdot f(\mathbf{y}_k)] - 2E[y_k \cdot n_k] + E[n_k^2] \tag{13.20}$$

The last two terms are independent of $f(\cdot)$. Thus the optimal $f(\cdot)$ can be found by minimizing the alternative cost,

$$\min_f \{E[(y_k - f(\mathbf{y}_k))^2] + 2E[n_k \cdot f(\mathbf{y}_k)]\} \tag{13.21}$$

The advantage of this formulation is that the first term only depends on the observed noisy speech, whereas the expectation in the second term can be evaluated using only knowledge of the noise statistics. The clean speech is not needed. The first term can also be viewed as the cost associated with filtering the noisy signal to itself (training the network with noisy target data). The second term corresponds to the expected product between the noise and the neural-network output, and acts to regularize the weights of the network to prevent the filter $f(\cdot)$ from simply becoming the identity map. The resulting estimator is referred to as a noise-regularized adaptive filter (NRAF) [54].

The evaluation of the regularization term cannot be performed analytically. Instead, an approximate solution is found using the unscented transformation (UT), a method for calculating the statistics of a random variable that undergoes a nonlinear transformation [55]. This method involves propagating a set of M vectors (based on the first- and second-order statistics of the signals) through the network at each time step, and then forming a weighted sample mean. The cost function (including the regularization term) can still be minimized with respect to the weights of the network by gradient based descent using standard back-propagation. The only assumption made by this approach is that the second-order approximation to the regularization term using the UT is sufficiently accurate to allow convergence of the network to the true minimum MSE (this issue requires further investigation).

As in the dual EKF approach, to address the nonstationarity of speech, the noisy data is windowed into short overlapping frames and a new filter is designed for each frame. The method also requires estimation of the noise statistics.

We report results using samples from the OGI Speech Enhancement Assessment Resource (SpEAR) [56]. The experimental setup consisted of 8 kHz speech, 600-point frames (overlap of 4), filter window length of 25, and two-layer feedforward networks with 11 hidden nodes. In addition, the raw noisy speech window was first embedded using a fixed Karhunen-Luéve transform (KLT) [50] with embedding dimension of 19. These values were found empirically by cross-validation. Figure 13.11 summarizes the performance on a sample speech sentence for a number of different noise sources. The nonlinear NRAF algorithm is compared to linear NRAF (neural network replaced by linear FIR filter) as well as standard spectral subtraction. The nonlinear NRAF filter shows a clear improvement.

One appealing aspect of the method is that it unifies a number of traditional approaches while extending them to the nonlinear domain. In the case where the neural network is replaced by a linear filter, the resulting estimator reduces to a simple time-domain implementation of

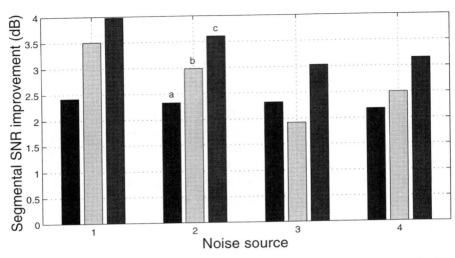

Figure 13.11 Comparison of segmental SNR performance for different noise sources 1) white (input SNR = 6.08, segmental SNR = 1.55), 2) pink (input SNR = 4.34, segmental SNR = 0.3), 3) factory (input SNR = 5.16, segmental SNR = 1.07), 4) F16 (input SNR = 4.61, segmental SNR = .46). Algorithms: a) standard implementation of spectral subtraction with 256 point frames (Duke Speech Processing Toolkit), b) linear NRAF with KLT, c) nonlinear NRAF. Note for these experiments a 3 dB improvement in segmental SNR corresponds to approximately 5dB improvement in SNR (Segmental SNR has been shown to correlate more closely with subjective quality evaluations [4]).

spectral subtraction. For four-layer networks with multiple outputs (block estimation), the number of hidden neurons can be made less than the number of outputs forcing an embedding of the input space. In this case, the method has the potential to provide nonlinear component analysis in contrast to the linear embedding used in traditional signal subspace approaches. An additional research direction would be to use larger networks that have been pretrained using the clean training set scenario, and then use the NRAF approach to tune these networks online. In summary, the NRAF approach has only recently been proposed, and while promising, further investigation is still necessary to fully characterize its potential.[13]

---

12. This is developed by Lee et al. [61] for the linear case.

13.6 Summary and Conclusions

In this chapter we have provided an overview of a number of different neural-network approaches to speech enhancement. We can summarize the techniques and their assumptions as follows:

- *Time-Domain Filtering:* The direct mapping approach trains a neural network using noisy inputs and clean targets. The implicit assumptions are that the statistics of both the speech and noise encountered will be the same as those of the training set, and that the statistics are constant (stationary) throughout.

  The extended Kalman filtering approach uses a predictive neural model (trained on clean speech) in a state-space framework in order to produce approximate maximum-likelihood estimates of the speech. The assumptions are that the signals are stationary, and the statistics of the speech to be enhanced will be the same as those of the training set. The speech signal is assumed to be well modeled by an autoregressive process driven by white Gaussian noise, and the additive noise is also assumed to be Gaussian (but can be colored); estimates of both the process and additive noise variances are assumed to be available. Compensation for channel effects is only possible when a model of the channel is available.

- *Transform-Domain Mapping:* Neural networks are trained from noisy input features to clean target features in a transformed domain. Domains are usually chosen for their desirable perceptual or recognition properties. The SNR or some other measure of the joint signal-noise statistics is typically used as an additional input to the network. Assumptions are that the SNR estimate is accurate, that the signal is stationary, and that the speech and noise statistics are representative of the training set.

- *State-Dependent Model Switching Methods:* These methods use a variety of different models trained on different classes of speech and noise signals in an attempt to better reflect the nonstationarity of the data. Two switching methods were discussed. The first was a classifier-based approach for choosing the appropriate neural mapping for the current signal. The second was a hidden Markov model approach which allows for modeling of state transition probabilities and which is typically used in conjunction with the extended Kalman filtering method of speech enhancement.

The assumption is that the number of states used is sufficient to model all the different regions of speech and noise statistics. In addition, the HMM (or classifier) must be able to identify the appropriate class given only the noisy signal.

- *Online Iterative Methods:* These methods adapt online to the specific noisy signal of interest. No assumptions are made about generalization because they do not rely on a training set. On the other hand, they do not make use of any possible prior knowledge available from a training set, and require significantly more computation during actual enhancement.

The simplest approach to training a speech enhancer online is to build an adaptive predictor of the speech. However, model-order constraints must be introduced to prevent prediction of the noise. It is assumed that the correlation length of the noise is less than that of the speech signal.

The second approach uses the EKF speech estimator in conjunction with a second EKF parameter estimator. Using short windows of the noisy speech and running these two estimators in parallel results in the dual EKF algorithm. This method provides an efficient approximate maximum-likelihood estimation for both the speech and the model parameters. Assumptions are the same as in the basic EKF approach, with the exception that stationarity is assumed only over short-term windows.

The last approach, referred to as noise-regularized adaptive filtering, utilized a novel expansion of the MSE cost function to allow training of direct time-domain filters online; in the linear case, this approach is closely related to a time-domain implementation of spectral subtraction and signal subspace embedding. This approach does not assume a model for the speech production. However, it is assumed that the second-order approximation to a necessary noise regularization term is sufficient to allow correct MSE training of the network. Other assumptions are the same as in the dual EKF approach.

While considerable progress has been made with these techniques, a number of key areas must still be addressed before we can expect widespread acceptance. Most important is the establishment of consistent evaluations to allow proper benchmarking between different approaches. Standardized databases should be used, with a variety of noise sources that include real-world examples and go beyond the simple white Gaussian noise assumption. Performance should be determined from established metrics (improvement in SNR, segmental SNR, Itakura distance, weighted spectral slope measures, mean-opinion scores, recognition accuracy, etc.). In addition, the basic techniques presented here must evolve to better incorporate perceptually relevant metrics for optimization. This is an area in which research on neural networks still lags considerably behind the traditional speech community.

Finally, the accurate estimation of the corrupting noise statistics remains a weak link in the algorithms that require these estimates as inputs. Research must be conducted to improve these estimates, or new techniques developed which avoid the need for explicit knowledge of the noise statistics.

In spite of the decades of work that has gone into understanding speech signals and issues in speech enhancement, the seemingly simple task of removing noise remains a formidable challenge. While it is still too early to draw definite conclusions, neural networks appear to offer an appropriate and powerful tool for further progress in this field.

Acknowledgments

This work was sponsored in part by the NSF under grant IRI-9712346.

References

[1] van Compernolle, D., "DSP Techniques for Speech Enhancement," *Proc. ESCA Workshop on Speech Processing in Adverse Conditions*, 1992, pp. 21–30.

[2] Kaneda, Y., and J. Ohga, "Adaptive Microphone-Array System for Noise Reduction," *IEEE Trans. on Acoustics Speech and Signal Processing*, Vol. 34, No. 6, 1986, pp. 1391–1400.

[3] Widrow, B., and S. D. Stearns, *Adaptive Signal Processing*, Englewood Cliffs, N.J.: Prentice-Hall, 1985.

[4] Deller, J. R., Jr., J. G. Proakis, and J. H. L. Hansen, *Discrete-Time Processing of Speech Signals*, New York: Macmillan, 1993.

[5] Lim, J. S., *Speech Enhancement*, Englewood Cliffs, N.J.: Prentice-Hall, 1983.

[6] Dahl, M., and I. Claesson, "A Neural Network Trained Microphone Array System for Noise Reduction," *IEEE Neural Networks for Signal Processing*, Vol. 6, 1996, pp. 311–319.

[7] Le, T. T., and J. S. Mason, "Artificial Neural Networks for Nonlinear Time-Domain Filtering of Speech," *IEEE Proc. Vis. Image Signal Process*, Vol. 143, No. 3, 1996, pp. 149–154.

[8] Moakes, P. A. and S. W. Beet, "Radial Basis Function Networks for Noise Reduction of Speech," *Proc. 4th Int. Conf. on Artificial Neural Networks*, 1995, pp. 7–12.

[9]   Tamura, S., "An Analysis of a Noise Reduction Neural Network," *Proc. ICASSP 87*, 1987, pp. 2001–2004.

[10] Ephraim, Y., and H. L. Van Trees, 'A Signal Subspace Approach for Speech Enhancement," *IEEE Trans. on Speech and Audio Processing*, Vol. 3, no. 4, 1995, pp. 251–266.

[11] DeMers, D., and G. Cottrell, "Non-linear Dimensionality Reduction," in *Advances in Neural Information Processing Systems 5* (Eds. Giles, Hanson, and Cowan), Morgan Kaufmann (1993).

[12] Oja, E., "Data Compression, Feature Extraction, and Autoassociation in Feedforward Neural Networks," *Artificial Neural Networks*, 1991, pp. 737–745.

[13] Tishby, N., "A Dynamical Systems Approach to Speech Processing," *Proc. ICASSP90*, Vol. 1, 1990, pp. 365–368.

[14] Lewis, F. L., *Optimal Estimation*, New York: Wiley, 1986.

[15] Paliwal, K. K., and A. Basu, "A Speech Enhancement Method Based on Kalman filtering," *Proc. ICASSP'87*, 1987, pp. 177–180.

[16] Lim, J. S., and A. V. Oppenheim, "All-Pole Modeling of Degraded Speech," *IEEE Trans. on Acoustics, Speech, and Signal Processing*, Vol. 26, No. 3, 1978, pp. 197–210.

[17] Wan, E. A., and A. T. Nelson, "Neural Dual Extended Kalman Filtering: Applications in Speech Enhancement and Monaural Blind Signal Separation," *IEEE Neural Networks for Signal Processing* Vol. 7, 1997.

[18] Dempster, A., N. M. Laird, and D. B. Rubin, "Maximum-Likelihood from Incomplete Data via the EM Algorithm," *Journal of the Royal Statistical Society*, Vol. B39, 1977, pp. 1–38.

[19] S. F. Boll, "Suppression of Acoustic Noise in Speech Using Spectral Subtraction," *IEEE Trans. on Acoustics Speech and Signal Processing*, Vol. ASSP-27, 1979, pp. 113–120.

[20] Hirsch, H. G., "Estimation of Noise Spectrum and Its Application to SNR-Estimation and Speech Enhancement," International Computer Science Institute, TR-93-012, 1993.

[21] Cohen, J., "Application of an Auditory Model to Speech Recognition," *J. Acoust. Soc. Am.*, Vol. 85, No. 6, 1989.

[22] Van Compernolle, D., "Noise Adaptation in Hidden Markov Model Speech Recognition Systems," *Computer Speech and Language*, Vol. 3, 1989, pp. 151–167.

[23] Berouti, M., R. Schwartz, and J. Makhoul, "Enhancement of Speech Corrupted by Additive Noise," *Proc. ICASSP'79*, 1979, pp. 208–211.

[24] Feit, A., "Intelligibility Enhancement of Noisy Speech Signals," M.Sc. Thesis, Israel Institute of Technology, 1973.

[25] Lim, J. S., and A. V. Oppenheim, "Enhancement and Bandwidth Compression of Noisy Speech," *Proc. of the IEEE*, Vol. 67, No. 12, 1979, pp. 1586–1604.

[26] McAulay, R. J., and M. L. Malpass, "Speech Enhancement Using a Soft-Decision Noise Suppression Filter," *IEEE Trans. on Acoustics Speech and Signal Processing*, Vol. 28, 1980, pp. 137–145.

[27] Oppenheim, A. V., and R. W. Schafer, *Digital Signal Processing*, 2nd ed. Englewood Cliffs, N.J.: Prentice-Hall, 1989.

[28] Xie, F., and D. V. Campernolle, "A Family of MLP Based Nonlinear Spectral Estimators for Noise Reduction," *Proc. ICASSP'94*, 1994, pp. 53–56.

[29] Sorensen, H. B. D., "A Cepstral Noise Reduction Multi-Layer Neural Network," *Proc. ICASSP'91*, 1991, pp. 933–936.

[30] Dawson, M. I. and S. Sridharan, "Speech Enhancement Using Time Delay Neural Networks," *Proc. 4th Australian Int. Conf. Speech Science and Technology*, 1992, pp. 152–155.

[31] Moon, S., and J.-N. Hwang, "Coordinated Training of Noise Removing Networks," *Proc. ICASSP'93*, Vol. 1, 1993, pp. 573–576.

[32] Wan, E., H. Hermansky, C. Avendano, and M. Kumashikar, "Application of Neural Networks to Speech Enhancement in Cellular Telephony," Machines That Learn Conference, Snowbird, Utah, 1995.

[33] Hermansky, H., N. Morgan, and H. G. Hirsch, "Recognition of Speech in Additive and Convolution Noise Based on RASTA Spectral Processing," *Proc. ICASSP'93*, Vol. 2, 1993, pp. 83–86.

[34] Kasper, K., H. Reininger, and D. Wolf, "A Neural Network Based Adaptive Noise Reduction Filter for Speech Recognition," *Proc. 7th European Signal Processing Conf.*, EUSIPCO-94, Vol. 3, 1994, pp. 1701–1704.

[35] Jones, M. B., and S. Sridharan, "Improving the Effectiveness of Existing Noise Reduction Techniques Using Neural Networks," *Proc. 4th Int. Symp. Signal Processing and Its Applications*, 1996, Vol. 1, pp. 387–388.

[36] Drucker, H., "Speech Processing in a High Ambient Noise Environment," *IEEE Trans. on Audio Electroacoustics*, Vol. AU-16, No. 2, 1968, pp. 165–168.

[37] Tamura, S., "Improvements to the Noise Reduction Neural Network," *Proc. ICASSP 90*, Vol. 2, 1990, pp. 825–828.

[38] Lee, K. Y., S. McLaughlin, and K. Shirai, "Speech Enhancement Based on Extended Kalman Filter and Neural Predictive Hidden Markov Model," *Neural Networks for Signal Processing*, Vol. 6, 1996, pp. 302–310.

[39] Ephraim, Y., "Speech Enhancement Using State Dependent Dynamical System Model," *Proc. ICASSP'92*, Vol. 1, 1992, pp. 289–292.

[40] Rabiner, L. R., "A Tutorial on Hidden Markov Models and Selected Applications in Speech Recognition," *Proceedings of the IEEE*, Vol. 77, No. 2, 1989, pp. 257–285.

[41] Rabiner, L., and Juang, B.-H., *Fundamentals of Speech Recognition*, Englewood Cliffs, N.J.: Prentice-Hall, 1993.

[42] Sambur, M. R., "Adaptive Noise Cancelling for Speech Signals," *IEEE Trans. Acoust. Speech Signal Process.*, Vol. ASSP-26, No. 5, 1978, pp. 419–423.

[43] Gao, X-M., S. J. Ovaska, and I. O. Hartimo, "Speech Signal Restoration Using an Optimal Neural Network Structure," *Proc. ICNN'96*, 1996, pp. 1841–1846.

[44] Musicus, B. R., and J. S. Lim, "Maximum Likelihood Parameter Estimation," *Proc. ICASSP'79*, 1979, pp. 224–227.

[45] Weinstein, E., A. V. Oppenheim, M. Feder, and J. R. Buck, "Iterative and Sequential

Algorithms for Multisensor Signal Enhancement," *IEEE Trans. on Signal Processing*, Vol. 42, No. 4, 1994, pp. 846–859.

[46] Lee, B.-G., K. Y. Lee, and S. Ann, "An EM-based Approach for Parameter Enhancement with an Application to Speech Signals," *Signal Processing:*, Vol. 46, 1995, pp. 1–14.

[47] Gannot, S., D. Burshtein, and E. Weinstein, "Iterative-Batch and Sequential Algorithms for Single Microphone Speech Enhancement," *Proc. ICASSP'98*, 1998, pp. 1215–1218.

[48] Gibson, J. D., B. Koo, and S. D. Gray, "Filtering of Colored Noise for Speech Enhancement and Coding," *IEEE Trans. on Signal Processing*, Vol. 39, No. 8, 1991, pp. 1732–1741.

[49] Hansen, J. H. L., and M. A. Clements, "Constrained Iterative Speech Enhancement with Application to Speech Recognition," *IEEE Trans. on Signal Processing*, Vol. 39, No. 4, 1991, pp. 795–805.

[50] Haykin, S., *Adaptive Filter Theory*, Upper Saddle River, N.J.: Prentice-Hall, 1996.

[51] Wan, E. A. and A. T. Nelson, "Dual Kalman Filtering Methods for Nonlinear Prediction, Estimation, and Smoothing," *Advances in Neural Information Processing Systems*, Vol. 9, 1997.

[52] Nelson, A. T., and E. A. Wan, "A Two-Observation Kalman Framework for Maximum-Likelihood Modeling of Noisy Time Series," *Proc. Int. Joint Conf. Neural Networks*, 1998.

[53] Avendano, C., H. Hermansky, M. Vis, and A. Bayya, "Adaptive Speech Enhancement Based on Frequency-Specific SNR Estimation," *Proceedings IVTTA*, 1996.

[54] Wan, E., and R. van der Merwe, "Noise-Regularized Adaptive Filtering for Speech Enhancement," *Proc. EUROSPEECH'99*, 1999.

[55] S. J. Julier, and J. K. Uhlmann, "A General Method for Approximating Nonlinear Transformations of Probability Distributions," Technical Report RRG, Dept. of Engineering Science, University of Oxford, 1996.

[56] SpEAR: Speech Enhancement Assessment Resource, Oregon Graduate Institute (http://ece.ogi.edu/NSEL/data).

[57] Thomas, I. B., and A. Ravindran, "Intelligibility Enhancement of Already Noisy Speech Signals," *Journal of the Audio Engineering Society*, Vol. 22, No. 4, 1974, pp. 234–236.

[58] Nelson, L. W., and Edwin Stear, "The Simultaneous On-Line Estimation of Parameters and States in Linear Systems," *IEEE Trans. on Automatic Control*, 1976, pp. 94–98.

[59] Matthews, M. B., and G. S. Moschytz, "The Identification of Non-linear Discrete-Time Fading-Memory Systems Using Neural Network Models," *IEEE Trans. on Circuits and Systems-II*, Vol. 41, No. 11, 1994, pp. 740–751.

[60] Goodwin, G. C., and K. S. Sin, *Adaptive Filtering Prediction and Control*, Englewood Cliffs, N.J.: Prentice-Hall, 1994.

[61] Lee, K. Y., B.-G. Lee, I. Song, and J. Yoo, "Recursive Speech Enhancement Using the EM Algorithm with Initial Conditions Trained by HMM's," *Proc. ICASSP'96*, 1996, pp. 621–624.

Index

design goal, 351
design optimality in practical
    situations, 326–27
development, 318
development concept, 322
discriminant function, 324
discriminative training, 336
distance classifier, 319–21
emulation of decision process, 321–24
formalization fundamentals, 319–27
form extension, 343
MCE, 198–99, 202
modular recognizer trained with, 312
optimization, 337
overview, 317–19
segmental, 329–31
selection of loss functions, 324–26
significance of, 342
smooth error count loss, 318, 325
smoothness, 327
speaker recognition with, 340–42
training, 326, 341
Genetic algorithms, 14
    application, 318
    defined, 14
Global restriction, 79
Gradient descent training, 227–28
    defined, 227–28
    weight change, 228
    See also Training
Gradient methods, 89
Gradient search algorithms, 14

Harmonic vector excitation coding
    (HVXC), 152
Hebbian learning rule, 443
Hidden Markov models (HMMs), 64, 160,
    218, 309, 358
3-state structure, 105
acoustic word models, 113
ANNs as front-ends for, 294–96
ANNs as postprocessors for, 296–97
classifiers, 112
competitive, 298
continuous, 106–9
defined, 103
design based on MCE, 204–12
discrete, 106

discriminant function, 206–7
formalization, 106
HMM-LVQ, 190–91
independence assumptions, 250
left-to-right, 104
LVQ-HMM, 189–90
mixed Gaussian, 347, 348
principles, 103–6
probabilistic acoustic modeling based
    on, 103–12
probability, 105
speaker identification, 341
state transition probability, 347
STLVQ interpretation, 187
Hidden neural networks (HNNs), 295
Hierarchical-mixtures-of-experts (HME),
    266, 288
architecture, 266
defined, 288
Hierarchical speech production processes,
    30–32
Hierarchical temporal decomposition
    (HTD), 145
Hierarchies of neural networks (HNNs),
    272–75
ACID clustered, for hybrid NN/HMMs,
    276
clustering, 274–75
defined, 272
experiments, 275–76
manually constructed, 273
results, 276–77
Human auditory peripheral system, 93
Human auditory process, 67
Hybrid ANN-HMM system, 294–96
    defined, 294
    structure, 294
Hybrid connectionist/stochastic networks,
    293–99
ANNs as front-ends for HMMs, 294–96
ANNs as postprocessors of HMMs,
    296–97
position of the problem, 293
solutions, 293–98
summary, 298–99
unified models, 297–98
Hybrid HME/HMM system, 268–69
Hybrid NN/HMM, 246

Recent Titles in the Artech House
Signal Processing Library

For further information on these and other Artech House titles,
including previously considered out-of-print books now available
through our In-Print-Forever® (IPF®) program, contact:

Artech House
685 Canton Street
Norwood, MA 02062
Phone: 781-769-9750
Fax: 781-769-6334
e-mail: artech@artechhouse.com

Artech House
46 Gillingham Street
London SW1V 1AH UK
Phone: +44 (0)171-973-8077
Fax: +44 (0)171-630-0166
e-mail: artech-uk@artechhouse.com

Find us on the World Wide Web at:
www.artechhouse.com